Interdisci
Engineering

T0093401

Interdisciplinary Engineering Sciences

Concepts and Applications to Materials Science

Ashutosh Kumar Dubey
Amartya Mukhopadhyay
Bikramjit Basu

CRC Press
Taylor & Francis Group
Boca Raton London New York

CRC Press is an imprint of the
Taylor & Francis Group, an **informa** business

Contents

Section I Fundamentals

Section II   Applications

Foreword

In the field of Science and Technology, we are used to read discipline-specific textbooks. In this context, I would say that this book is unique because it breaks away from the above 'degeneracy'. Here, the focus has not been just on the fundamentals or on the technological aspects, but on the ways the fundamental understandings of various disciplines/sub-disciplines can be integrated and finally translated to technology.

In order to provide a detailed commentary and understandings on the above, the authors have primarily selected two technological aspects of great societal impact as examples, *viz.*, healthcare and electrochemical energy storage. In the contexts of these two applications, this book logically and systematically covers the various associated fundamental aspects with sufficient depth. At each stage, the correlation between the fundamental and applied aspects has been brought to the fore, which also immediately provides an interdisciplinary flavor to the discussion. The second part of the book focuses on how to further integrate the concepts discussed in the first half and take the same forward toward the successful build-up of the related technologies. The major objective is to point out the ways and means of translation toward the same from the associated fundamental knowledge base. The discussion also indicates the technological and social benefits arising from such developments, again linking it to the associated scientific aspects.

To sum up, this book points out, in explicit terms, the lacuna of the age-old practice of compartmentalizing science and engineering disciplines, while successfully bringing to the fore the need to promote interdisciplinary science and engineering to keep pace with the technological requirements of society. The necessity toward the creation of collaborative centers of excellences and achievements is introduced to the discussion toward the closure of the book. I am convinced that this book will inspire senior undergraduate or graduate students in engineering disciplines to proceed toward a more interdisciplinary approach in academia or in industry.

Ashutosh Sharma
Secretary, Department of Science and Technology
Ministry of Science and Technology, Government of India
New Delhi

Preface

In general, the academic curricula in any university/academic institution are designed within a specified domain of a particular discipline. In the modern technological era, the research and development to address the existing/upcoming challenges of reasonable societal impact requires synergistic interaction, that is, cross-talk of various disciplines. To address such an important aspect, the necessity of interdisciplinary educational programs has been realized over the last few decades and is emerging at an accelerated pace across the globe. The academic programs focusing on a particular discipline provide an in-depth knowledge of that discipline and its role, importance, and implication in exercising diverse problems. The interdisciplinary programs reduce the barrier height among various disciplines, which facilitates the smooth transition from one area of study to another. Such smooth transitions will have a major implication in revolutionizing the next-generation technological advancement in various sectors such as energy, public health, security, artificial intelligence, and the like. As far as the nature of existing/upcoming challenges of major societal impact are considered, a particular discipline can be regarded as an island of knowledge, in which the interaction of fundamental concepts of diverse disciplines becomes important. In view of the above, the demands of interdisciplinary academic and research programs are in continuous demand.

This book endorses a fairly unique idea, which is pertinent to and important for the present-day needs of society as a whole. In general, education benefits society in various ways, with one of the more important aspects being the development of technologies needed for present and future generations of human beings. The technologies, which serve human society directly, are outcomes of scientific-cum-engineering discoveries and advancements (often interdisciplinary in nature), which are, in turn, born out of basic scientific and engineering principles pertaining to different disciplines. Currently, a large number of textbooks and handbooks are available, which provide sufficient detailing on different scientific and engineering principles at various levels, usually in isolation. However, lacking are books and literature which discuss aspects associated with bringing scientific and engineering concepts/principles together to evolve methodologies toward the development of efficient technologies.

Against this backdrop, the basic scientific and engineering principles relevant to health care and energy storage are discussed in Section I of this book. This is followed by showing how principles translate into actual technological developments (Section II). The challenges associated with the developments are discussed as are the importance of each aspect toward efficient build-up and application of the concerned technologies. The interdisciplinary nature of the principles directly associated with technological development and the need for collaborative efforts between experts in different areas are highlighted.

In this context, the need for interdisciplinary approach is discussed in Chapter 1; Chapter 2 describes the basic bonding characteristics, which dictate the correlation between bonding and some of the inherent properties of materials, in general. This paves the way toward the discussion, in Chapter 3, on aspects associated with advanced bonding characteristics and distribution of electrons in various sub-orbitals when forming ionic–covalent bonds, which, in turn, cast significant influence toward the electrochemical behavior of various materials in advanced batteries and supercapacitors. Chapter 4 covers the fundamentals of mechanical properties of materials. It also discusses the relevance of various numerical approaches in predicting the response of materials. Chapter 5 introduces various conventional and advanced manufacturing techniques, including additive manufacturing to produce three-dimensional structures with the help of machine readable/controllable design. Considering that batteries and supercapacitors are basically electrochemical systems, the thermodynamic aspects, firstly in a generic sense and then relevant to electrochemistry, have been discussed in Chapter 6. This introduces some of the very important electrochemical concepts such as equilibrium voltage, polarization, overpotential, and so forth. The same chapter also briefly discusses about some electroanalytical techniques which are used more often for investigation of the functionalities and performances of electrochemical energy storage systems. Continuing from the thermodynamic aspects, Chapter 7 discusses the kinetic aspects of electrochemical reactions and stresses on the importance of the same toward application in electrochemical energy storage systems. Chapter 8 aims to provide a foundation of biological systems to a non-biologist which primarily focuses on the structural characteristics as well as properties of various elements of biological systems such as protein, cell, bacteria, tissue, and so on. In view of the fact that these biological systems are electrically active in nature and play a key role in regulating various metabolic processes in living beings, Chapter 9 provides the fundamental origin of bioelectricity along with its consequences in the living system. Chapter 10 covers the basic aspects and underlying principles behind solar-thermal energy harvesting.

Following the discussions on the fundamental aspects described above, Section II opens by stressing present day demands of energy being harvested from renewable sources (like solar and wind energy), as governed by the concerns over increasing environmental pollution and depleting fossil fuel reserves (Chapter 11). This, in turn, stresses the need for the development and integration of efficient energy storage systems. Chapter 12 covers the technological aspects of material development to utilize the direct solar energy. Borrowing the principles discussed in the relevant chapters in Section I, Chapter 13 presents considerable detailing on the science-cum-technologies involved in the development and functioning of supercapacitors and advanced batteries, including the steps-cum-challenges associated with the fabrication of efficient devices of such electrochemical energy storage systems. At each stage, focus has been directed toward the interdisciplinary aspects of technology development and the importance of collaborative efforts. Chapter 14 provides insight of the critical issues and approaches in developing high-performance functional materials for armor applications. Chapter 15 covers the challenges and implications in the design and development of functionally graded materials which can provide a bone-mimicking electro-active response without altering the surface chemistry of the base material. Chapter 16 discusses the various aspects of design, development, and testing of components of thermal protection systems in hypersonic space vehicles. Finally, Chapter 17 summarizes the various aspects associated with the understanding of integrated interdisciplinary approach toward translational research in various sectors.

Ashutosh Kumar Dubey
Indian Institute of Technology (BHU)
Varanasi, Uttar Pradesh, India

Amartya Mukhopadhyay
Indian Institute of Technology, Bombay
Maharashtra, India

Bikramjit Basu
Indian Institute of Science, Bangalore
Karnataka, India

Acknowledgments

This book is the outcome of several years of teaching undergraduate and postgraduate level courses in the respective expertise areas of the authors, at IIT Kanpur, IISc, Bangalore, IIT Bombay, and IIT (BHU), Varanasi. In brief, this book consists of two sections. Section I (up to Chapter 10) covers the fundamental aspects of various science/engineering disciplines and Section II deals with their practical/technological aspects. As part of Section II, a few of the chapters summarize the extensive research outcomes from authors' groups, which have been/are being supported by the Council of Scientific and Industrial Research (CSIR), Department of Biotechnology (DBT), Department of Science & Technology (DST), Ministry of New and Renewable Energy (MNRE), Indo-US Science and Technology Forum (IUSSTF), UK-India Education Research Initiative (UKIERI). The financial support from these agencies is gratefully acknowledged.

I (BB) would also like to acknowledge the recent multi-institutional research program "Translational Centre on Biomaterials for Orthopedic and Dental Applications," supported by Department of Biotechnology, Government of India under "Centres of Excellence and Innovation in Biotechnology" scheme. The author also acknowledges the ongoing collaboration and interaction under the umbrella of this center with several colleagues, including Drs. Vamsi Krishna Balla, Amit Roy Chowdhury, Debasish Sarkar, R. Joseph Ben Singh, Biswanath Kundu, Manoj Komath, Sivaranjani Gali, Vibha Shetty, A. Sabareeswaran, Aroop Kumar Dutta, Tony McNally, B. Ravi, Michael Gelinsky, Jonathan Knowles, C. P. Sharma, S. Bose, D. C. Sundaresh, and B. Vaidhyanathan.

Present and past group members, who deserve special recognition, include Greeshma, T., B. Sunil Kumar, Ravi Kumar K., Yashoda Chandorkar, B. V. Manoj Kumar, G. B. Raju, Indu Bajpai, Shekhar Nath, Subhodip Bodhak, D. Sarkar, Atiar R. Molla, Naresh Saha, Shouriya Dutta Gupta, Garima Tripathi, Alok Kumar, Shilpee Jain, Shibayan Roy, Ravi Kumar, Prafulla Mallik, R. Tripathy, U. Raghunandan, Divya Jain, Nitish Kumar, and Sushma Kalmodia. The author also acknowledges the past and present research collaboration with a number of researchers and academicians, including Drs. Omer Van Der Biest, Jozef Vleugels, K. Lambrinou, M. Chandrasekaran, Dileep Singh, M. Singh, T. Goto, Sanjay Mathur, T. J. Webster, G. Sundararajan, K. Chattopadhyay, R. Gupta, Mira Mohanty, P. V. Mohanan, Ender Suvaci, Hasan Mondal, Ferhat Kara, S. J. Cho, Doh - Yeon Kim, J. H. Lee, Arvind Sinha, and Animesh Bose. BB is grateful for constant encouragement received from a number of international colleagues, including Robert S Langer, Cato Laurencin, Barry Carter, Samir Mitragotri, Nicholas Peppas, Serena Best, Sarah Cartmell, Philip J Withers, Abhay Pandit, Surya Mallapragada, Amit Bandyopadhyay, Susmita Bose, and many others.

I am also grateful for the suggestions from several colleagues, including Professors M. S. Valiathan, Dipankar Das Sarma, David Kaplan, S. C. Koria, David Williams, G. Padmanaban, Indranil Manna, Kamanio Chattopadhyay, G. K. Ananthasuresh,

K. K. Nanda, N. Ravishankar, Abhishek Singh, Prabeer Barpanda, A. M. Umarji, Vikram Jayaram, Dipankar Banerjee, and N. K. Mukhopadhyay and express further gratitude to my long-time friend and collaborator, Dr. Jaydeep Sarkar, for his constant inspiration during the writing of this book.

Finally, I am grateful to my parents, Manoj Mohan Basu and Chitra Basu, my wife, Pritha Basu, and son, Prithvijit Basu, for their constant moral support and encouragement.

AKD is grateful to his parents, Ashok Kumar Dubey and Sushila Dubey, wife, Richa Dubey, and son, Shivansh Dubey.

AM is grateful to his parents, Sonali Chakrabarty and Shankar Mukhopadhyay, and wife, Madhurima Barman. The author is further thankful for the great help rendered by Prerana S., Gopinath N K, Asish Kumar Panda, Rahul Mitra, Vignesh, Rea Johl, Nihal Kottan, Nitu Bhaskar, Deepa Mishra, Swati Sharma, Soumitra Das, Vidushi, Gowtham N H, Nandita Keshavan, Ranjith Kumar, Atasi Dan, Srimanta Barui, Subhadip Basu, Abhinav Saxena, Urvashi Kesharvani, Angraj Singh, Deepak Khare, Alok Singh Verma, Priya Singh, Maneesha Pandey, Richa and Priyanka Gupta during the writing of this book. AM would also like to acknowledge all past and present graduate students, who, as part of his research group, are always keen on having discussions on various aspects, right from fundamentals to applications to technology development, which has, in turn, been a major driving force for him to contribute toward a book having a very similar theme.

The authors would like to place on record their sincere gratitude to Professor Ashutosh Sharma to agree to write the foreword of this book.

Authors

Ashutosh Kumar Dubey is an assistant professor and Ramanujan Fellow in the Department of Ceramic Engineering at Indian Institute of Technology (BHU), Varanasi, UP, India. He is currently a Visiting Faculty at Nagoya Institute of Technology, Nagoya, Japan. His research interests include external electric field and surface charge-mediated biocompatibility evaluation of electro-bioceramics, piezoelectric toughening of bioceramics, functionally graded materials, nanoporous bio-ceramics, orthopedic biomaterials, and analytical computation. As recognition of his research work, Professor Dubey has received prestigious awards/fellowships including the Young Scientist Award by Indian Science Congress Association (2011–2012) as well as by Indian Ceramic Society (2015; Dr. R. L. Thakur Memorial Award), Japan Society for the Promotion of Sciences (JSPS) Fellowship for Foreign Researchers (2012–2014), and Ramanujan Fellowship by Department of Science and Technology, Government of India (2015–2020).

Amartya Mukhopadhyay is presently an associate professor at the Department of Metallurgical Engineering and Materials Science, Indian Institute of Technology-Bombay, Mumbai, India. His research interests include materials for electrochemical energy storage and engineering ceramics. Among his major recognitions, he has been awarded the INAE Young Engineer Award 2016, ASM-IIM North America Visiting Lectureship 2016, IIT Bombay Young Investigator Award 2014, and Dr. R. L. Thakur Memorial Award (as Young Scientist) by the Indian Ceramic Society in 2013 and has also been recognized by the *Royal Society of Chemistry* (UK) journals as one of the 2019 Emerging Investigators.

Bikramjit Basu is a Professor at the Materials Research Center and an Associate faculty at the Center for BioSystems Science & Engineering as well as at Interdisciplinary Center for Energy Research at Indian Institute of Science (IISc), Bangalore, India. He is currently a Visiting Professor at School of Materials, University of Manchester, UK and at Wuhan University of Technology, China. His research integrates the concepts of engineering, physical, and biological sciences to develop novel materials science-based solutions for solar energy harvesting and human health care. A recipient of India's highest Science & Technology award, the Shanti Swarup Bhatnagar prize (2013), Professor Basu is an elected Fellow of the American Ceramic Society (2019), American Institute of Medical and Biological Engineering (2017), National Academy of Medical Sciences (2017), Indian National Academy of Engineering (2015), and National Academy of Sciences, India (2013).

1

Introduction

Science, engineering and, in essence, human thought have been profoundly impacted by the developments of the past century. The essence of these changes can be perceived in many different ways; complexity, to start with. It includes complexity in the range of issues considered in the tools employed, and in profound insights consequently derived. It is now clearly understood that complexity originates from inter-connectedness of different disciplines, processes, and techniques, increasing as their inter-dependence grows. The network connecting basic units weaves a web, whose response to stimuli can be as varied as can be imagined. The development of different materials/components/technologies for a variety of applications and their associated equipment, or in a broader sense, the fundamental knowledge base and research facilities that are used/employed in research and development, display this awesome complexity and perform in ways that could not have been imagined at all. Computers have gained capabilities to analyze systems in their entirety. Algorithms process data on multiple scales, applying laws of nature as they appear at every level of hierarchy. Aggregating these technologies, the tone of "research and development" itself has undergone a vast change in the last few decades.

Nowadays, most development that benefits the human society necessitates interdisciplinary science and research. The researchers in the field of interdisciplinary engineering science face challenges stemming from intrinsic complexity of scientific ideas and the means to bring them together, leading to a sustainable development. At one level, colleagues with complementary backgrounds have formed short-term partnerships, thus creating elements of an inter-connected human network. This web-of-life is a key to understand sophisticated interactions that abound around us. It is also the key to exploit such interactions to address the numerous societal needs—be it energy, security, economy, food, or public health. In the above context, the creation of "Centers of Excellence" in the field of interdisciplinary engineering sciences within an academic or research institute should be appreciated in the context of scientific and technological upheaval happening around the globe. The human brain, as the original battleground (and of course, the original network that helps us function in multiple dimensions), has given way to human clusters as the basic unit that can maneuver a meaningful idea, if at all.

Over the last five decades, the research philosophy at a number of institutes around the world has evolved from the classical way of research being carried out as theses/dissertations of graduate students to individual research groups, sponsored projects, inter-departmental collaboration, and inter-institutional partnerships. Another parallel trend has been from the conventional way of conducting research within the boundary of a specific discipline to adopting a more interdisciplinary approach, thereby crossing the scientific boundaries of the existing disciplines. The latest in the series of evolutionary steps involves researchers from multiple disciplines and multiple institutes, ranging from humanities to "pure"/biological sciences to engineering to biomedical sciences/engineering, with a variety of knowledge base and skill sets. Inter-disciplinary research as a matter of necessity and multi-disciplinary research as a matter of enlightenment appeared on the stage. Apart from research topics and projects, the *research program*, with a broader perspective in problem definition as well as implementation, is now being followed. It is increasingly believed that "Departments" in their present form are just degree-granting entities, with concomitant responsibilities such as teaching, curriculum definition, developing teaching/research laboratories, and skill training. By contrast, research is rapidly getting concentrated in multi-faculty initiatives and centers, which are endowed with state-of-the-art facilities.

A common perspective is that complexity arises at the juncture of multiple disciplines and associated fundamental knowledge/skills. The interdisciplinary approach can potentially reduce the barrier height from one discipline to another and smooth transition will play a major role to revolutionize the next generation of scientific and technological advancement. A re-organized approach becomes absolutely necessary to face challenges stemming from the intrinsic complexity of various scientific disciplines. The inter-connected network can significantly harmonize the sophistications by considering and dealing with the individual disciplines, as independent. It is also the key to exploit such interactions to address our many societal needs, as mentioned above. This is largely enabled from the interaction of concepts and ideas drawn from multiple disciplines, including physical, chemical, biological, and engineering sciences (see Figure 1.1). This has been possible by seamlessly integrating the fundamental laws of a specific discipline with another discipline.

Interdisciplinary science is often considered synonymous with convergent science, which is defined as "an approach to problem solving that integrates expertise from life sciences with physical, mathematical, and computational sciences as well as engineering to form comprehensive frameworks that merge areas of knowledge from

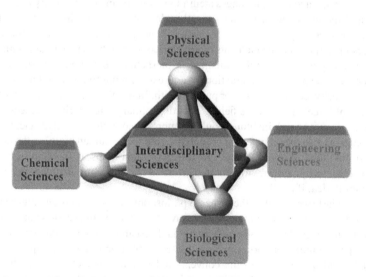

FIGURE 1.1 A pyramid depicting the contribution of the various scientific disciplines toward the interdisciplinary sciences, which, in turn, serve as the building block toward technological developments necessitated by the present and future societal needs.

multiple fields to address specific challenges." Independent of definitions, interdisciplinary science or convergent science, has made a significant impact on society. In other words, a particular discipline appears as an island of learning/ knowledge, while the cross-disciplinary approach gives it an ocean's magnitude. Overall, the interdisciplinary approach brings creativity. It has been often seen that the "immigrants" from a particular discipline provide new pathways for different disciplines, which depend on the interdisciplinary knowledge. The mistake of short sightedness by a person of a particular discipline can be best viewed and analyzed by a person with the knowledge of a different or multiple disciplines. Also, a person from a particular discipline can provide relatively new/novel approaches/methodologies for the benefit of other disciplines. The demand for interdisciplinary approach is in continuous thrust for the solution of a number of intellectual, social, and practical problems. Accordingly, the present concept of getting expertise/specialization within a narrow domain of a particular discipline needs to be looked into and reframed by adopting an approach with cross-linked multiple specializations.

> Interdisciplinary science is at the crossroads of traditional science disciplines. The innovative aspects and societal impact of interdisciplinary science can be better utilized if science is taught in a compartmentless manner.

Multidisciplinary approach involves more than one discipline in parallel. As per the perspective provided by Nissani, a single discipline is "self-contained and isolated domain of knowledge and experience. However, interdisciplinarity brings many isolated domains together."

In the above context, interdisciplinary science can therefore be defined as blending of two or more science disciplines toward solving a complex research problem, with the end objective being successful and efficient realization of a targeted engineering application by coupling engineering aspects with the combined knowledge base, which otherwise cannot be perceived using a single science-cum-engineering discipline alone in a conventional approach.

> Interdisciplinary science does not depend on the length scale involved in scientific research (e.g., nanoscience and nanotechnology), and that way it is unique in scientific approach.

Such interdisciplinary basic sciences are ultimately the building blocks toward engineering sciences, which, at some point, get directed toward technological developments and advancements, as per the needs of the present and future society.

We can substantiate the above with a few examples. For electronic devices, the laws of physics and advanced manufacturing techniques can be integrated to develop functional devices for multitude of engineering applications. For devices meant for energy harvesting and storage, the laws of physics, principles of chemistry, fundamentals in materials, and also the principles involved in developing efficient electronic devices get integrated. Thus, the knowledge combined can harvest and store the "green" (or environmentally friendly) solar energy for electric grids and also allow vehicles to be run by using such stored "green" energy, instead of on gasoline (the "non-green" form). Overall, for the society to progress further, the demand for energy in the useful form (i.e., electricity) is bound to go up very rapidly and needs to be accompanied by preserving the environment in the quest for satisfying this demand. This is where the integration of knowledge toward efficient, as well as "green," energy harvesting/storage is expected to lead toward. Similarly, the principles and methods of chemical sciences, as well as biological sciences, are used to develop life-saving drugs for biomedical applications. Here, drugs can be suitably loaded into biodegradable polymers for sustained release *in vivo*, which needs expertise from biomedical engineering. Clearly, the design of a disease-specific drug-delivery system demands the effective integration of knowledge from chemical, biological, and engineering sciences.

It may be worthwhile to mention that traditionally each of the above-mentioned science disciplines is further compartmentalized into multiple sub-disciplines. For example, chemical sciences can be divided into physical chemistry, inorganic chemistry, organic chemistry, solid-state and structural chemistry, electrochemistry, etc. Physical sciences can be divided into condensed matter physics, high energy physics, mechanics, and so on. Similarly, biological sciences have sub-disciplines such as neurosciences, biochemistry, cell and molecular biology, genetics, and so on. Engineering science has also very well-defined sub-disciplines such as materials, mechanical, chemical, electrical, civil, instrumentation, manufacturing, and others (see Figure 1.2).

It has been well-perceived in scientific community that many laws of physical and chemical sciences are efficiently utilized in various sub-disciplines of engineering sciences. There has been significant attention in the recent past to bridge the gap between biological and engineering sciences and consistent effort in this direction has resulted in the conceptual evolution of "bioengineering" or "biomedical engineering" or "clinical engineering" as a distinct discipline. It has also been recognized in the community that while chemical and biological sciences have close linkages, similar linkages also exist between engineering and physical sciences. For example, in the case of energy storage, final engineering and technology development needs very

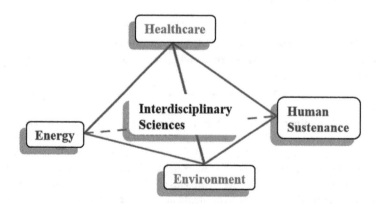

FIGURE 1.2 Quadrilateral depicting the contribution of the interdisciplinary sciences, which in turn gets contributed from the various scientific disciplines (as shown in Figure 1.1) toward the different engineering/technological domains that ultimately lead to human sustenance.

close integration of the concepts in physical sciences and chemical sciences; ultimately necessitating efficient instrumentation and manufacturing. In the above contexts, it is recommended that, starting with basic academics, students should be offered interdisciplinary courses and be trained on how the concept of one discipline can be used in dealing with the problems of other disciplines. Such training will be helpful in solving the real-life problems by cutting across the boundary of a particular discipline.

In the above context, some of the successful research programs launched in the field of healthcare, such as by the National Academy of Medicine, USA, include "Human Gene Editing," "The BRAIN initiative" (Brain Research through Advancing Innovative Neurotechnologies), "The Precision Medicine Initiative," "National Cancer Moonshot Initiative," "National Cancer Institute: Nanotechnology alliance," "Clinical Proteomics Tumor Analysis Consortium," "Defense Advanced Research Projects Agency (DARPA)," and "Biological Technology Office." Many of these programs are expected to transfer human longevity and address the challenges of human aging. Some of the above-mentioned programs are also initiated by the National Academy of Engineering, USA as part of their grand challenges program since 2008. Some of the health-related challenges include "Reverse Engineer the Brain," "Engineer Better Medicines," and "Advance Health Informatics." On a similar note, some of the programs, such as TAPSUN (Technologies and Products for Solar Energy Utilization through Networks) by CSIR, India, Indian Government's "Energy Storage Mission" and "National Electric Mobility Mission," BATT (Batteries for Advanced Transportation Technologies) Program by the United States, have the motive toward bringing together the relevant knowledge bases and technological know-hows toward efficient development and integration of technologies meant for harvesting renewable energies and storage of the same, with the application in electric vehicles deservingly getting some additional look into. The above, which are just a very few examples, are direct initiatives from think-tanks and policy makers that necessitate integration of various disciplines of knowledge, science, engineering, and technologies. Overall, the initiatives by the governments themselves of various countries to bring technologists, engineers, scientists, and students of various disciplines together to achieve goals and developments necessary for the progress and sustenance of the human society are justification enough toward a look into the translation at various stages/levels and for different applications of fundamental scientific knowledge toward engineering and technological solutions.

Section I

Fundamentals

2

Bonding and Properties of Materials

The physical, mechanical, electrical, and thermal properties of materials are decided by the nature of the bond, which combines atoms into molecules/crystals. The concept of molecules/crystals can be considered to be originated from the tendency of atoms to attain the inert gas configuration. This tendency of atoms binds them together to form a stable configuration/structure. The force that holds the atoms together is, in general, referred to as the chemical bond. The formation of molecules/crystals can be represented as the possible lowest energy form of their free entities. In view of the fact that the bonding plays an important role in understanding the properties of materials, this chapter discusses the fundamental aspects of chemical bonding and its correlation with the various properties of materials.

2.1 Introduction

The simplest approach to understand the bonding forces is to consider two isolated atoms at an infinite distance and study their interaction behavior, while bringing them close. Of course, there is no interaction at a large distance. The interaction between the atoms begins at a certain finite distance. Up to a minimum finite distance, the forces are attractive in nature. However, at quite close distance, the outer electron clouds of the atoms start to overlap, as a result of which a repulsive force acts to separate them. Therefore, the resultant force is a function of interatomic separation.

A state of equilibrium exists for the counterbalance of attractive (F_A) and repulsive forces (F_R), that is, $F_A + F_R = 0$. In such situations, the distance between the atoms corresponds to the minimum potential energy, that is, stable configuration. The magnitude of this energy is called the bond energy.

Let us discuss the problem in terms of potential energy. The potential energy of attraction is negative as the work is carried out by the atoms. However, the repulsive energy is positive as external work is needed to make the atoms come close.[1]

The total potential energy,

$$V(r) = V_{att} + V_{rep}$$

$$V(r) = -\frac{A}{r^n} + \frac{B}{r^m} \tag{2.1}$$

where A and B are constants and m varies between 9 and 15.[2]

Mathematically, the force is related to the potential energy as,

$$F_{(r)} = -\frac{dV}{dr} = -\frac{nA}{r^{n+1}} + \frac{mB}{r^{m+1}} \tag{2.2}$$

At some equilibrium separation (say r_e), the force of attraction counter balances the force of repulsion, which is characterized by the minimum potential energy. For many atoms, $r_e \sim 0.3$ nm.[3,4]

At equilibrium separation, $F(r) = 0$, at $r = r_e$,

$$F_{(r)} = -\frac{dV}{dr} = -\frac{nA}{r_e^{n+1}} + \frac{mB}{r_e^{m+1}} = 0 \tag{2.3}$$

$$r_e = \left(\frac{mB}{nA}\right)^{1/(m-n)} \tag{2.4}$$

If the total energy of the molecule/solid/crystal is lower than the energy of free atoms (before bonding), then the crystal will be stable. The difference in the crystal energy and their constituent free atom's energy is a measure of energy of cohesion.

Similarly, the stability of the structure is proportional to the amount of energy released. For the most stable structure, a maximum amount of energy should be released.

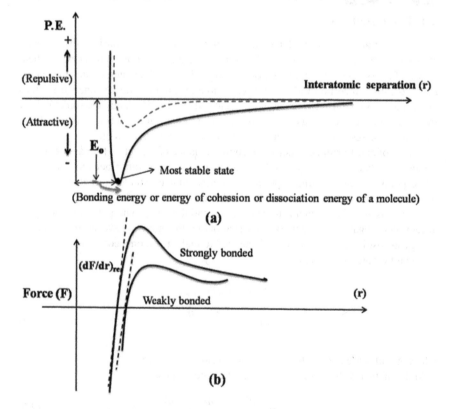

FIGURE 2.1 Variation of (a) potential energy, and (b) interatomic forces as a function of distance between two atoms.[3]

This suggests that the energy of free atoms is released during the formation of a chemical bond.

A number of material properties depend on E_0, materials with large bonding energies generally possess high melting temperatures (Figure 2.1a).

At room temperature, $(E_o)_{solid} > (E_o)_{liquid} > (E_o)_{gas}$

The shape of the force versus interatomic separation curve (Figure 2.1b) provides the idea about the mechanical stiffness (or modulus of elasticity) of any material.[15]
As

$$E \alpha \left. \frac{dF}{dr} \right|_{r=r_e} \tag{2.5}$$

The stiffer material consists of a quite steep slope at $r = r_e$, whereas the flexible material can be suggested to have shallower slopes. The fundamental differences in the elastic modulus of metals, ceramics, and polymers are due to their different types of bonding characteristics.[1,4]
Now,

$$V_{(r=r_e)} = -\frac{A}{r_e^{\,n}} \left[1 - \frac{B}{A} r_e^{\,n-m} \right] \tag{2.6}$$

$$V_{min} = -\frac{A}{r_e^{\,n}} \left[1 - \frac{n}{m} \right] \text{(Cohesive energy)}$$

Now,

$$\left. \frac{dV}{dr} \right|_{r=r_e} = \frac{nA}{r_e^{\,n+1}} - \frac{mB}{r_e^{\,m+1}}$$

$$\left. \frac{d^2V}{dr^2} \right|_{r=r_e} = -\frac{n(n+1)A}{r_e^{\,n+2}} + \frac{m(m+1)B}{r_e^{\,m+2}}$$

$$\left. \frac{d^2V}{dr^2} \right|_{r=r_e} = -\frac{nA}{r_e^{\,n+2}} (n-m) \tag{2.7}$$

For V to be minimum,

$$\left. \frac{d^2V}{dr^2} \right|_{r=r_e} > 0 \quad ; \quad m > n \tag{2.8}$$

If $m < n$, the bond is not formed.[15]

2.2 Bonding in Solids

Depending upon the bond strength, the chemical bonds can be categorized into primary and secondary types. In general, the primary bonds are interatomic bonds whereas secondary bonds are intermolecular bonds. Covalent, ionic, and metallic bonds are grouped as primary types. Secondary bonds possess an order of magnitude (>1) lower bonding energy as compared to those of the primary bonds. Van der Waals and hydrogen bonds are common examples of secondary bonding. Many of the common materials form via mixed bonding. These materials are classified on the basis of the dominating bond type. In general, the properties and response of materials are predicted on the basis of dominant bonding characteristics. However, such a prediction is not accurate for every case.

The line joining the centers of atoms, involved in bond formation, is a measure of the bond length. The stronger bonds are characterized by relatively smaller length (≤ 2 Å). The nature of the bond depends upon the configuration of the outermost electronic shell. In general, the primary bonds form due to the tendency of atoms to attain the stable inert gas configuration. In addition to strong primary bonds, weak secondary bonds or physical forces also exist in a number of solids.

> Despite being quite weak, the secondary bonds dominate in deciding the properties of materials, which consist of both, primary as well as secondary bonds.

2.2.1 Ionic Bonding

Ionic bonds are among the most important electro-static bonds, which are formed due to Coulombic interactions. To achieve a stable electronic configuration, the metallic atoms loose their valence electrons, whereas, the non-metallic atoms gain electrons. In this practice, atoms attain the inert gas configuration and are converted into ions. The interaction between these cations and anions helps to form a stable molecule or structure. For example, NaCl is a classical ionic material. A Na atom (Na^{11}) acquires the electronic configuration of Ne (Ne^{10}) by transferring one electron to Cl (Cl^{17}), which attains the stable configuration of Ar (Ar^{18}). In this practice, Na and Cl atoms become ions with positive and negative charges, respectively.

Overall, an ionic bond is formed via the electrostatic force between electropositive and electronegative ions.

To illustrate the energies of ionic bonding, consider the bond formation between Na and Cl atoms. The ionic bond is formed due to Coulomb-type interactions between two oppositely charged ionic species. Let us see the amount of energy released during the formation of NaCl molecules. Consider the free Na and Cl atoms at an infinite distance.

The ionization energy for Na is 5.14 eV and the electron affinity of Cl is 3.61 eV. Ionization energy is the energy required to remove an electron from the neutral Na atom and electron affinity is the change in the energy of Cl atom after gaining an electron. Therefore, a net energy (5.14 − 3.61 eV) is spent in creating Na^+ and Cl^- ions from their respective neutral atoms.[4]

$$Na + 5.14 \text{ eV(Ionization energy)} \rightarrow Na^+ + e \qquad (2.9)$$

This released electron occupies the vacant electronic state of Cl as,

$$Cl + e \rightarrow Cl^- + 3.61 \text{ eV (Electron affinity)} \qquad (2.10)$$

For NaCl, the equilibrium separation distance r_e is 2.36 Å, where the potential energy is minimum and the crystal is stable. The energy released in the formation of NaCl molecule is called the bond energy.[4]

$$Na + Cl^- \rightarrow Na^+Cl^- + 7.9(\text{Cohesive energy}) \qquad (2.11)$$

The energy for one molecule of NaCl is (7.9 − 5.1 + 3.6) = 6.4 eV lower than the energy of the neutrally separated Na and Cl atoms.[4]

The change in energy in bringing the ions upto a distance r is given as,

$$E_{pot} = -\frac{Z_1 Z_2 e^2}{4\pi\varepsilon_0 r}(q = z.e) \qquad (2.12)$$

It is now clear that the energy is released when the ions are taken together to form molecules. Consider the situation, if the distance between the ions is taken as zero, an infinite amount of energy will be released. It can, therefore, be suggested that the repulsive forces must come into action after a minimum finite distance between the ions. The consequences of both, the attractive and repulsive forces, result in the formation of stable molecules/crystals. From the above equation (2.12), it is clear that the amount of released energy is proportional to the charge (z) on the ions. Therefore, the ionic crystals consisting of multivalent ions (e.g., Al_2O_3; $Z_1 = 3$, $Z_2 = 2$), generally, form quite a strong bond.

> In a similar line, the crystals with multivalent ions have higher melting points than those with monovalent ions.

Therefore, Al_2O_3 and MgO are used as refractories, which have melting points of 2072°C and 2852°C, respectively. For example, common molten metals can be easily kept in an Al_2O_3 crucible.[2]

As the ionic bonds are formed via electrostatic forces, they are non-directional in nature. This non-directionality facilitates the bonding of a maximum number of anions to a particular cation in three-dimensional space and vice versa. In a solid, the effective Coulomb attractive/repulsive energy term consists of the contribution of all such ions, involved in the bond formation, that is, the Coulomb energy depends on the mutual attraction and repulsion of a particular ion with the rest of the ions in the crystal lattice. This idea came into picture after the pioneer work by Madelung. The energy of any particular ion pair in the crystal can be obtained by multiplying the Madelung constant into Coulomb energy term of that isolated ion pair. The calculation of Madelung constant is straightforward.

2.2.1.1 Calculation of Madelung Constant

Let us consider a linear arrangement of Na^+ and Cl^- ions and Na^+ ions at O as a reference ion (Figure 2.2).

This Na^+ ion has two Cl^- ions as nearest neighbors at distance r_0. The Coulomb energy term due to attractive interaction with both the Cl^- ions can be given as,

$$-\frac{e^2}{4\pi\varepsilon_0 r_0} + \left[-\frac{e^2}{4\pi\varepsilon_0 r_0} \right] = -\frac{2e^2}{4\pi\varepsilon_0 r_0}$$

Now, let us consider the next-nearest neighbors (Na^+) at a distance of $2r_0$. The Coulomb energy term due to repulsive interactions between Na^+ ions can be given as $\frac{2e^2}{4\pi\varepsilon_0 (2r_0)}$. Similar consideration for next neighbors at a distance of $3r_0$ provides the Coulomb energy term as $-\frac{2e^2}{4\pi\varepsilon_0 (3r_0)}$ and so on.

Therefore, the total energy due to entire linear arrangement can be given as,

$$\begin{aligned}
E_{pot} &= -\frac{2e^2}{4\pi\varepsilon_0 r_0} + \frac{2e^2}{4\pi\varepsilon_0 (2r_0)} - \frac{2e^2}{4\pi\varepsilon_0 (3r_0)} + \cdots \\
&= -\frac{e^2}{4\pi\varepsilon_0 r_0} 2 \left(1 - \frac{1}{2} + \frac{1}{3} - \frac{1}{4} + \cdots \right) \\
&= -\frac{e^2}{4\pi\varepsilon_0 r_0} \left[2\ln(1+1) \right] \\
&= -\frac{e^2}{4\pi\varepsilon_0 r_0} 2\ln 2
\end{aligned} \tag{2.13}$$

The term [2ln 2] is known as Madelung constant for the molecule in an ionic solid. As far as the binding energy of ionic crystal is concerned, the primary contribution comes from electrostatic interaction, which is called as Madelung energy.[4] Let us calculate the Madelung constant for a real NaCl crystal structure (Figure 2.3). Let us consider point A (center) as the reference point, that is, Na^+ at the center as the

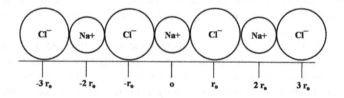

FIGURE 2.2   Linear arrangement of Na^+ and Cl^- ions.[4]

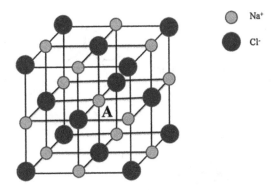

FIGURE 2.3 Crystal structure of NaCl.[4]

reference ion. This ion is coordinated by 6 Cl^- ions as nearest neighbors, twelve Na^+ ions at a distance of $\sqrt{2}r_0$ as next-nearest neighbors, and eight Cl^- ions at $\sqrt{3}r_0$, and so on.[4]

The total potential energy can be calculated as,

$$E_{pot} = \frac{-e^2}{4\pi\varepsilon_0 r_0}\left[6 - \frac{12}{\sqrt{2}} + \frac{8}{\sqrt{3}} - \cdots\right] \qquad (2.14)$$

where,

$$\alpha = \left[6 - \frac{12}{\sqrt{2}} + \frac{8}{\sqrt{3}} - \cdots\right]$$

$\alpha = 1.747565$ (Madelung constant).

2.2.1.2 Calculation of Cation to Anion Radius Ratio

In ionically bonded solids, the entire crystal can be considered to be made up of cations and anions. Owing to the non-directional nature of such solids, a cation can bond as many anions as possible and vice versa, depending upon the relative sizes of the cations and anions, which also helps to determine the structures of the ionic crystals.[2,3]

Let us consider the ionic radii of cations and anions to be r and R, respectively. As the cation is formed by the atom which loses the electron, it is usually smaller in size than the anions, which is formed by gaining the electron by respective atoms. The number of anions surrounding a cation is called the coordination number. As far as the formation of ionic crystals is concerned, only the coordination numbers 3, 4, 6, 8, and 12 are observed. The arrangement of cations and anions for coordination numbers 3, 4, and 6 is represented in Figure 2.4.

Let us consider the case of coordination number 3. As a limiting case, let us consider that all the anions are touching the central cation and anions are just in contact with each other (Figure 2.5).

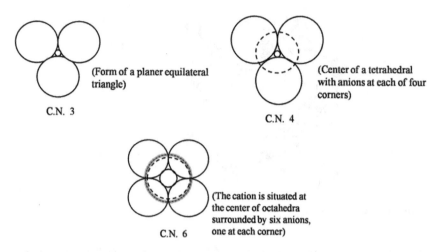

C.N. 3 — (Form of a planer equilateral triangle)

C.N. 4 — (Center of a tetrahedral with anions at each of four corners)

C.N. 6 — (The cation is situated at the center of octahedra surrounded by six anions, one at each corner)

FIGURE 2.4 Schematic illustrating the arrangement of anions and cations for coordination number 3, 4, and 6, respectively (as limiting case).[2,3]

From Figure 2.5,

$$OP = r + R, \; QP = R$$

$$\cos 30° = \frac{R}{r + R}$$

$$\frac{r}{R} = \left(\frac{2 - \sqrt{3}}{\sqrt{3}} \right)$$

$$\frac{r}{R} = 0.155$$

Therefore, the radius ratio of the cation to anion for coordination number 3 is 0.155 (in limiting case). This value is called as the critical radius ratio (for CN = 3).

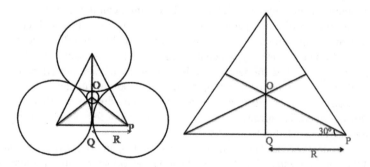

FIGURE 2.5 Geometrical consideration for the evaluation of critical radius ratio for coordination number 3.[3]

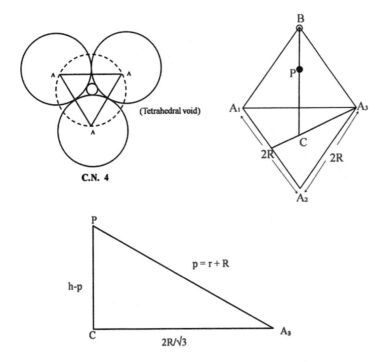

FIGURE 2.6 Geometrical consideration for the evaluation of critical radius ratio for coordination number 4.[3]

Now, considering the case of coordination number 4 (Figure 2.6), one can obtain the radius ratio as,

$$BC = h, \; BP = p, \; PC = h - p$$

$$A_3C = \frac{2R}{\sqrt{3}}$$

From triangle PCA_3 (Figure 2.6),

$$h^2 - 2hp + \frac{4R^2}{3} = 0 \tag{2.15}$$

$$h^2 = \frac{8R^2}{3} \tag{2.16}$$

Putting this value in Equation (2.15), one can obtain

$$h^2 - 2hp + \frac{h^2}{2} = 0, \; p = \frac{3h}{4}$$

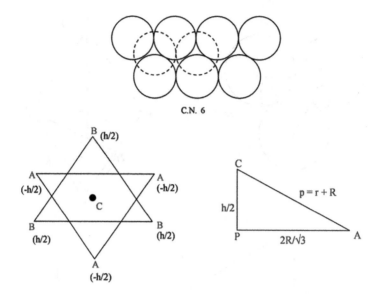

FIGURE 2.7 Geometrical consideration for the evaluation of critical radius ratio for coordination number 6.[3]

that is,

$$r + R = \frac{3h}{4}$$

$$\frac{r}{R} = \left(\frac{\sqrt{3} - \sqrt{2}}{\sqrt{2}}\right) \text{(Using the value of } h \text{ from Equation 2.16).}$$

$$\frac{r}{R} = 0.225$$

Therefore, for coordination number 4, one can have $0.225 \leq \frac{r}{R} \leq ?$

For coordination number 4, the value of $\frac{r}{R} \geq 0.225$, which also represents the upper case for coordination number 3. However, in order to minimize the energy, the coordination number 4 is favored for $\frac{r}{R} = 0.225$ as compared to coordination number 3.

Now, let us consider the case of coordination number 6 (octahedral void) (Figure 2.7). From triangle PCA (Figure 2.7),

$$p^2 = \left(\frac{h}{2}\right)^2 + \left(\frac{2R}{\sqrt{3}}\right)^2, \, p = \frac{\sqrt{3}}{2}h \tag{2.17}$$

$$r + R = \frac{\sqrt{3}}{2} \times \frac{2\sqrt{2}}{\sqrt{3}} R$$

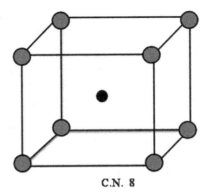

C.N. 8

FIGURE 2.8 Schematic arrangement of cations and anions for coordination number 8.[3]

$$\frac{r}{R} = (\sqrt{2} - 1) = 0.414$$

$$C.N. = 6; 0.414 \leq \frac{r}{R} \leq \cdots$$

The coordination number 8 represents a cube (Figure 2.8). For coordination number 8, the critical radius ratio can be obtained as,

Along <111> (Figure 2.8),

$$2r + 2R = \sqrt{3}a, \text{ that is, } 2r + 2R = \sqrt{3} \times 2R \tag{2.18}$$

$$\frac{r}{R} = 0.732$$

$$C.N. = 8, \ 0.732 \leq \frac{r}{R} \leq \cdots$$

For coordination number 12, anions and cations are of the same size, that is, when $\frac{r}{R} \geq 1$. Therefore, in a summarized way, we have,

$$C.N. = 2, \ \frac{r}{R} < 0.155$$

$$C.N. = 3, \ 0.155 \leq \frac{r}{R} \leq 0.225$$

$$C.N. = 4, \ 0.225 \leq \frac{r}{R} \leq 0.414$$

$$C.N. = 6, \ 0.414 \leq \frac{r}{R} \leq 0.732$$

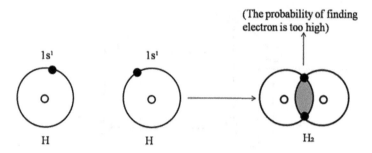

FIGURE 2.9   Covalently bonded H_2 molecules.[2]

$$C.N. = 8, \ 0.732 \le \frac{r}{R} \le 1$$

$$C.N. = 12, \ \frac{r}{R} \ge 1$$

2.2.2 Covalent Bonding

In contrast to the ionically bonded solids, covalent bonds are formed by the sharing of electrons of incomplete outermost shells of the atoms. Such sharing is facilitated by the overlapping of orbitals, which effectively reduces the potential energy of the system. For a significant reduction in potential energy, that is, stability of the structure, good overlapping between the orbitals is required. In most such cases, the outermost electronic orbitals are not spherical in shape. The overlapping of non-spherical orbitals (e.g., *p*-orbitals) provides the directional nature to the covalent bond.

To satisfy Pauli's exclusion principle, the shared electron pairs must have opposite spins. For a strong bonding to occur, the spins of the bonding electrons must be opposite to each other. The formation of hydrogen molecules is the simplest example of covalent bonding (Figure 2.9).

Due to the similar polarity of atoms, involved in bonding, the covalent bonds are also known as homo-polar bonds.

In H_2 molecules, the shared electrons are more probably found between two nuclei.

For minimizing the potential energy, a good overlap of orbitals is required, that is, the shared electrons should be in close proximity to the nucleus.

These bonds are primarily observed in organic and some inorganic molecules.

As mentioned above, if the overlapping orbitals are not spherical in nature, the requirement of significant overlapping of non-spherical orbitals (e.g., *p*-orbitals) to minimize the potential energy gives rise to the directionality to the covalent bond.

The formation of a covalent bond usually occurs by the sharing of un-occupied *p*-orbital electrons. Owing to the directional nature of *p*-orbitals, significant overlapping occurs in the direction of maximum electron density.

2.2.2.1 Valence Shell Electron Pair Repulsion (VSEPR) Theory

Sidgwick and Powell (1940) proposed a model to obtain a molecular geometry or shape based on valence electron pairs.[5] This theory was modified by Nyholm and Gillespie (1957)[6] as valence shell electron pair repulsion (VSEPR) theory, which is based on the repulsion between the electron pairs (bond and/or lone pairs [lp]) of the outermost (valence) shell of a central atom. These electron pairs try to orient in a position having minimum repulsion or maximum distance which provides a definite shape to the molecule. The lone pairs (lp) exist on the central atom, while unpaired electrons shared with the unpaired electrons of another atom for bond formation. However, lone pairs (lp) occupied more space than bond pairs (bp).[7] Therefore, lp–lp repulsion > lp–bp repulsion > bp–bp repulsion.[6,7]

Example; BeF_2 (Figure 2.10a), $BeCl_2$, O_2, N_2, and CO_2 (linear shape), CCl_2, SO_3, and BF_3 (Figure 2.10b) (trigonal planer), CH_4 and SiH_4 (tetrahedral), NH_3 (trigonal pyramidal), SF_6 (octahedral), H_2O, SO_2, and O_3 (bent shape), and so on.[6,7]

The VSEPR theory could not explain the concept of chemical bond formation in a polyatomic molecule, which can be explained with the help of valence bond theory and hybridization.[8,9] The overlapping between atomic orbitals (*s–s*, *s–p*, *p–p*, etc.) is responsible for the strength of covalent bonds; a good overlap results in larger bond strength. The *s* and *p* orbitals possess spherical and dumbbell shapes, respectively (Figure 2.11a,b). The overlapping between *s–s*, *s–p* and end-to-end *p–p* orbitals form sigma (σ) bonds (Figure 2.11c–e). However, side by side or parallel overlapping between *p* and *p* orbitals produces pi (π) bonds (Figure 2.11f).[8,9] The overlap between the two orbitals having nucleus on the same line results in maximum overlapping as compared to angular axes overlapping, and therefore, the strength of σ bonds is greater than that of π bonds. In the case of σ bonds, electron clouds can be observed along the axis of the orbitals, which are absent along the axis of parallelly overlapped orbitals of π bonds. It should be noted that multiple bonds like single and double bonds, formed by the overlapping of σ and π bonds. The single bond must be σ bond, however, in case of double or triple bonds, the other bond/bonds except the σ bond will be π bond/ bonds. For example, a double bond will be composed of one σ and one π bond and a triple bond can be produced by the overlap of one σ and two π bonds. The bond strength of single, double, and triple bonds can be compared as

$$[C-C](347 \text{ kJ/mol}) < [C=C](614 \text{ kJ/mol}) < [C≡C](839 \text{ kJ/mol})^{8,9}$$

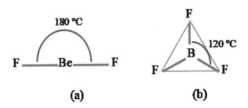

(a)   **(b)**

FIGURE 2.10   Molecular geometry of (a) BeF_2 and (b) BF_3 molecules.[6,7]

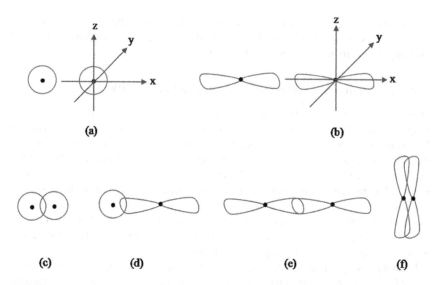

FIGURE 2.11 Schematic illustration of shape and orientation of (a) spherical shaped s orbital, (b) dumbbell-shaped p orbital, (c) s–s sigma bond, (d) s–p sigma bond, (e) p–p sigma bond, and (f) p–p π bond.[8,9]

2.2.2.2 Hybridization and Hybrid Atomic Orbitals

The concept of overlapping (valence bond theory) failed to explain the stable geometrical shapes of polyatomic molecules. For example, the actual molecular geometry of H_2O molecules with bond angle H–O–H as 104.5° can be explained by a combined model of valence bond theory and hybridization. In hybridization, s and p orbitals are intermixed or combined to produce sp^n orbitals, where n is the number of p orbitals taking part in the hybridization ($n = 1$, 2, or 3).[3,8,9]

Few important aspects of hybridizations are as follows:

1. Hybridization takes place between the orbitals having almost similar energies and belonging to the same atom. For example, a $2s$ orbital will only hybridize with $2p$ orbital and not with the $3p$ orbital.[3,8]

2. The d orbitals often do not undergo hybridization due to their large size and higher energies than s or p orbitals. However, it can be possible in case when bonding atom is more electronegative than central atom, then electronegative atoms are attracted towards the central atom and the size of d orbital of central atom reduces. For example, PCl_5 and SF_6 correspond to sp^3d and sp^3d^2 hybridization, respectively.[3,8,10]

3. The overlapping of hybrid and non-hybrid orbitals produce σ and π bonds, respectively.[8]

4. The number of hybrid orbitals must be equal to the number atomic orbitals which produce hybrid orbitals. For example, in sp^2 hybridization, three sp^2 hybrid orbitals form through the hybridization of three atomic orbitals (one s and two p).[3,8,9]

5. The electron-pair geometry inferred from VSEPR theory is responsible for the type of hybridization that takes place.[6,7]

6. All hybrid orbitals tend to reach in position of maximum distance between each orbital to minimize the repulsive force and maximize stability.[6]

7. Hybridization results in the formation of orbitals having equivalent energy and therefore, hybrid orbitals are more stable than non-hybridized atomic orbitals.[6-8]

8. Hybridization can take place in the central atom having a single valence electron and a lp of electron. Hence, hybridized orbitals may have both bp as well as lp.[6-8]

2.2.2.3 Types of Hybridization

2.2.2.3.1 sp Hybridization

In this type of hybridization, one valence s orbital and one valence p orbital intermix to form two equivalent half-filled sp hybrid orbitals, which can be utilized for bond formation.[3,8,9] For example, in BeF_2 molecule, the electronic configuration of central atom at lower energy level can be given as $1s^2 2s^2 2p^0$ (Figure 2.12a).[9] However, at the higher energy state, one electron from the 2s orbital occupies the vacant p orbital (Figure 2.12b). Now, one s and one p orbitals with valence electrons hybridized and generate a set of two sp hybrid orbitals with a single electron (Figure 2.12c). These two half-filled sp hybrid orbitals overlapped with the unpaired electron of $3p$ orbital of F atoms for bond formation.[8,9] Example, BeF_2, $BeCl_2$, $HgCl_2$, etc.

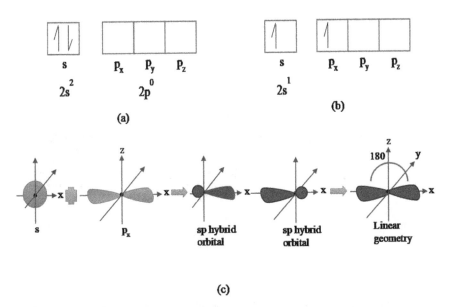

FIGURE 2.12 In the covalent bond formation in BeF_2 molecules, (a) ground-state electronic configuration of the valence shell of the central Be atom, (b) excited state electronic distribution of the valence shell of the central Be atom. (Adapted from J. D. Lee, *Concise Inorganic Chemistry*, 4th edn, Chapman & Hall, 1991, ISBN 0 412 40290 4), (c) intermixing of valence electrons of s and p orbitals to produce two equivalent half-filled sp orbitals which is used to form bonds with two unpaired electrons of the valence shell of fluorine (Adapted from P. Flowers, *Chemistry*, Rice University, ISBN-10 1-938168-39-9).

2.2.2.3.2 *sp² hybridization*

In this hybridization, one valence *s* orbital intermixes with two valence *p* orbitals to form three equivalent half-filled *sp²* hybrid orbitals which are available to overlap with the valence electron of other atoms for bond formation.[3,8,9] For example, in the formation of BF_3 molecules, the electronic configuration of central B atom in the ground state is $1s^2 2s^2 2p^1$ (Figure 2.13a).[9] At the excited state, one electron of $2s$ orbital occupy the space of one vacant $2p$ orbital and electronic configuration becomes $1s^2 2s^1$ $2p^2$ (Figure 2.13b).[9] Now, these one valence $2s$ orbital and two valence $2p$ orbitals were combined together to form three identical half-filled hybrid *sp²* orbitals (Figure 2.13c). These three half-filled *sp²* orbitals overlapped with three valence electrons of $2p$ orbitals of fluorine for bond formation.[8,9] Example, BCl_3, BF_3, BH_3, etc.

2.2.2.3.3 *sp³ hybridization*

In this type of hybridization, one *s* orbital with valence electron/lp combines with three *p* orbitals having unpaired electrons without/with lp and generates four identical *sp³* hybrid orbitals, which are used for bond formation.[3,8,9] The *sp³* hybridization can be divided into two categories, one is *sp³* hybridization without a lp and the other as *sp³* hybridization with a lp. The example of *sp³* hybridization without a lp is CH_4 molecule. Here, the electronic configuration of central carbon atom at lower energy

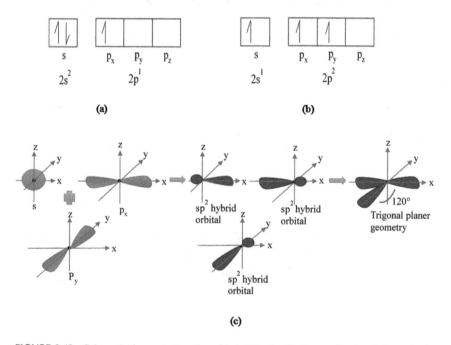

FIGURE 2.13 Schematic demonstrating the *sp²* hybridization during covalent bond formation in BF_3 molecules. Electron distribution of central B atom at (a) ground state, (b) excited state. (Adapted from J. D. Lee, *Concise Inorganic Chemistry*, 4th edn, Chapman & Hall, 1991, ISBN 0 412 40290 4), (c) intermixing of valence electrons of one *s* and two *p* orbitals to produce three equivalent half-filled *sp²* orbitals, which can be utilized for bond formation with three unpaired electrons of the valence shell of fluorine atoms. (Adapted from P. Flowers, *Chemistry*, Rice University, ISBN-10 1-938168-39-9).

level is $1s^2\,2s^2\,2p^2$ (Figure 2.14a). However, at the higher energy level, one electron of $2s$ orbital occupies the vacant space of $2p$ orbital and electronic configuration of the central carbon atom becomes $1s^2\,2s^1\,2p^3$ (Figure 2.14b).[9] Now, one valence s orbital and three valence p orbitals combine together and form four identical half-filled hybrid sp^3 orbitals which further overlapped with unpaired electrons of the valence shell of the s orbital of hydrogen atom for bond formation (Figure 2.14c). The sp^3 hybridization with lp can be observed in H_2O molecules. In this case, the electronic configuration of the central oxygen atom is $1s^2\,2s^2\,2p^4$. Here, two unpaired electrons of $2p$ orbitals form a bond with unpaired valence electron of $1s$ orbital of two different hydrogen atoms, according to the valence bond theory (Figure 2.15a). In this case, the bond angle H–O–H should be 90° due to the mutually perpendicular axes of two p orbitals (Figure 2.15b). However, the experimentally estimated H–O–H bond angle is 104.5°. It can be explained by the principle of hybridization.[3,8,9] According to this principle, in the formation of H_2O molecules, one s and three p orbitals of the valence shell of central oxygen atom undergo hybridization and produce four equivalent sp^3 hybrid orbitals with a tetrahedral geometry (bond angle − 109.5° from the VSEPR theory) (Figure 2.15c).[6,7] Among these four hybrid orbitals, two p orbitals contain an unpaired electron each, which combined with the unpaired electron of the valence shell (one s orbital) of two H atoms to form a bp. However, one $2s$ and one $2p$ orbitals of oxygen

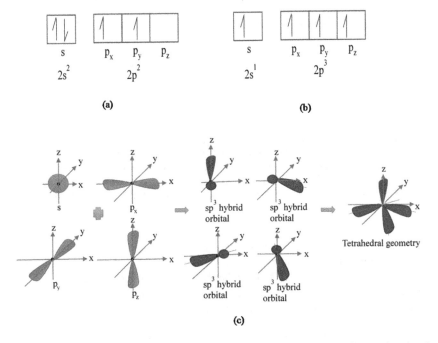

FIGURE 2.14 Schematic illustrating the mechanism of sp^3 hybridization for covalent bond formation in the CH_4 molecule. Electron distribution of the central carbon atom at (a) ground state and (b) excited state, (Adapted from J. D. Lee, *Concise Inorganic Chemistry*, 4th edn, Chapman & Hall, 1991, ISBN 0 412 40290 4), and (c) intermixing of valence electrons of one s and three p orbitals of the central carbon atom to produce four equivalent half-filled sp^3 orbitals, which can be utilized for bond formation with three unpaired electrons of the valence shell of the hydrogen atom. (Adapted from P. Flowers, *Chemistry*, Rice University, ISBN-10 1-938168-39-9).

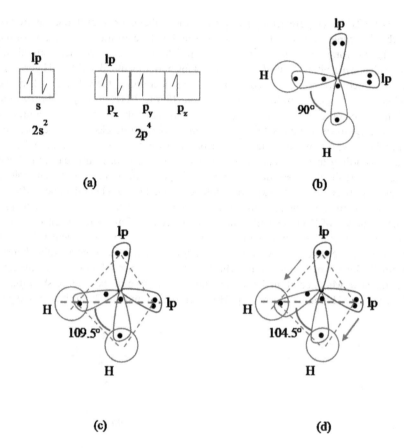

FIGURE 2.15 Schematic demonstrating (a) half-filled and completely filled orbitals in the valence shell of the oxygen atom, (b) bond angle H–O–H is 90° due to the mutually perpendicular axes of p_y and p_z orbitals, (c) according to VSEPR theory, the central oxygen atom is surrounded by two bp and two lp and forms a tetrahedral structure and consequently, the bond angle H–O–H is 109.5°, (d) the bond angle H–O–H becomes 104.5° due to the greater repulsion between lp and lp as compared to lp–bp.[7–9]

atom contain two electrons for each, known as lp. The lp domains contain greater space as compared to the bp domains and therefore, the bond angle is reduced to 104.5° from 109.5° (Figure 2.15d).[7–9] On the other hand, as described earlier, the repulsion between lp and lp is greater than the repulsion between lp-bp and consequently, the lp push the bp in the opposite side and the bond angle H–O–H becomes 104.5°. Example: CH_4, H_2O, NH_3, etc.

The molecules having d orbitals in the valence shell produce five sp^3d (one s, three p, and one d orbitals) hybrid orbitals or six sp^3d^2 (one s, three p, and two d orbitals) hybrid orbitals.[8,11]

For example, the molecules PCl_5, SF_4, ClF_3, etc. exhibit sp^3d hybridization and SF_6, IF_5, and XeF_4 exhibit sp^3d^2 hybridization.

Most of the compounds are formed by the combination of ionic as well as covalent bonding. In the situation, where the bonding is partially ionic and partially covalent, the characteristics of the compound are generally decided by the dominating bond type. The formation of either bond type primarily depends on the electro-negativity difference between the constituent atoms.

A smaller difference in electro-negativity leads to the formation of a covalent bond as the dominating bond type. However, ionic bonding characteristics dominate in the case of a larger electro-negativity difference.

Let us consider two elements A and B, where the electro-negativity of A is very high as compared to B. The ratio of ionic character of the bond in the compound can be evaluated using the expression,

$$\% \text{ Ionic character} = \{1 - \exp[-(0.25)(X_A - X_B)^2]\} \times 100 \qquad (2.19)$$

where X_A and X_B represent the electro-negativities of elements A and B, respectively.[3]

2.2.3 Metallic Bonding

Similar to the covalent bonding, metallic bonds are formed by the sharing of electrons. The difference between these two types of bonding is that the bonding electrons in the covalent bond are "localized" whereas, in metallically bonded solids, the bonding electrons are "delocalized" and are available to drift throughout the solid. It is because of the fact that the valence electrons are insufficient to provide the inert gas configuration for each atom. Therefore, the metallic bonds are non-directional in nature.

The sharing of electrons between two particular atoms is momentary, that is, only for part of time, which results in an incomplete covalent bond. The metallic bonds are also known as unsaturated covalent bond.

Overall, a metallic crystal can be thought of as a distribution of positive ions (nuclei), shielded by the mutual electrostatic forces via free electrons. The free electrons act as a "glue" to hold the entire ion core of opposite nature (Figure 2.16).

In general, all elemental metals consist of metallic bonding.

Because of the availability of free electrons, metals are good conductors of heat and electricity. In contrast, ionically and covalently bonded solids are insulators due to the presence of localized electrons.

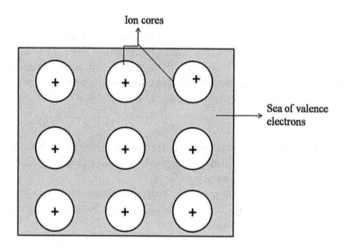

FIGURE 2.16   Schematic representing the formation of metallic bonds.[3]

2.2.4 Secondary Bonds

As discussed above, the primary bonds are interatomic bonds, whereas the secondary bonds are intermolecular bonds.

These bonds are quite weaker than primary bonds. The secondary bonding generally occurs as a result of dipolar interactions. These dipoles are usually formed due to the asymmetric distribution of electron clouds around the nucleus or uneven distribution of charges in asymmetric molecules. These dipoles can be permanent or temporary.

The intermolecular bonds are primarily of two types:

1. Van der Waals bond
2. Hydrogen bond

2.2.4.1 Van der Waals Bond

Consider the case of atoms of the inert gases, which consist of stable and symmetric electronic configuration with respect to positively charged nucleus. They form monoatomic gases. This stable configuration negates the possibility of any interaction with other atoms. However, at quite low temperature, most of the inert gases can be solidified. This is due to the momentary dipole formation, owing to a small fluctuation in the electronic configuration around the nucleus. Such momentary fluctuation is sufficient to induce the formation of dipoles in the neighboring atom. Such interactions can be classified as dipole-induced-dipole interaction.

Overall, the electrostatic interactions among the atoms of inert gases are governed by the momentary non-uniformity in spherical distribution of electrons around the nucleus (Figure 2.17). These atoms interact via a weak Van der Walls bond.

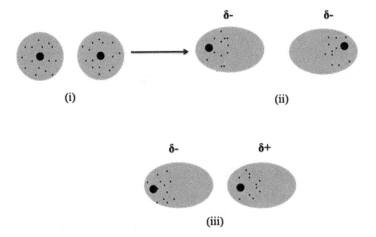

(i)   (ii)

(iii)

FIGURE 2.17 Schematic illustration representing the dipole-induced-dipole interactions due to the momentary shift of electron clouds around the nucleus.[2]

2.2.4.2 Hydrogen Bonding

Let us consider the case of a single water molecule, where the oxygen atom is bonded with two hydrogen atoms via a covalent bond. This means that the atoms in a single H_2O molecule are strongly bonded. Now, consider the case of a number of water molecules. Initially, the bond in a single water molecule is formed by sharing of electrons between oxygen and hydrogen atoms. Because of the higher electronegativity of oxygen than that of hydrogen, the shared electron is pulled towards the oxygen atom. In this practice, the oxygen attains a net negative charge, followed by the corresponding positive charge on hydrogen which gives rise to a permanent dipole moment to water molecules. Now, the bond formed between the negatively charged oxygen of one water molecule to the positively charged hydrogen of another water molecule is called the hydrogen bond (Figure 2.18). These bonds are directional in

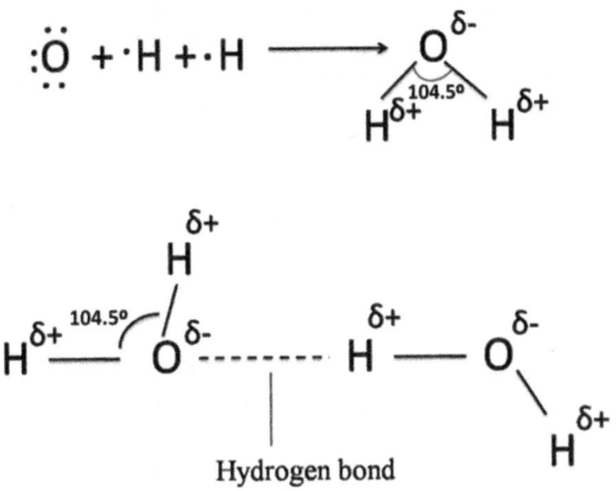

Hydrogen bond

FIGURE 2.18 Formation of the hydrogen bond due to the electrostatic interaction between two water molecules.

nature. Although, these dipolar bonds are relatively weak, they are strong enough to exist in a liquid state.

It is interesting to note that H_2O is liquid while H_2S is gas because of the higher electro-negativity of O than S.

2.3 Correlation between Bonding and Properties of Materials

The response of materials toward external stimuli such as temperature, pressure, mechanical stress, electric field, magnetic field, light, heat and so on is dependent on the respective properties of the materials. The properties of any material depends on a number of factors such as bonding, crystal structure, chemical composition, microstructure, etc.[12] The correlation between bonding and some important properties of materials including mechanical properties, thermal properties, and electrical properties is summarized, as below.

2.3.1 Mechanical Properties

Mechanical properties deal with the behavior of material under the applied load or force. Depending upon the magnitude of the load and characteristics of the material, it may deform plastically or undergo elastic deformation. The bonding affects a number of fundamental mechanical responses such as stiffness, strength, hardness, ductility, and toughness.[3,13]

2.3.1.1 Elastic Behavior/Stiffness

It is known that the elastic deformation is reversible where the stress is proportional to the strain. The slope of the linear region of stress–strain defines the modulus of elasticity.[3] At the atomic level, the basis of the modulus of elasticity comes from the interatomic distance.[14] Microscopic elastic strain results from the stretching of the interatomic bonds and small changes in separation between two adjacent atoms.

The modulus of elasticity (Young's modulus) is directly proportional to the interatomic bonding force.[3]

Higher is the interatomic bonding force, more resistance will be offered in the path of elastic deformation and consequently, higher elastic modulus. Figure 2.1b shows the interatomic bonding force–distance curve for both strongly bonded (primary bond) and weakly bonded (secondary bond) materials.[3] At equilibrium separation r_e, the slope of the curve provides the estimate of elastic modulus of the materials as per Equation (2.5).

From Figure 2.1,[3] it is clear that strongly bonded materials have a sharp curvature which provides a high elastic modulus. Similarly, the shallow curvature of weakly bonded materials provides low value of elastic modulus, as discussed above.

> Materials with a high value of elastic modulus possess high melting point.

2.3.1.2 Strength

The theoretical strength of any material can be determined by the ease of its dislocation movement under shear stress.[15] A strong bonding restricts the dislocation motions which provide strength to the material.[2]

> Strength and modulus of elasticity are correlated to each other on the basis of bond strength. The material with high elastic modulus possesses high strength.

The strongly bonded materials such as ceramics have high strength. Polymers with low elastic modulus possess very low strength.[14] The strength of the metals is the intermediate of the strength of ceramics and polymers.

2.3.1.3 Ductility and Brittleness

Ductility is a measure of the ability of a material to be deformed plastically without fracture under an applied load.[16] A material that has very little or no plastic strain is called a brittle material. The concept of ductility and brittleness for the differently bonded materials can be explained on the basis of dislocation. The extent of relaxation and width of dislocation decide the amount of Peierls–Nabarro stress to move the dislocation. In case of covalent bonding, the dislocation is narrow in nature due to small relaxation. Consequently, Peierls–Nabarro stress is high and the material experience brittle fracture before the movement of dislocation when tensile load is applied.[2] Therefore, ceramic materials possess high strength but very low ductility.[14] The metals are softer than ceramics and bonding is non-directional. In this case, the dislocations are wide and Peierls–Nabarro stress is low.[2] Therefore, the stress required to move the dislocation is low and consequently, the metallic crystals possess good ductility.[14]

2.3.1.4 Hardness

Hardness can be used as a qualitative measurement of strength of the material,[16] as the plastic deformation in materials occurs by motion of dislocation. Any mechanism that tends to make the dislocation motion difficult will aid hardness to the material. Due to localized bonding, covalent and ionically bonded solids make the dislocation movement difficult which provides hardness to such a material. In addition, the availability of a limited number of slip systems also aids hardness.[3] However, the metallically bonded crystals are not as strong as covalently bonded, which makes the slip easier.

2.3.2 Thermal Properties

2.3.2.1 Melting Point

The melting temperature is directly related to the characteristics of the interatomic bond, that is, bonding energy. Atoms must have sufficient amount of thermal energy to break this bond. For higher bonding energy, more thermal energy is required to break the bond and consequently, the melting temperature is higher.[2] In case of primary bonding, covalent and ionic bonds have higher bonding energies than metallic bonds, and therefore, their melting temperature is higher. The materials, in which atoms are interconnected by secondary bonds, have the lowest melting and boiling points, as the lowest amount of thermal energy is required to break a secondary bond.[14]

> The types of bonding in the decreasing order of melting point can be arranged as: covalent bond > ionic bond > metallic bond > van der Walls bond.[2,14]

2.3.2.2 Thermal Expansion

In solids, thermal expansion is the change in the dimensional parameters as a function of temperature.[17] Generally, it quantifies in terms of coefficient of thermal expansion[15] which is used to predict various intrinsic properties of solid materials such as binding force, band and crystal structure, degree of crystallinity, phase transition, and lattice dynamics.[18] In the mathematical form, thermal expansion coefficient can be expressed as follows,[16]

$$\alpha_L = \frac{\varepsilon_L}{\Delta T} \tag{2.20}$$

where ε_L represents the linear strain $\left(\varepsilon_L = \frac{\Delta l}{l_0} \right)$

l_0 and Δl represent the initial length and change in length with respect to change in temperature ΔT, respectively, and the parameter α_L depicts the linear coefficient of thermal expansion.[3] Dilatometer is the most common method to measure the linear coefficient of thermal expansion.[18]

The volume coefficient of thermal expansion (α_v) can also be defined as the change in volume with temperature as,

$$\alpha_v = \frac{\varepsilon_v}{\Delta T} = \frac{\Delta V}{V_0 \Delta T} \tag{2.21}$$

In crystals, thermal expansion is often anisotropic in nature, that is, different in different crystallographic directions.[19] Therefore, linear expansion coefficient is preferred over the volume expansion coefficient. For an isotropic material, $\alpha_v = 3\alpha_L$.[16]

A potential energy curve (Figure 2.19) is the best way to understand the possible mechanism of thermal expansion of solid from an atomistic point of view.

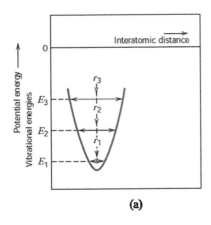 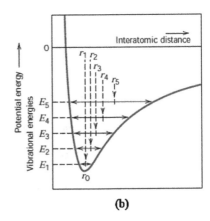

(a)   (b)

FIGURE 2.19 (a) Symmetric potential energy curve as a function of interatomic distance, (b) real potential energy versus interatomic separation plot for a typical molecule. (Reproduced with permission from Callister W. D., D. G. Rethwisch. *Materials Science and Engineering: An Introduction*, John Wiley & Sons, New York, 2007.)

Thermal expansion is the result of an increase in interatomic separation.[16]

In solids, atoms vibrate about their mean position. Upon heating, atoms absorb thermal energy which leads to an increase in the vibrational amplitudes of neighboring atoms. (Figure 2.19a) shows the symmetric intermolecular potential energy curve in which the shape of the curve is parabola. In this case, no net change in interatomic distances will occur with the variation of temperature due to the symmetric shape of parabola and therefore, no thermal expansion will occur.[3,15] In practice, most of the materials expand with heat treatment. Therefore, we must go further to explain the material expansion mechanism.

In reality, potential energy versus interatomic separation curve is not parabolic in nature (Figure 2.19b). The shape is only periodic at very bottom of the curve but when we go upward from the well, the shape tends to depart from the parabolic behavior.[15] When the material is heated from temperature T_1 to T_5, the vibrational energy of atoms increases from E_1 to E_5 (Figure 2.19b). The amplitude of atomic vibration corresponds to the width of the curve at each temperature and average interatomic separation increases, with successive temperature, from r_1 to r_5.[3,12]

The magnitude of linear coefficient of thermal expansion is directly proportional to the temperature. The materials, which have higher bonding force or higher bonding energy, represent narrow and deeper potential energy trough.[12] For such materials, the increase in interatomic separation with an increase in temperature is smaller, and consequently, the lower thermal expansion coefficient. For example, covalent crystals such as, diamond and other ceramic materials have very strong bonding and therefore, they possess lower coefficient of thermal expansion than the metallic crystals. For the materials having lower bonding energy, the increase in the interatomic distance with an increase in temperature is larger, yielding higher thermal expansion (Figure 2.20). For example, polymers show quite large thermal expansion upon heating due to the presence of secondary bonds.

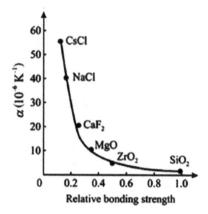

FIGURE 2.20 Variation of the thermal expansion coefficient with relative bonding strength for different classes of materials. (Reproduced with permission from Poplavko Y. M. *Electronic Materials: Principles and Applied Science*, 2019, Chapter 3, pp. 95–120.)

2.3.2.3 Thermal Conductivity

The rate, at which heat transferred through the material, is measured by thermal conductivity.[16] The transfer of heat takes place due to non-uniform distribution of temperature.[19] It is a microstructure-sensitive and temperature-dependent property and can be defined as,[3]

$$q = -k \frac{dT}{dX} \tag{2.22}$$

where q indicates the heat flow or heat flux per unit time per unit area (perpendicular to the direction of flow), the parameter k is called the thermal conductivity, dT/dX denotes the temperature gradient, and the minus sign indicates that heat is flowing from the hotter part to the cooler part in the material. This equation is only valid for steady-state heat flow.

The mechanism of thermal conductivity is different for each class of material, namely, metals, ceramics, polymers, and so forth. In metals, there are two mechanisms for transfer of heat (i) free electrons and (ii) phonons. The total thermal conductivity is the addition of electronic and photonic contributions as follows,

$$k = k_e + k_p \tag{2.23}$$

where k_e and k_p indicate the electron and phonon thermal conductivity.[15] As the metal is heated, the valance electrons take thermal energy and migrate toward the cooler area to achieve the thermal equilibrium and transfer their kinetic energy to other atoms, as a result of collision between moving electrons and phonons. The amount of energy transfer is dependent on the free electron concentration and their mobility. In addition, phonons also give their contribution during the transfer of heat through metals.[16] Metals have a large number of free electrons and their electronic contribution is predominant over the phononic contribution because electrons have a higher mobility and are not easily scattered in comparison to phonons. Therefore, metals usually have higher thermal conductivity.[3]

In the case of ceramics, there are fewer free electrons due to strong covalent bonding. Hence, phonons play a major role in conducting heat in ceramic materials. For perfect harmonic lattice, phonons transfer the heat without any resistance. In this case, thermal conductivity would be infinite. Since, phonons are scattered easily by lattice imperfections, ceramics must have some heat flow resistance. In this case, the thermal conductivity is not infinite.[15] The thermal conductivity of ceramics is much smaller than that of metals due to the absence of electronic contribution. On increasing the temperature, the thermal conductivity diminishes because the mean free path decreases due to the increased probability of scattering of phonons. This occurs at relatively low temperature. But at higher temperature, the conductivity increases due to the transfer of radiant heat.[3,15]

In the case of polymers, thermal conductivity is quite low, that is, of the order of 0.3 W/m K for most of the polymers. This type of material transfers thermal energy by vibration and rotation of polymeric chains.[3] A high level of degree of crystallinity is responsible for high thermal conductivity because the crystalline state enhances the vibration of polymer chains. Moreover, higher degree of polymerization, lower branching, and greater cross-linking are also preferred to achieve the high thermal conductivity.[16]

2.4 Closure

The fundamental difference between various class of materials such as metals, ceramic, and polymers comes from their bonding characteristics. In this perspective, this chapter initially described the conditions for the formation of bonds between two atoms. Further, the basic mechanisms, involved in the formation of bonds, have been elaborately discussed. Towards the end, a correlation between the various properties of materials, such as, mechanical, physical, thermal, and so on, and bonding has been established.

REFERENCES

1. Verma A. R., O. N. Srivastava, *Crystallography Applied to Solid State Physics*, 2nd edn, New Age International (P) Ltd., 1991.
2. Raghavan V., *Materials Science and Engineering*, 5th edn, Prentice-Hall of India Pvt. Ltd., 2004.
3. Callister W. D., D. G. Rethwisch, *Materials Science and Engineering: An Introduction*, John Wiley & Sons, New York, 2007.
4. Kittel C., *Introduction to Solid State Physics*, 8th edn, John Wiley & Sons, Inc, 2005.
5. Sidgwick N. V., H. M. Powell, Bakerian lecture. Stereochemical types and valency groups, *Proc. Roy. Soc. A.*, 1940, 176(965), 153–180.
6. Gillespie R. J., R. S. Nyholm, Inorganic stereochemistry, *Q. Rev. Chem. Soc.*, 1957, 61, 339.
7. Gillespie R. J., E. A. Robinson, Models of molecular geometry, *Chem. Soc. Rev.*, 2005, 34, 396–407.

8. Flowers P., *Chemistry*, Rice University, ISBN-10 1-938168-39-9.

9. Lee J. D., *Concise Inorganic Chemistry*, 4th edn, Chapman & Hall, 1991, ISBN 0 412 40290 4.

10. Jain V. K., The concept of hybridization: A rapid and new innovative method for prediction of hybridized state of an atom in a very short time, *Indian J. Appl. Res.*, Mar 2014, 4, 3, ISSN - 2249-555X.

11. Galbraith J. M., On the Role of d Orbital Hybridization in the Chemistry Curriculum, *J. Chem. Educ.*, 2007, 84, 5, 783. https://doi.org/10.1021/ed084p783

12. Padmavathi D. A., Potential energy curves; Material properties, *Mat. Sci. Appl.*, 2011, 2, 97–104.

13. Courtney T. H., *Mechanical Behavior of Materials*, 2nd edn, Waveland Press, Inc., Long Grove, Illinois, 2000.

14. Banerjee S., J. K. Chakravartty, H. B. Singh, R. Kapoor, Super-strong, super-modulus materials, in *Functional Materials*, eds. S. Banerjee, A. K. Tyagi, Elsevier, Waltham, Massachusetts, 2012, pp. 467–505.

15. White M. A., *Physical Properties of Materials*, 2nd edn, CRC Press, Taylor & Francis Group, 2011.

16. Askeland D. R., P. P. Phule, W. J. Wright, *The Science and Engineering of Materials*, Cengage Learning, Boston, Massachusetts, 2011.

17. Collins J. G., G. K. White, Chapter IX thermal expansion of solids, *Prog. Low. Temp. Phys.*, 1964, 4, 450–479.

18. Ventura G., L. Risegari, Measurement of thermal expansion, in *The Art of Cryogenics*, Elsevier, Cambridge, Massachusetts, 2008, Chapter 13, pp. 289–296.

19. Poplavko Y. M., Thermal properties of solids, in *Electronic Materials: Principles and Applied Science*, 2019, Chapter 3, pp. 95–120.

3

Advanced Bonding Theories for Complexes

In continuation to our earlier discussion on electronic configurations and bonding, this chapter discusses the theoretical advances made to understand the nature of bonding in complex molecular systems. The modifications in the electronic orbital structure due to the formation of cation–ligand complexes often involve considerable electrostatic interactions between the negatively charged ligands and the electrons of the central cationic species. Such fundamental structural changes, the origin of which is primarily the lowest energy configuration (with respect to the distribution/occupancy of electrons), have a significant impact on technologically important properties (such as, optical, magnetic, electrochemical).

3.1 The Valence Bond Theory

The discussion on the types of bonding involved in the formation of cation–ligand complexes has to be initiated with a brief recap of the "valence bond theory;" some preliminary aspects and rules of which were previously discussed in Section 2.2.2.2, while introducing the concept of hybridization. The "valence bond theory" was postulated and subsequently refined by a number of scientists in the early part of the nineteenth century. As mentioned earlier, the formation of covalent bonds via sharing of electron pairs was put forward by Gilbert N. Lewis in the year 1902. Later, in 1927, Walter Heitler and Fritz London put forward more advanced theories and models that provide the concept toward the stability of covalent molecules via sharing electron pairs. The Heitler–London model thus, eventually, established the foundation of the "valence bond theory."

For a more refined and widely accepted version of the valence bond theory, the primary credit goes to Linus Pauling, who put forward the idea behind the overlap of the associated atomic orbitals or in more precise terms, "mixing" of the atomic orbitals, to form what is known as the "hybrid orbitals" (as has been introduced in Section 2.2.2.2). The concept of "hybridization" will be further elaborated here from slightly different perspectives, with the aim being to move forward toward more advanced theories concerning the bonding characteristics and the resultant orbitals of cation–ligand complexes.

> It may be mentioned upfront here that these hybrid orbitals, in turn, allow the formation of bonds for various "cation–anion" complexes; with the more electropositive atom ("cation") being usually (but not always) denoted as the central atom. The type of hybridization also influences the shape of the as-formed molecule.

The typical hybridized orbitals include sp, sp^2, sp^3, dsp^3, and d^2sp^3 orbitals, which need not be the "pristine" atomic orbitals (viz., $1s$, $2s$, $2p$, $3s$, $3p$, $3d$, $4s$, $4p$, and so on). Rather, the concerned atomic orbitals or the hybridized/hybrid orbital may have combinatorial characteristics of the different kinds of "pristine" atomic orbitals. In fact, the hybridized orbitals are usually at a lesser energy level, as compared to some of the associated "pristine" atomic orbitals (see Figure 3.1). Here it must be noted that an atomic orbital is manifestation of the energy state of electrons, which are bound to just one (or the associated) atomic nucleus. This can change when the same electron is under the force field of another atomic nucleus, that is, when the concerned atom gets bonded to another. Hence, one needs to look beyond the "pristine" energy states or the atomic orbitals to understand the theory of bonding.

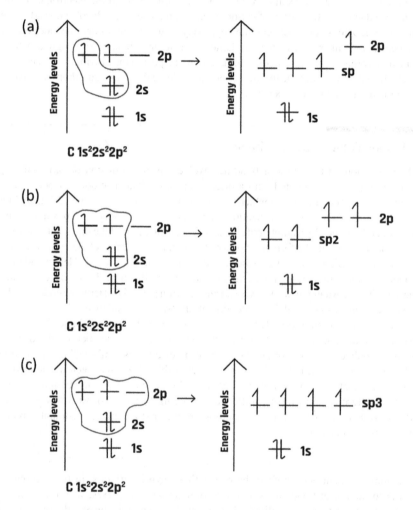

FIGURE 3.1 The concept and associated energy levels of the hybridized orbitals (as compared to the energy levels of the pristine atomic orbitals in the ground state) for (a) sp hybridization, as in acetylene (or C_2H_2), (b) sp^2 hybridization, as in ethylene (or C_2H_4) and (c) sp^3 hybridization, as in methane (or CH_4). (Adapted from 'The Molecular Orbital Model' – Chemistry LibreTexts').

To better explain the above considerations or the valence bond theory, let us first take the example of the simplest case of *sp* hybrid orbital. As the name immediately suggests, it is a combination of *s* and *p* atomic orbitals. In more specific terms, usually, two *sp* hybridized orbitals are formed by the combination of the $2s$ and one of the $2p$ "pristine" atomic orbitals with each of the *sp* orbitals consisting of 50% *s* and 50% *p* character. The energy levels of the *sp* hybridized orbitals are typically above the energy level of the pristine *s* orbital, but below the pristine *p* orbital (see Figure 3.1a). These hybridized *sp* orbitals possess a region of high electron density, with the two orbitals centered on the concerned atom and located almost exactly opposite to each other (i.e., 180° apart w.r.t. the central atom) in the form of lobes (see Figure 2.12). As denoted, this leads to a linear configuration for a molecule developed by forming bonds with *sp* hybrid orbitals.

Let us take a specific example of Be (atomic number: four), which is the first element in the periodic table, that forms *sp* hybridized bonds. The ground state or "pristine" configuration of Be is $1s^2 2s^2$. It has to form bonds with two H atoms to form BeH_2. Since in the ground state, the $2s$ orbital is fully filled, it cannot accommodate another electron (from H). Hence, to form the bond, the $2s$ pristine orbital of Be will get hybridized with one of the $2p$ orbitals (i.e., p_x) of the same (as mentioned above) to form two *sp* hybridized orbitals, now each having one unpaired electron from Be to host another electron. The single $1s$ electron in each H will get shared with the 1 *sp* electron in Be to form two Be–H bonds, thus forming a linear H–Be–H covalent molecule. The case of $BeCl_2$ is also similar, with the single valence unpaired $3p$ electron of Cl (having ground-state electronic configuration of $1s^2 2s^2 2p^6 3s^2 3p^5$) being shared with each of the single *sp* electron of Be, thus forming the Cl–Be–Cl linear molecule (or $BeCl_2$).

Now let us consider *sp²* hybridization. In this case, one $2s$ orbital and two of the three $2p$ orbitals hybridize to form a total of three hybridized orbitals. Each of the three hybridized orbitals has 67% of *p* and 33% of *s* character; hence, the name *sp²*. In contrast to the linear character of the *sp* hybrid orbitals, the frontal lobes of the *sp²* hybrid orbitals point toward the corners of an equilateral triangle from the central atom, making an angle of 120° to each other. This minimizes the electron repulsion and hence reduces the energy for such configuration. Hence, *sp²* hybridization leads to a trigonal planar shape of the molecule. With regard to the energy level, once again, the *sp²* hybrid orbitals are at a level only slightly above the pristine $2s$ orbital, but at a lower level as compared to the pristine $2p$ orbitals (see Figure 3.1b). In order to substantiate further, let us take a simple case as an example, namely, now that of BF_3 (i.e., the next element after Be in the periodic table forming the central atom). The electronic configuration in the ground state of B (having atomic number 5) is $1s^2 2s^2 2p^1$. As mentioned above, the $2s$ and two of the $2p$ orbitals (i.e., p_x^1 and p_y) will hybridize to form the *sp²* hybrid orbital, each having one unpaired electron. These unpaired electrons in each of the three *sp²* orbitals get shared with one electron each from the unpaired fifth electron in the $2p$ orbital of three F atoms (viz., p_z^1; having ground-state electronic configuration of $1s^2 2s^2 2p^5$); thus forming the BF_3 covalent molecule having a trigonal planar structure around the central B atom (see Figure 2.13).

To explain the *sp³* orbital, we can consider the very next atom (i.e., after B) in the periodic table, namely, C, having atomic number 6 and ground-state electronic configuration of $1s^2 2s^2 2p^2$. It is to be noted that this has two unpaired electrons in the *p* orbital and a total of four valence electrons available for bonding, unlike the cases of Be and B shown earlier. Based on the picture of hybridization, as provided above, now four hybrid orbitals are needed to place four electrons (unpaired state) at the same energy level (i.e., degenerate). This is achieved by hybridization between $2s$ orbital and all the three $2p$ orbitals (i.e., p_x^1, p_y^1, and p_z) to form four sp^3 hybrid orbitals, each having 75% p and 25% s character and energy slightly lower than $2p$ orbitals in the ground state (see Figure 3.1c). Again in the simplest and very common case of C–H bonding, this sp^3 hybridization of C is the basis for the formation of CH_4 or methane. In the case of sp^3 hybridized orbitals, the unpaired electron of C in each of the four degenerate sp^3 lobes extend an angle of 109.5° with each other. This results in a tetrahedral configuration (see Figure 2.14), thus forming four covalent C–H single bonds with four H atoms around the one C atom. It needs to be mentioned here that in the case of sp^3 hybridization, some molecules may be formed by utilizing all the four hybridized orbitals of the central atom for bonding purposes. For example, in the case of H_2O, that is, H–O–H, only two of the four sp^3 hybrid orbitals of O, having altogether six electrons in the L level (ground-state atomic configuration: $1s^2\ 2s^2\ 2p^4$), are used, with two lone pairs not being used. These non-bonding electrons in O (viz., in the "3rd and 4th" unused sp^3 orbital) force the two O–H bonds to make ~105°, instead of the typical sp^3 bond angle of 109.5° due to lone pair–bond–pair repulsion. On a similar note, even ammonia or NH_3 with three covalent bonds with H also has sp^3 hybridization in the central N atom. However, with five electrons in the L level (ground-state atomic configuration: $1s^2\ 2s^2\ 2p^3$), one lone electron pair (or non-bonding electrons) remains in the "4th" unused sp^3 orbital.

In the context of the C–H bond, now let us consider ethylene, which has a chemical formula of C_2H_4. It is interesting to note that the same carbon atom, with the same number of valence electrons, now seems to be bonding with two, instead of three H (as in methane), per carbon atom. In more specific terms, the bonding in ethylene can be represented in simple terms as $H_2C=CH_2$. Nevertheless, even in this structure, each C appears to be still forming four bonds, namely, two with two H atoms and two with the other C atom in the molecule. The difference here is that, instead of all the three pristine $2p$ atomic orbitals of C, only two of them get hybridized with the $2s$ of C, forming sp^2 hybridization. The third p orbital still contains the 4th unpaired electron, but at a slightly elevated energy level as compared to the degenerate hybridized sp^2 orbitals (see Figure 3.1b). Therefore, the unpaired electrons in the three sp^2 orbitals of each of the C atoms form three covalent σ bonds spaced at 120° with respect to each other around the C atom. Again, two of them have the unpaired $1s$ electrons of two H atoms and the third with the unpaired third sp^2 electron of the other C atom. This makes the overall shape of ethylene planar, as opposed to tetragonal for methane, with the unpaired (lone) electrons in the p orbitals of each of the C atoms forming π bond with each other in between the two C atoms. If we consider that the $sp^2\sigma$ bonds lie in the x–y plane, the π bonds will be on an orthogonal plane, with the plane having the z-axis. Now, with a further reduction in the H:C ratio to 1:1, namely, in the case of acetylene (C_2H_2), each C atom will be having sp hybridized orbitals forming two σ bonds, one with H and the other with the other C, with the two remaining unpaired electrons in the pristine p_y and p_z orbitals forming two π bonds between the C atoms. Typical of sp hybridization, this leads to a purely linear structure.

Going beyond the three basic hybridizations (as mentioned above and also discussed to some extent in Chapter 2) and now involving d atomic orbitals, let us now look at the concept of dsp^3 hybridization. In this case, as the name suggests, one $3d$, one $3s$, and three $3p$ atomic orbitals hybridize to form five dsp^3 hybrid orbitals. For example, the first element in the periodic table having one electron in each of the $3p$ orbitals, namely, P($1s^2\ 2s^2\ 2p^6\ 3s^2\ 3p^3$) forms a compound $PClF_4$ by bonding with the one unpaired

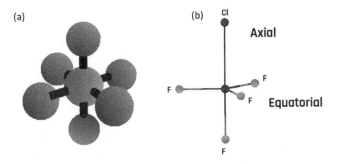

FIGURE 3.2 Ball and stick representations of the geometries for (a) d^2sp^3 and (b) dsp^3 hybridizations.

electron in $3p_z$ orbital of one Cl atom and one unpaired electron in the $2p_z$ orbital in each of the four F atoms via the five dsp^3 hybrid orbitals in a "trigonal bi-pyramidal" configuration around the central P atom (see Figure 3.2).

The next higher level of hybridization will be d^2sp^3, which involves hybridization between two $3d$, one $3s$, and three $3p$ atomic to form six hybrid orbitals forming an octahedral configuration around the central atom upon bonding to form a molecule. The first element in the periodic table that is known to form d^2sp^3 hybridized orbitals is S ($1s^2$ $2s^2$ $2p^6$ $3s^2$ $3p^4$). For example, in the case of compound SF_6, the six d^2sp^3 hybridized orbitals of each S will form a bond each by sharing the one unpaired electron in the $2p_z$ orbital in each of the six F atoms around the central S atom (see Figure 3.2).

Let us now look at a couple of examples for more complex situations. These involve transition metals as the central atom that bond with non-elemental ligands, necessitating "Valence bond theory" or hybridization. In the case of the coordination complex, $Co(NH_3)_6^{3+}$ ion, the six NH_3 molecules bind, as ligands, around the central Co^{3+} having ground-state electronic configuration of $1s^2$ $2s^2$ $2p^6$ $3s^2$ $3p^6$ $4s^0$ $3d^6$ $4p^0$. If we consider that the six $3d$ electrons get paired in the d_{xy}, d_{xz}, and d_{yz} orbitals, that leave two $3d$ orbitals (viz., d_{x2-y2}, d_{z2}), one $4s$ orbital, and three $4p$ orbitals available to form six d^2sp^3 hybridized orbitals. Note that in this case, unlike the simpler system discussed above, all the pristine atomic orbitals involved in the d^2sp^3 hybridization do not have the same principle quantum number. Nevertheless, the six d^2sp^3 hybridized orbitals in the case of the $Co(NH_3)_6^{3+}$ coordination complex will form six bonds having an octahedral geometry around the central Co^{3+} cation by each hosting a pair of nonbonding electrons (see above) from one neutral sp^3 bonded NH_3 molecule (see Figure 3.3a). Such complexes are known as "inner shell" complexes, because of the usage of an inner shell (viz., two $3d$) for the hybridization.

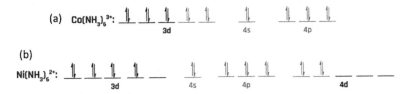

FIGURE 3.3 Representations of the (a) d^2sp^3 hybridization scheme for $Co(NH_3)_6^{3+}$ and (b) sp^3d^2 hybridization scheme for $Ni(NH_3)_6^{3+}$ complexes; with the lighter shade denoting the hybridized orbitals in both the cases. (Adapted from the 'Bodner Research Web', hosted by Purdue University, USA).

We can look at a still more complex system, namely, that of $Ni(NH_3)_6^{2+}$ ions, the same six NH_3 molecules bind, as ligands, now around a central Ni^{2+} ion having a ground-state electronic configuration of $1s^2\,2s^2\,2p^6\,3s^2\,3p^6\,4s^0\,3d^8\,4p^0$. Therefore, in the $3d$ orbital in the ground state, now there are already eight electrons (instead of six, as in the previous case), which will not make two $3d$ orbitals available for the d^2sp^3 hybridization. The way around this is to invoke the usage of only N (i.e., principle quantum number of 4) orbitals for the hybridization, namely, the two $4d$ (i.e., $4d_{x2-y2}$ and $4d_{z2}$), one $4s$, and three $4p$ (i.e., $4p_x$, $4p_y$, and $4p_z$) orbitals for forming the six d^2sp^3 hybrid orbitals (or, in more correct terms, sp^3d^2), each of which accept the lone electron pair from each of the six NH_3 molecules to form the six bonds having octahedral geometry (see Figure 3.3b). Note that, unlike in the previous case, "this complex with unavailable inner shell" for hybridization may be termed as "inner shell" complex.

3.2 The Molecular Orbital Theory

As has been discussed thoroughly above and previously, the electrons in atoms are placed in atomic orbitals. In contrast, for the molecules, which were formed by some combination of the individual atoms, the electrons are placed in molecular orbitals, which are, in turn, formed by the combination of the corresponding atomic orbitals. If the atomic orbitals are considered as wave functions ψ_{ao}, that are solutions of the Schrödinger wave equation, as per "Linear Combination of Atomic Orbitals (or LCAO)," the wave functions for the molecular orbitals (ψ_{mo}) can be obtained by a linear combination of the wave functions corresponding to the associated atomic orbitals (viz., by addition and subtraction of the ψ_{ao}s). In other words, the ψ_{mo}s for a molecule formed by the combination of one atomic orbital, each atom A and B can be represented as $\psi_{aoA} \pm \psi_{aoB}$.

> Hence, the combination of two atomic orbitals (viz., one from each atom) leads to the formation of two molecular orbitals.

The molecular orbital formed upon addition of the wave functions of the two atomic orbitals (viz., $\psi_{aoA} + \psi_{aoB}$) is known as the "Bonding Molecular Orbital" (BMO; represented as σ or π, as the case may be), whereas, the molecular orbital formed upon subtraction of the wave functions of the two atomic orbitals (viz., $\psi_{aoA} - \psi_{aoB}$) is known as the "Anti-bonding Molecular Orbital" (ABMO; represented as σ^* or π^*, as the case may be). The bonding (BMO) and anti-bonding (ABMO) orbitals are at different energy levels, with the bonding and anti-bonding orbitals having lower and higher energy, respectively, with respect to the corresponding atomic orbitals (see Figure 3.4).

> Hence, during filling-up of the molecular orbitals (electrons being added one at a time), the BMOs get filled-up before the ABMOs, in accordance with Aufbau Principle.

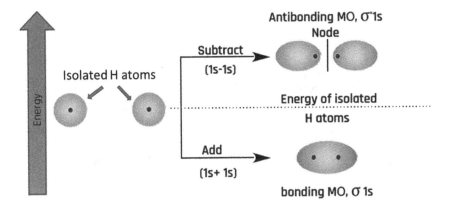

FIGURE 3.4 Energy levels of the bonding and anti-bonding molecular orbitals (MO) with respect to the energy level of the corresponding atomic orbitals of the free atoms in the ground state. (Adapted from 'Bonding and antibonding orbitals' – Chemistry LibreTexts).

The reason for the lower energy level of the BMO is that an increase in the probability of finding the electrons in between the two atomic nuclei leads to enhanced attraction. In contrast, the electrons in ABMO try to move away from in-between the two nuclei, causing greater probability toward the other sides; hence leading to repulsion between the two atoms (see Figure 3.4). Hence, electrons in BMO favor the formation of bonds between the two orbitals, whereas those in ABMO make bonding unfavorable; as also can be inferred based on the relative energies (with BMO having greater stability even as compared to the associated atomic orbitals). If, for reasons involving symmetry or energy aspects, the "mixing" of an atomic orbital with an atomic orbital of the other associated atom does not take place, a "Non-bonding Molecular Orbital" (NBMO) is created. The energy level of a NBMO is usually unaffected by the bonding and is at a level similar to that of the corresponding atomic orbital; namely, in between that of the valence shell BO and the corresponding ABMO. The occupation of NBMO by electrons does not alter the bond order (and hence, stability) among the associated atoms. In many cases, a NBMO containing electrons becomes the highest occupied molecular orbital (HOMO) in the concerned molecule. Some of the examples for non-bonding electrons include electrons in the oxygen ($2p$) non-bonding states and also, sometimes, those in the non-bonding d orbitals of the transition metal compounds (primarily, transition metal oxides), which are used for many important technological applications, including in electrochemical energy storage. Overall, the anti-bonding and non-bonding orbitals of transition metal compounds influence the characteristics of the bonds, and, concomitantly, their structure and properties. A few examples of these effects, in the context of application in electrochemical energy storage, will be mentioned in Chapter 13.

The molecular orbitals formed by combinations of atomic s orbitals are known as σ (or σ^*) molecular orbitals (or σ bonds) because they look more or less like an s orbital, when viewed along the bond axis (see Figure 3.4). For similar reasons, one of the molecular orbitals formed by combining one of the sets of p atomic orbitals (say, p_z orbitals) is also called σ (or σ^*) molecular orbital (see Figure 3.5a). In contrast, combinations of the other two orthogonal sets of p atomic orbitals (viz., p_x and p_y) appear somewhat like the p orbitals and hence are known as π (or π^*) molecular orbitals (see Figure 3.5b).

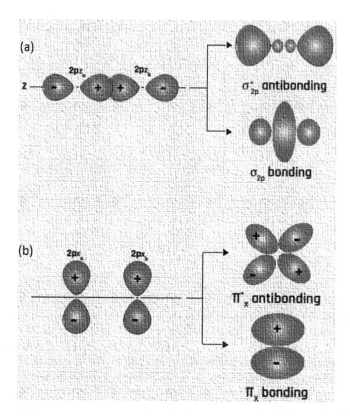

FIGURE 3.5 Schematic representation of the geometries of the molecular orbitals formed upon combination of the atomic (a) p_z and (b) p_x orbitals, if molecular axis lies along the Cartesian z-axis. (Adapted from the 'Bodner Research Web', hosted by Purdue University, USA).

Unlike σ (and σ^*) orbitals, which concentrate electrons along the bond axis, π (and π^*) orbitals concentrate the electrons either above or below the axis on which the atomic nuclei lie (i.e., bond axis). For any given principle quantum number, σ molecular orbitals formed due to the favorable combination of the atomic s orbitals have the lowest energies compared to those formed due to the combination of the atomic p orbitals. Now among the molecular orbitals that form due to the combination of the atomic p orbitals, since p_z orbitals meet head-on, they also lead to the formation of σ molecular orbitals. As a consequence, the interaction is stronger, as compared to the interaction between the otherwise degenerate p_x or p_y orbitals, which meet edge-on (forming the π molecular orbitals). The π (or π^*) molecular orbitals formed due to the combination of the $2p_x$ or $2p_y$ orbitals are themselves degenerate. Hence, in many cases (especially, diatomic molecules, such as O_2, F_2, etc.), the energy levels of the molecular orbitals, or the sequence of filling of electrons in the molecular orbitals, follow the following sequence (i.e., from left to right): σ_{1s}, σ^*_{1s}, σ_{2s}, σ^*_{2s}, σ_{2pz}, $\pi_{2px} = \pi_{2py}$, $\pi^*_{2px} = \pi^*_{2py}$, σ^*_{2pz}, and so on. In a few cases (like for B_2, C_2, N_2), the interaction between the $2s$ orbital of one atom and $2p_z$ orbital of the other also introduces s–p hybridization into the concept of molecular orbitals. Therefore, the sequence is a bit different, namely, the degenerate π_{2px} and π_{2py} have lower energy, as compared to the σ_{2pz}.

> The relative stabilities of molecules, along with whether formation of a molecule from concerned atoms are favorable or not, can be inferred from the "Bond order" (BO).

Numerically the "Bond Order" of a molecule is one half of the difference between the number of electrons present in the bonding molecular orbitals and the antibonding molecular orbitals in a given molecule (viz., BO = ½ [no. of electrons in BMO − no. of electrons in ABMO]). Here, a positive BO, meaning more number of electrons located in BMO, as compared to ABMO, signifies a stable molecule (w.r.t. the associated atoms) or the fact that the formation of the molecule is energetically favored. In contrast, both zero and negative values for the BO implies that the concerned molecule is not stable and is not likely to form from the concerned atoms. Additionally, a greater positive value of the bond order indicates greater bond dissociation energy and a shorter bond length.

To start, let us take a very simple example of the formation of a H_2 molecule. Each atomic H has only one electron in the $1s$ orbital. The combination of $1s$ orbital of each of the atomic H will create two molecular orbitals, namely, σ_{1s} and $\sigma*_{1s}$. Again, going by the Aufbau principle of filling orbitals, the two electrons contributed by the two H atoms will take up only the σ_{1s} molecular orbital, with the $\sigma*_{1s}$ remaining vacant. Hence, in molecular H_2, the electrons reside at even lower energy level (viz., σ_{1s}), as compared to $1s$ in the atomic hydrogen (see Figure 3.4). This indicates that the formation of molecular H_2 is energetically favorable. The same is also inferred numerically from the bond order, which in this case is ½ (2 − 0) = 1. We apply the same concept toward forming a hypothetical di-atomic He_2 molecule. Two He atoms, each having ground state electronic configuration of $1s^2$, will lead to filling of the σ_{1s}, as well as the $\sigma*_{1s}$ molecular orbitals. This will lead to a bond order of ½ (2 − 2) = 0, indicating that the formation of diatomic He_2 molecule is not energetically favorable; which indeed is the case.

In the case of di-atomic O_2 molecule (see Figure 3.6), with the atomic orbital of each O atoms having a ground state electronic configuration of $1s^2\,2s^2\,2p^4$, we need to consider only the valence shell electrons ($2s^2\,2p^4$ electrons). On combining the two O atoms, the $2s^2$ electrons in each of the atoms will go into the σ_{2s} and $\sigma*_{2s}$ molecular orbitals. The four $2p$ electrons on each of the O atoms first fill-up the σ_{2pz} orbital and then the degenerate π_{2px} and π_{2py} orbitals (viz., the sequence of increasing energy; as mentioned above). After that, each electron takes up the degenerate antibonding $\pi*_{2px}$ and $\pi*_{2py}$ orbitals (singly occupied). Note that the electrons in these molecular orbitals are unpaired (viz., $\pi*_{2px}$ and $\pi*_{2py}$ orbitals are singly occupied), because the filling of orbitals at the same energy level follows the Hund's rule (i.e., electrons are first placed singly in degenerate orbitals and pairing takes place only after each degenerate orbital consists of one electron). In this case, the bond order is ½ (8 − 4) = 2, indicating that formation of diatomic O_2 molecules is energetically favorable; again, which indeed is the case. Another point to note here is that, due to the presence of unpaired electrons in the molecular orbitals of O_2 (i.e., in the $\pi*_{2px}$ and $\pi*_{2py}$ orbitals), O_2 is paramagnetic; namely, attracted to magnetic field (which liquid O_2 indeed is). In contrast, N_2 (electronic configuration in atomic state: $1s^2\,2s^2\,2p^3$), having molecular orbital electronic configuration of $\sigma_{2s}^2\sigma*_{2s}^2\pi_{2px}^2 = \pi_{2py}^2, \sigma_{2pz}^2$, has a bond order of 3 (viz.,½ [8 − 2]) and is diamagnetic (i.e., repelled by magnetic field), because the molecular orbitals do not have any unpaired electron.

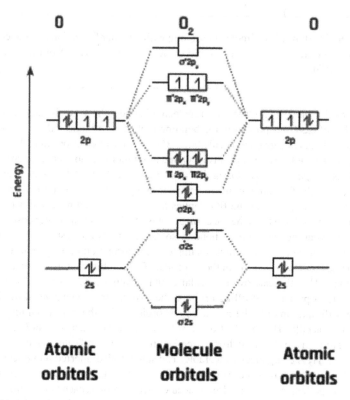

FIGURE 3.6 Molecular orbitals of di-atomic O_2 molecule, as filled with the associated electrons. (Adapted from the web of Institute for Advanced Studies [IFAS], India).

3.3 The Crystal Field Theory

Before discussing the crystal field theory in more specific terms, it may be recalled here that the formation of a complex is one of the manifestations of Lewis acid–base reactions (i.e., $A + :B \rightarrow A - B$); namely, the formation of a covalent bond between a species that can supply an electron pair (i.e., Lewis base) and a species that can accept the electron pair (i.e., Lewis acid). In the case of transition metal complexes, such as $[M(Lig)_6]^{n+}$, the central metal ion (viz., M; a transition metal) may be considered as the Lewis acid and the Lig (viz., ligand) as the Lewis base, possessing a lone pair of electrons.

Usually, metal ions are capable of acting as electron acceptors (i.e., as Lewis acid) to several ligands, with the number of ligands that may attach to a central metal ion being partly controlled by the efficacy of spatial packing arrangement of the ligands around the metal ion.

Now returning to the main topic here, that is, crystal field theory (CFT), it may be simply stated that CFT provides a formulation concerning the breaking of degeneracy of the non-bonding *d* orbitals of the concerned transition metal in such transition metal complexes under the influence of the ligands surrounding the same. Such breakage of degeneracy of the *d* orbitals affects the strength of the metal–ligand bonds, the overall energy/stability of the complex, and also the spin state (viz., pairing) of the electrons occupying the *d* orbitals.

> With respect to the properties, such an influence of the ligands on the degeneracy of the transition metal *d* orbitals (and concomitantly on the electron pairing/spin states), in turn, impacts the magnetic properties, electrochemical properties, and the color.

In a more generic sense, when ligands approach a transition metal ion, their electrons experience varying "opposition" from the *d*-orbital electrons based on the structure of the concerned ligand molecule. In other words, there is interaction between the electrons in the *d* orbitals of the transition metals and those on the ligand molecules. Now, considering that the ligands approach from different directions (perhaps symmetrically around the central metal ion), due to the non-uniform spatial distribution/arrangement, of the transition metal *d* orbitals, not all of them, interact directly with the ligands. Hence, these varying levels of electrostatic interactions between the electrons in the ligand molecules and in the *d* orbitals of the transition metals break the erstwhile degeneracy of the *d* orbitals and cause them to "split" in terms of the energy levels.

To understand this better, let us first take the example of octahedral coordination geometry (such as in most transition metal−oxide ion complexes as MnO_2, $LiCoO_2$, $LiMn_2O_4$, CoF_2, etc.), namely, six ligands surround the central metal cation forming an octahedral around the same (similar to the d^2sp^3 hybridization valence band theory). Considering the Cartesian coordinate system, six ligands can be visualized to approach the central transition metal cation along *x*, *y*, and *z* axes. Among the five *d* orbitals of the transition metal, the d_{z^2} and $d_{x^2-y^2}$ orbitals directly lie along these axes, with the other three (viz., d_{xy}, d_{yz}, and d_{zx}) lying between these axes. Hence, d_{z^2} and $d_{x^2-y^2}$ orbitals directly point toward the ligands, whereas d_{xy}, d_{yz}, and d_{zx} lie in between the ligands (see Figure 3.7). Hence, a greater repulsion between the electrons on the ligand and those present (or supposed to be filled) in the d_{z^2} and $d_{x^2-y^2}$ orbitals increases their energy, as compared to other three *d* orbitals, namely, d_{xy}, d_{yz}, and d_{zx} (see Figure 3.8a). The d_{z^2} and $d_{x^2-y^2}$ orbitals are also known as e_g orbitals, while the other three are known as t_{2g} orbitals, as per the terminology used in the crystal field theory. This

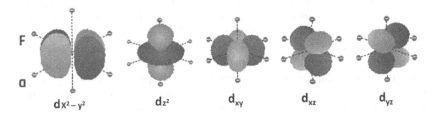

FIGURE 3.7 Spatial arrangement of the five *d* orbitals of the transition metal ion with respect to the ligands in a coordination complex having octahedral arrangement of the ligands. (Adapted from 'Crystal Field Theory' – Chemistry LibreTexts').

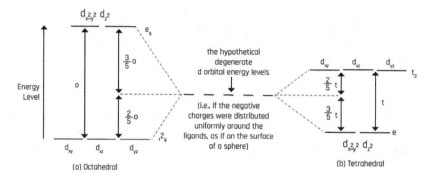

FIGURE 3.8 The split, in terms of difference in energy levels of the *d* orbitals of the central transition metal ion, under the influence of the fields from (a) octahedral and (b) tetrahedral coordination/ geometry of the surrounding ligands. (Adapted from 'Crystal Field Theory' – Chemistry LibreTexts').

splitting of the otherwise degenerate *d* orbitals of the free transition metal into e_g and t_{2g} orbitals having different energies, under the influence of the ligands, is known as "crystal field splitting" and the difference in energy between the two sets (i.e., the e_g and t_{2g}) is known as "crystal field splitting energy" or "crystal field stabilization energy" or CFSE (the usual symbol used being Δ). Usually, the e_g orbitals are raised with respect to the energy level of the supposedly degenerate *d* orbitals (if existed) by a magnitude equal to $(3/5)\Delta$, while those are lowered with respect to the energy level of the hypothetical degenerate *d* orbitals (viz., if the negative "ligand charges" were distributed uniformly around the orbitals) by $(2/5)\Delta$.

With regard to filling of the electrons in these split e_g and t_{2g} orbitals, in a more general perspective, electrons are supposed to fill the lower energy orbitals prior to filling-up the higher energy orbitals (as per the Aufbau principle discussed earlier). In the context of the present split orbital energy levels for the octahedral coordination (as discussed above), the electrons are supposed to fill-up the three "degenerate" d_{xy}, d_{xz}, and d_{yz} (i.e., t_{2g}) orbitals first, followed by the two "degenerate" d_{z2} and d_{x2-y2} (i.e., e_g) orbitals. Now, as per Hund's rule (discussed earlier), electrons are expected to fill-up the orbitals in a way that will lead to the highest possible number of unpaired electrons. In the context of these basic rules, now let us look at what would happen in the case of the technologically important transition metal complexes, especially for those that have up to three electrons in the *d* orbitals in the period 4 of the periodic table, namely, for Sc, Ti, V (or the corresponding M^{2+} ion). In these cases, the electrons would be present as unpaired electrons only in the lower energy t_{2g} orbitals. The addition of one or more electrons in the *d* orbital, starting with transition metal ions like Cr^{2+}, Mn^{2+}, and so on (in the same period), two possible situation arises, namely, either the additional electron(s) goes to one of the higher energy (d_{z2} or d_{x2-y2}) orbitals or pair with an already present electron in the lower energy t_{2g} orbitals. As per the discussion in the preceding paragraph, for going to one of the higher energy e_g orbitals, energy equivalent to the crystal field splitting energy (Δ) is needed, while pairing of electrons requires "spin pairing energy" (SPE). Now, if the SPE is less than the Δ, the next or additional electron will pair up with one of the electrons residing in the t_{2g} orbitals and vice versa. The former situation results in the presence of the least number of unpaired electrons, and accordingly, the transition metal ion is said to be in low spin (LS) state. In the case of the latter, that is, if the $\Delta < $ SPE, it results in the next electron filling-up one of the higher energy e_g orbitals instead of pairing up with an already present electron the t_{2g} orbital; a high spin (HS) state results due to the presence of enhanced number of unpaired electrons, as compared to the former case.

The possibility of occurrence of both LS and HS configurations occurs only for transition metal ions having four to seven electrons in the d orbitals (i.e., only for d^4, d^5, d^6, and d^7 configurations), with only a single arrangement type for the d electrons possible for the other configurations (viz., d^1, d^2, d^3, d^8, d^9, and d^{10}) (see Figure 3.9).

Overall, if it takes lesser energy to pair electrons in the t_{2g} orbitals (i.e., $\Delta > $ SPE), the corresponding transition metal–ion complex is in the LS state. By contrast, if it takes lesser energy to excite the next electron to the e_g orbitals (i.e., $\Delta < $ SPE), the complex is in HS state. For example, let us take the case of CoF_2 (having a rutile structure), where Co^{2+} has 7 electrons in the d orbital (i.e., d^7), with the Δ being 10 Dq, where Dq stands for differential of quantum; implying energy. Upon the crystal field splitting induced by the F^- ligands forming octahedral around Co^{2+}, the t_{2g} of Co^{2+} gets stabilized (i.e., energy lowered) w.r.t. the degenerate d orbitals of the free ion by 4 Dq, while the e_g orbitals get destabilized by 6 Dq. If the Co^{2+} is in the LS state (viz., $t_{2g}^6 e_g^1$), the net CFSE is [6 × 4 Dq − 1 × 6 Dq=] 18 Dq. By contrast, in the case of the HS state of Co^{2+} (viz., $t_{2g}^5 e_g^2$), the net CFSE becomes [5 × 4 Dq − 2 × 6 Dq =] 8 Dq. Now in the case of d^5 transition metal ion complexes (such as Mn^{2+}, Fe^{3+}, etc.), LS configuration (viz., $t_{2g}^5 e_g^0$) leads to net CFSE of [5 × 4 Dq − 0 × 6 Dq =] 20 Dq. By contrast, HS configuration (viz., $t_{2g}^3 e_g^2$) leads to net CFSE of [3 × 4

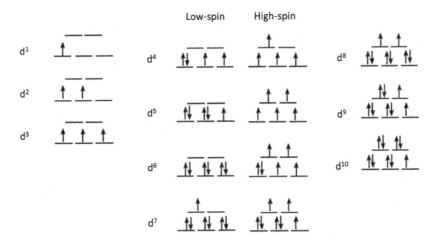

FIGURE 3.9 Representation of the LS and HS electron configurations in the transition metal ion *d* orbitals, split under the crystal field splitting. This also shows that both the states (viz., LS and HS) are possible only for the d^4, d^5, d^6 and d^7 configurations, with only one arrangement type being possible for the other *d* orbital configurations. (Adapted from the 'Bodner Research Web', hosted by Purdue University, USA).

Dq − 2 × 6 Dq=] 0 Dq. In other words, the crystal field stabilization provided by the electrons in the lower energy t_{2g} orbitals gets canceled out by the electrons in the higher energy e_g orbitals. Hence, the LS configuration appears to be more energetically favorable for the complexes based on d^5 transition metal ions. It may be mentioned here that compounds that contain unpaired electron(s) are paramagnetic and the force of attraction to a magnetic field is proportional to the number of unpaired electrons.

> In principle, it is therefore possible to determine whether the transition metal in a complex is in HS or LS state by measuring the strength of interactions between the same and a magnetic field.

The feasibility of LS or HS configuration depends on the transition metal under consideration, the charge/valence on the same (usually Δ increases with the oxidation state) and, more importantly, the type of ligands that form the complex. In this context, the ligands which lead to a relatively smaller value for the crystal field splitting energy (CFSE or the Δ) are known as "weak field ligands," which usually leads to HS configuration. In contrast, the ligands that produce a larger crystal field splitting (or value of Δ) are known as "strong field ligands," which usually leads to LS configuration. The ligands are arranged in the increasing order of CFSE; they are expected to form a complex with a given transition metal (and given transition metal oxidation state) in a series known as the "spectrochemical series" absorption spectra of transition-metal complexes). A part of the "spectrochemical series" series (which is usually included in the more complicated ligand field theory), listing ligands from small to large Δ is: $O_2^{2-} < I^- < Br^- < S^{2-} < SCN^-$ (for bonding with S) $< Cl^- < N_3^- < F^- < NCO^- < OH^- < \underline{O}NO^- < C_2O_4^{2-} \approx H_2O < NCS^-$ (for bonding with N) $< CH_3CN < NH_3 < NO_2^- < PPh_3 < CN^- < CO$. The transition metal ions can also be tentatively arranged in a form similar to the above "spectrochemical series" of the ligands, namely, in the order of increasing CFSE or Δ (for a fixed ligand type); as $Mn^{2+} < Ni^{2+} < Co^{2+} < Fe^{2+} < V^{2+} < Cu^{2+} < Fe^{3+} < Cr^{3+} < V^{3+} < Co^{3+} < Rh^{3+} < Ir^{3+} < Pt^{4+}$. Overall, the range of Δ is quite significant, with Δ for octahedral complexes (more specifically denoted as Δ_o) is ~100 kJ/mol for $Ni(H_2O)_6^{2+}$, ~104 kJ/mol for Co^{2+} in CoF_2, and as high as ~520 kJ/mol in the $Rh(CN)_6^{3-}$ ion (just as a few examples).

In the case of tetrahedral complexes (such as in CuCl), the relative energy levels of the t_{2g} (i.e., d_{xy}, d_{yz}, and d_{zx}) and e_g (i.e., d_{x2-y2}, d_{z2}) orbitals of the associated transition metal ions are just the opposite to that for the octahedral complexes, namely, the t_{2g} orbitals are at a higher energy level as compared to the e_g orbitals (say, by Δ_t). Again based on the Cartesian coordinate system and symmetry considerations, if we can consider that the four ligands forming the tetrahedra approach, the central cation along the four corners of a cube, the d_{x2-y2} and d_{z2} orbitals on the metal ion at the center of the cube, lie now in between the ligands rather than pointing towards the ligands (see Figure 3.10). Hence, due to similar influence (i.e., interaction and electrostatic repulsion) of the electrons on the ligands, the energy level of the d_{xy}, d_{yz}, and d_{zx} orbitals now gets accrued with respect to the "hypothetical" degenerate d orbitals (by $0.4\Delta_t$), whereas the energy level of the d_{x2-y2} and d_{z2} orbitals gets lowered (by $0.6\Delta_t$ w.r.t. "hypothetical" degenerate d orbitals) (see Figure 3.8b). In comparison with the CFSE

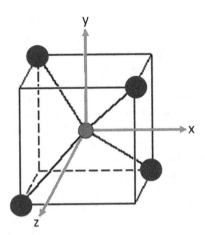

FIGURE 3.10 Spatial arrangement of the ligands in a coordination complex having tetrahedral arrangement, indicating that in this case the d_{z2} and d_{x2-y2} would not be directly pointing toward the ligands, but now with the other three d orbitals (viz., the d_{xy}, d_{yz}, and d_{zx}) doing so. (Adapted from the 'Bodner Research Web', hosted by Purdue University, USA).

for octahedral coordination, the splitting energy in the case of tetrahedral coordination is usually lesser (viz., $\Delta_t \sim 0.4\ \Delta_o$).

3.4 Jahn–Teller Distortions

The crystal field theory can be used to successfully explain the distortion that some of the octahedral complexes develop while coordinating some of the transition metal ions, especially those having d^4 (HS; like Mn^{3+}; $t_{2g}^3\ e_g^1$), d^7 (LS; like Ni^{3+}; $t_{2g}^6\ e_g^1$) and d^9 (like Cu^{2+}, Cr^{2+} etc.; $t_{2g}^6\ e_g^3$), electron configurations. This is known as "Jahn–Teller distortion," which either elongates or shortens two of the axial metal–ligand bonds with respect to the other four.

> As can be inferred from the d orbital electronic configurations associated "Jahn–Teller distortion" (mentioned above), odd number of electrons in the higher energy e_g orbitals usually lead to the same.

For example, consider the case of Cu^{2+} or d^9 configuration, in which one of the e_g orbitals, say d_{z2}, will have two electrons with d_{x2-y2} having only one (unpaired) electron. The different electron configuration/occupation and spin energy of the d_{z2} and d_{x2-y2} no longer allow them to stay degenerate, especially because these e_g orbitals directly point toward the ligands and face strong repulsion from the electrons on the ligands (say, in the p_z orbitals of the coordinating O^{2-} anion). Under the undistorted condition, this will lead to a greater repulsion for the doubly occupied d_{z2} orbital, as compared to the singly occupied d_{x2-y2} orbital. This makes the undistorted arrangement energetically unfavorable, especially since the orbital containing the greater number of electrons (in this case, d_{z2}) would become a higher energy orbital. To circumvent this situation, the bond lengths get enhanced along z-directions, namely, the two ligands approaching the Cu^{2+} along the z-directions get "pushed away" due to excess repulsion from the doubly occupied d_{z2} orbital, till the point that the d_{z2} orbital can lower its energy with respect to the singly occupied d_{x2-y2} orbital (see Figure 3.11). Hence, the distorted octahedral coordination, with two elongated axial bonds, becomes longer than other four in this case. This leads to a more energetically favorable (or stable) configuration. Overall, e_g orbitals no longer remain degenerate in the case of the "Jahn–Teller distortion," with sometimes even t_{2g} orbitals being also affected, but to a considerably lesser extent.

In the same light, one can consider the case of Mn^{3+} (d^4; HS; $t_{2g}^3 e_g^1$), as against Mn^{4+} (d^3 or t_{2g}^3). Mn^{4+} forms an undistorted octahedral structure, resulting in cubic symmetry for MnO_2. When another electron gets added to Mn^{4+}, thus forming Mn^{3+} (the "Jahn–Teller" active ion), this additional electron has to go the e_g orbital making (say) d_{z2} singly occupied (with no occupation of the other d_{x2-y2}). The situation is now similar to the case of Cu^{2+} with $d_{z2}^2 d_{x2-y2}^1$ (as above). Hence, to stabilize the configuration (and have the energy of d_{z2}^1 lower than that of d_{x2-y2}^0; as in Figure 3.11), the ligands (say O^{2-}) will be pushed away along the two bonds on the z-axis, thus elongating them with respect to the other four Mn^{3+}–O^{2-} bond sand distorting the

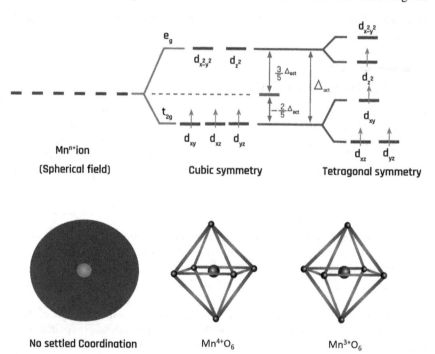

FIGURE 3.11 Schematic representation of the splitting of the $3d$ orbitals of Mn ions under the crystal field effect due to the presence of O^{2-} ions (as ligands) forming octahedral co-ordination around Mn ions and further splitting of the e_g and t_{2g} orbitals (with associated reduction of symmetry from cubic to tetragonal) upon addition of one electron to Mn^{4+} to form the "Jahn–Teller" active Mn^{3+} ion (having one, i.e., odd number of electrons in the e_g orbital). (Adapted from C. Liu et al., *Mater. Today*, 2016, 19, 109.)

structure (as per "Jahn–Teller distortion") of the corresponding oxide to tetragonal, rather than cubic symmetry. As will be discussed later, this particular phenomenon has immense significance toward electrochemical energy storage.

3.5 Closure

Theoretical concepts pertaining to bonding in metal–ligand complexes have been described in this chapter, with suitable examples. The discussions were initiated with the basic atomic orbital theory, followed by molecular orbital theory, and leading to the practically more relevant crystal field theory. The crystal field theory was, in turn, invoked to explain a particular type of distortion, namely, Jahn–Teller distortion, which has its origin in the "crystal field splitting/stabilization energy" and the occupation of the electrons in the molecular orbitals under the influence of the crystal field. On a more important note, the above theories and, in particular, distortion (such as the Jahn–Teller distortion), cast their influences on the performance and structural stability of the associated materials in various applications, including electrochemical energy storage, an important example of which will be highlighted in Chapter 13.

FURTHER READING

Liu C., Z. G. Neale, G. Cao, Understanding electrochemical potentials of cathode materials in rechargeable batteries. *Mater. Today*, 2016, 19, 109.

West A. R., *Basic Solid State Chemistry*, 2nd edn, 2000. John Wiley & Sons Inc., ISBN 0-471-98756-5.

4

Engineering Mechanics and Mechanical Behavior of Materials

For a number of engineering applications, such as bridges, aerospace, armors for defense, semiconductor chips, and electrode materials for batteries, it is important to design the materials to sustain loads or stresses during real applications. Understanding the response of materials toward applied loads and developing/designing structures/ components with the desired stability against such loads mandate the understanding of concepts related to mechanics. Accordingly, this chapter discusses the mechanics concepts with direct relevance to applications involving bulk structural parts and also thin films. In particular, the concepts and analytical methodologies involved in finite element analysis (FEA) of material properties are briefly discussed.

4.1 Introduction

Engineering materials are classified into three main categories, namely, metals and alloys, ceramics and glasses, and polymers. Among these categories, metals and alloys as well as polymers are the most used engineering materials for structural applications. Nevertheless, ceramics have become an attractive candidate material in the field of materials science in the last four decades for a variety of applications.[1-4] Ceramics possess high melting point, compressive strength, hardness, elastic modulus, high-temperature strength retention, and resistance to oxidation, corrosion, wear and abrasion, along with good refractoriness. The advancement in materials technology has enabled ceramicists to design advanced ceramics for use in various sectors such as aerospace, defense, automobile, and electronic industry. An emerging class of materials, known as composites, is developed based on the combination of advantageous properties possessed by the primary material classes (metals/ceramics/ polymers). Composite materials are further categorized based on the dominant presence of primary material class, as metal matrix composites (MMC), ceramic matrix composites (CMC), and polymer matrix composites (PMC).

The word "ceramics" has a Greek origin, referred to as "keramikos," meaning potter's earth. Ceramics are refractory, inorganic materials with covalent and ionic bonds and are processed/used at high temperatures. Ceramics are broadly classified as oxides (Al_2O_3, ZrO_2, etc.) and non-oxide ceramics (SiC, TiB_2, etc.). Some inter-metallic compounds, such as phosphides, antimonides, arsenides, aluminides, and beryllides are also considered as ceramics. Based on applications, ceramic materials can be grouped as traditional and advanced ceramics. As the name suggests, traditional ceramics are derived from raw materials, which are naturally occurring, such as sand, clay, quartz, etc. These find traditional applications as earthware, glassware, cement, etc. Advanced ceramics are being developed by tailoring the material properties by controlling the composition and internal structure, so as to satisfy the functional requirements for the system by making use of materials science and technology. These ceramics usually are carbides (SiC), oxides (Al_2O_3), nitrides (Si_3N_4), zirconia-toughened alumina, non-silicate glasses, etc. and find applications in automotive engines, electronic components, cutting tools, abrasives, high-temperature superconductors, and so on. With recent advances in materials technology, advanced ceramics show a 50-fold increase in specific strength as compared to primitive traditional ceramics. Advanced ceramics, which serve as structural members (subjected to mechanical loading) and demonstrate enhanced mechanical properties under demanding conditions, are generally known as structural ceramics. A special category of structural ceramics which has melting temperatures greater than 3000°C are referred to as ultra-high-temperature ceramics (UHTCs). In particular, non-oxide UHTCs are of interest for high-temperature structural applications, which include borides, carbides, and nitrides of transition metals. Borides as structural ceramics are potential candidate materials for many tribological and high-temperature applications, because of their unique combination of properties as low density, low thermal expansion, high melting temperature (~3000°C), high elastic modulus (~500 GPa), good abrasion, wear, creep and corrosion resistance, considerable chemical stability, and high hardness (>20 GPa) with metal-like characteristics as high thermal (60–120 W/mK) and electrical conductivity (~10^7 S/m). Higher elastic modulus provides these ceramics with good resistance to contact damage. All these characteristic properties make borides potential materials for various technological applications, as hypersonic aerospace vehicles, atmospheric re-entry vehicles, rocket nozzles, etc. However, despite these promising properties, the broader application of these monolithic borides is rather limited due to high machining costs, low-to-moderate fracture toughness, poor bending strength, poor thermal shock resistance, poor sinterability, and exaggerated grain growth at high temperature.[5] High-temperature structural intermetallics, such as transition metal silicides ($MoSi_2$, $TiSi_2$, Ti_5Si_3, etc.), have also gained considerable interest in the field of advanced high-temperature materials for use above 1000°C due to their high melting point, low density, and improved oxidation resistance at elevated temperatures.

Engineering ceramics are frequently exposed to high temperatures. Therefore, high-temperature stability, especially the oxidation behavior, is an important property to be considered.[6–13] The properties of various refractory ceramics are tabulated in Table 4.1. It can be noted here that HfB_2, ZrB_2, and TiB_2 can be used for very high-temperature applications compared to other borides. B_4C, on the other hand, is suitable for applications requiring low density and higher hardness. Over the years, it has been realized that an optimum combination of high toughness and environmental stability along with high hardness and strength is required for the majority of the current and future applications of structural borides. However, monolithic borides are not optimal for all structural applications. Therefore, ceramic composites are explored to develop structural components. The major classes of ceramic composites are particle-, fiber- and whisker-reinforced ceramic composites. Among these classes, particle-reinforced CMC offer the most cost-effective route to develop materials with improved and optimal combination of mechanical properties. Several refractory intermetallic compounds, such as $MoSi_2$, Ni_3Al, $FeAl$, Ti_5Si_3, etc. can be used as substitutes of composites/ceramics used for high-temperature applications above 1000°C, as they tend to form a protective impervious oxide layer over their surfaces

TABLE 4.1
Properties of Different Refractory Materials[18-23]

Material	Density (g/cm³)	Melting Point (°C)	Hardness (GPa)	Young's Modulus (GPa)	Fracture Toughness (MPa m$^{1/2}$)	Flexural Strength (MPa)	Thermal Conductivity (W/m/K)	Electrical Resistivity (μΩ cm)	Coefficient of Thermal Expansion (10^{-6}/K)
B$_4$C	2.52	2450	28.0–37.0	450–470	3.0–3.5	300	30–40	10^6	5.0
TiB$_2$	4.52	3225	25.0–35.0	560	5.0–7.0	700–1000	60–120	10.0–30.0	7.3
ZrB$_2$	6.10	3245	22.0–26.0	300–350	4.0–6.0	300	20–58	9.2	6.8
HfB$_2$	11.21	3380	21.0–28.0	500	....	350	104	11.0	6.3
EuB$_6$	4.99	2580	18.0–26.0	....	....	183	23	85.0	6.9
CrB$_2$	5.20	2200	11.0–20.0	211	....	600	32	30.0	10.5
MoSi$_2$	6.30	2050	13.0	384	2.0–2.5	....	50–221	21.0	8.4
TiSi$_2$	4.04	1540	8.7	250	....	....	....	<15.0	9.9
Ti$_5$Si$_3$	4.32	2130	9.6	160	....	....	....	....	7.1

at elevated temperatures, thus imparting necessary oxidation resistance to the material for high-temperature applications. Also, these compounds are lower in density, which reduces the net weight of the component.

4.1.1 Mechanical Properties: Principles and Assessment

Considering the participation of researchers from widely disconnected disciplines of science and engineering in the field of interdisciplinary research, the discussion in the following subsections starts with the basic definition of stress and strain as tensorial quantity. The difference in mechanical behavior, in terms of the stress–strain response, of materials under different modes of loading is also explained. This is then followed by a brief discussion on the deformation and strengthening mechanisms of metals. While metals are ductile (permanent deformation under stress), the ceramics are inherently brittle. The relevant theories to explain the brittle behavior of ceramics are explained with sufficient details. The concept of criticality under fracture conditions, leading to the definition of the fracture toughness, has been emphasized. Thereafter, the viscoelastic deformation of thermoplastic polymers is described in reference to the necking and crystallization during tensile deformation. Various experimental methodologies together with guidelines to conduct mechanical property evaluation are also summarized. Overall, the following discussion establishes a theoretical framework to understand and experimentally determine the mechanical properties of the materials used for various engineering applications. Some contemporary discussion can be found elsewhere.[14–16]

4.1.2 Conceptual Understanding of Stress and Strain

We will first explain the concept of stress and strain. When a complex shaped material is subjected to an external load, like the one shown in Figure 4.1, one may be interested

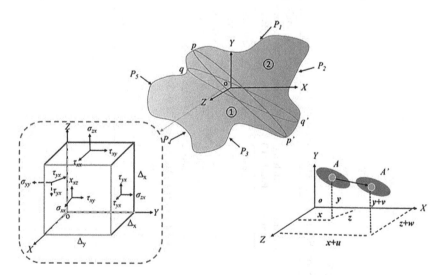

FIGURE 4.1 Conceptual definition of stress at a plane in a solid under equilibrium of external forces, depicting the stress acting at a point on an elemental unit cube (left) and strain due to the displacement of a point A to A′ in a solid continuum. (Adapted from Ref. 14).

to find out the stress at a plane and at a point "O." This demands that the definition of stress on a plane and that at a point to be differently expressed.

The stress at a given point is determined by the orientation as well as cross-sectional area of the hypothetical plane around it, with respect to the net force acting on the plane. This is possible to construct the free-body diagram around the region of interest. For example, one can consider two hypothetical planes, mm and nn around the point "O" in Figure 4.1. In defining stress, net force is to be determined from the equilibrium of the free-body diagram. In reference to Figure 4.1, the area of the pp' or qq' plane is different as well as the orientation with respect to the net force.

The above can be substantiated by defining stress at point O with the hypothetical plane pp' and then to determine the net force, under equilibrium of the external forces acting on the solid.

Mathematically, the stress at a given point can be defined as,

$$\sigma = \lim_{\Delta A \to 0} \frac{\Delta P}{\Delta A} \tag{4.1}$$

where ΔP is the resultant of all load components (P_3, P_4, P_5) that the solid experiences under equilibrium (see Figure 4.1). The stress at a point "O" is therefore a measure of net force on a hypothetical plane around the point of interest, when the area of the plane is assumed to be negligible or zero.

At a macroscopic level, the stability of any material is fundamentally determined by the equilibrium of mechanical forces acting on it from different directions and NOT by the equilibrium of stresses. This condition of equilibrium can be expressed as,

$$\sum_{x,y,z} P_i = 0$$

where P_i represents the components of the different forces acting on the solid along x, y, and z directions. The above discussion also leads one to arrive at the following condition,

$$\sum_{i,j} \sigma_{ij} \neq 0$$

In many engineering applications, a material experiences three types of stresses (see Figure 4.2). Two generic variants of stress are normal and shear stress. Normal stresses can be either tensile or compressive in nature. The tensile stress state is defined as when force of equal magnitude is applied in the opposite direction and under such loading, a cylinder will extend in length with a decrease in diameter (Figure 4.2a). The reverse scenario is the case of compressive force, when equal force is applied in the same direction (see Figure 4.2b). Under compression, a cylinder will increase in diameter, but will shorten in length. The shear loading involves the application of equal forces at two opposite faces of a cube. While both tensile and compressive

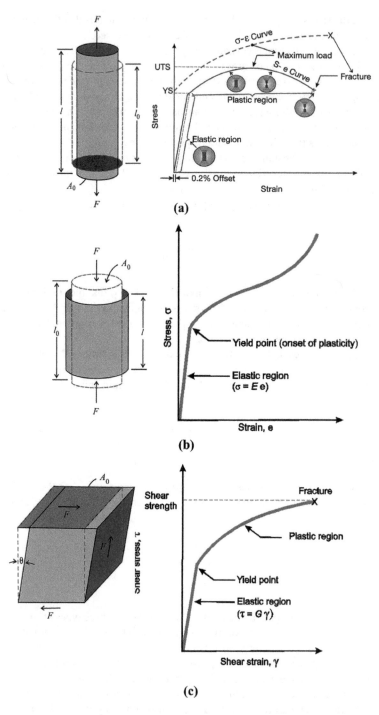

FIGURE 4.2   Type of loading and stress–strain response of metallic materials illustrating critical points and their significance on deformation behavior under (a) tension, (b) compression, and (c) shear. (Adapted from Ref. 14).

stresses will cause volume change, the shear stress leads to a shape change or angular distortion, measured by angle "θ" (Figure 4.2c). The compressive and tensile stress can be defined as the ratio of force (F) to the initial cross-sectional area (A_0) of the cylinder (F/A_0). Similarly, shear stress can be quantified as F/A_0, where A_0 is the area of one of the parallel faces of a cube. Note that tensile/compressive force is also applied perpendicular to the marked plane of a cylinder with area, A_0 (see Figure 4.2). It may be worthwhile to mention here that although Fig. 4.2a describes the generic tensile response of a ductile metal; the response will be largely different in case of ceramics and polymers, as will be explained later in reference to Figure 4.3.

> It should be clear that the stress is to be represented not only by magnitude and direction, but also by its orientation with respect to a plane.

It should therefore be realized by the reader that stress is a tensorial quantity. A scalar is represented only by magnitude, while a vector by magnitude and direction. In contrast, a tensor requires magnitude, direction, and a reference plane. To define stress, one needs to know the magnitude, direction as well as the plane across which it acts. Since stress is a tensor of rank 2, it needs a total of $3^2 = 9$ terms. With reference to Figure 4.1, the stress state at a point (O) can be defined when considering the stress components for different faces of a cube around point "O".

> Stress is not a vector, but a tensor of rank 2.

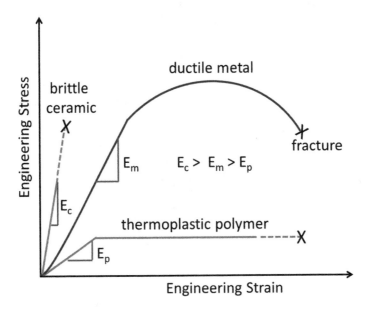

FIGURE 4.3 Schematic illustration comparing the stress–strain response of three primary classes of materials under uniaxial tension. (Adapted from Ref. 14).

The first subscript of a component of stress tensor (σ_{ij}) refers to the reference plane and the second subscript the direction. The normal stress component has double identical subscript, indicating that force/stress is acting on the planes along the plane normal.

In contrast, the shear stress has different subscripts. In Figure 4.1, the three stress components on the X-plane are σ_{xx}, τ_{xy}, and τ_{xz}. Similarly, specific stress components for Y-plane or Z-plane are also shown in Figure 4.1. While considering the small areas of the cube faces (change in stress over faces therefore negligible), the moments created by the shear forces would be equal. Thus, one would be able to realize that $\tau_{xy} = \tau_{yx}$; $\tau_{yz} = \tau_{zy}$; $\tau_{xz} = \tau_{zx}$. Therefore, the stress tensor, σ_{ij}, will have six independent stress components and is defined as,

$$\sigma_{ij} = \begin{pmatrix} \sigma_{xx} & \tau_{xy} & \tau_{xz} \\ \tau_{yx} & \sigma_{yy} & \tau_{yz} \\ \tau_{zx} & \tau_{zy} & \sigma_{zz} \end{pmatrix} \tag{4.2}$$

Likewise, the strain at a point in a material can be explained in the following manner. In high school textbook, one has learnt that, strain is the ratio of change in the dimension to the original dimension, when a solid experience a stress state. In figure 4.1, the strain has been described in terms of displacement of a point, A, in a solid continuum. While A has a coordinate (x, y, and z), the displaced point A' has a different coordinate ($x + u$, $y + v$, and $z + w$). Thus, the displacement component along X, Y, and Z direction is u, v, and w, respectively. According to the convention, the normal displacement gradients are defined as,

$$e_{xx} = \frac{\partial u}{\partial x}, \; e_{yy} = \frac{\partial v}{\partial y} \text{ and } e_{zz} = \frac{\partial w}{\partial z} \tag{4.3}$$

Similar to stress tensor, the entire displacement tensor can be defined as follows,

$$e_{ij} = \begin{pmatrix} e_{xx} & e_{xy} & e_{xz} \\ e_{yx} & e_{yy} & e_{yz} \\ e_{zx} & e_{zy} & e_{zz} \end{pmatrix} = \begin{pmatrix} \dfrac{\partial u}{\partial x} & \dfrac{\partial u}{\partial y} & \dfrac{\partial u}{\partial z} \\ \dfrac{\partial v}{\partial x} & \dfrac{\partial v}{\partial y} & \dfrac{\partial v}{\partial z} \\ \dfrac{\partial w}{\partial x} & \dfrac{\partial w}{\partial y} & \dfrac{\partial w}{\partial z} \end{pmatrix} \tag{4.4}$$

Like stress, strain is also a tensor of rank 2.

Following the continuum mechanics based treatment, the strain tensor is defined as,

$$\varepsilon_{ij} = \begin{pmatrix} \varepsilon_{xx} & \frac{1}{2}\gamma_{xy} & \frac{1}{2}\gamma_{xz} \\ \frac{1}{2}\gamma_{xy} & \varepsilon_{yy} & \frac{1}{2}\gamma_{yz} \\ \frac{1}{2}\gamma_{xz} & \frac{1}{2}\gamma_{yz} & \varepsilon_{zz} \end{pmatrix} \tag{4.5}$$

Similar to stress tensor, that strain tensor has six dependent components. Furthermore, the total volume change is zero during plastic or permanent deformation of a metal and this can be expressed as,

$$\Delta V = \varepsilon_{xx} + \varepsilon_{yy} + \varepsilon_{zz} = 0 \tag{4.6}$$

The strain tensor has five independent components with two normal and three shear components.

4.2 Stress–Strain Response of Metals under Different Loadings

Many materials in service encounters a variety of mechanical loadings, namely, tensile, compressive, shear, and at times, a combination of them (see Figure 4.2). It is worthwhile to reiterate that in pure tension mode, a material is subjected to equal and opposite forces, which typically leads to elongation of length along the loading axis. In case of pure compression mode, equal force is applied along the same direction to both the ends of the material, resulting in shortening of length along the loading axis. In contrast, shearing mode involves application of tangentially opposing force on the opposite surface, which leads to angular distortion.

The mechanical response of a ductile metal under tension, compression, and shear is schematically shown in Figure 4.2a–c. Clearly, a metal exhibits an initial linear stress–strain response, irrespective of the mode of loading (tensile/compressive/shear). In the linear elastic region, the stress is proportional to the strain (Hooke's law of elasticity). In tensorial notation, this can be expressed as

$$\sigma_{ij} = C_{ijkl}\varepsilon_{kl} \tag{4.7}$$

Alternatively, $\varepsilon_{ij} = S_{ijkl}\sigma_{kl}$ (4.8)

Where C_{ijkl} and S_{ijkl} are the stiffness and compliance tensor, respectively.

Unlike stress–strain, the elastic stiffness or compliance is a tensor of rank 4.

Clearly, C_{ijkl} or S_{ijkl} would require 81 terms to define them. However, more analysis to arrive at the number of independent terms of these two tensors is beyond the scope of this book. In all the modes of loading in metallic material namely, tension, compression, and shear, there exists a region of linear response in the initial portion of the stress–strain curve, which represents the linear elastic region. The slope of the curve in this region is used to determine the elastic modulus of the material. Beyond the elastic region, transition from the elastic to plastic regime is observed, which is denoted as the point of yielding. Beyond the yield point, the material exhibits a non-linear deformation trend, which continues till the point of fracture. The transition from linear to non-linear deformation is defined as the yield strength. In simplistic terms, linear deformation is known as elastic deformation and non-linear response is known as plastic deformation.

> The term "yield" means that a metal surrenders at yield stress and would start to deform permanently.

Prior to yield stress, the strain is proportional to stress and in the post-yield region, the response is characterised by a residual strain even after the complete removal of the load. It is worthwhile to mention that the non-linear stress–strain response qualitatively depends on the mode of loading. Although, the stress monotonically increases both in compression and shear, the increase is more significant in the last stage of compression loading. For shear loading, the stress increases but at a lower rate to reach almost a steady value. We shall now discuss more the tensile deformation in metals as crack propagation leading to failure is easier under tensile than under any other loading mode.

In materials science, a distinction is always made between engineering stress/strain and true stress/strain. In reference to Figure 4.2a, tensile deformation can be described in terms of both engineering stress (S)–strain (e) and true stress (σ)–strain (ε). These terms are defined as follows:

$$\text{Engineering stress, } S = P/A_0 \tag{4.9}$$

$$\text{True stress, } \sigma = P/A_i \tag{4.10}$$

$$\text{Engineering strain, } e = \Delta l/l_0 \tag{4.11}$$

$$\text{True strain, } \varepsilon = \ln l_i/l_0 \tag{4.12}$$

In the above definitions, P is the given load, A_0 and A_i are the original and instantaneous cross-sectional area, respectively Δl is the change in the length, and l_i and l_0 are the instantaneous length and original length of the test sample, respectively. It is important to recognise that length and area of the cross-section constantly changes with load during mechanical deformation (e.g., tensile loading), the "true" stress–strain realistically therefore should describe the dynamic changes in the stress state of a material under load. Since the strain in the elastic regime is usually very low or negligible with respect to plastic strain in case of metals, the engineering stress–strain relationship is normally used. However, the stress–strain response, when plotted in terms of true stress–strain, will be shifted upwards and will also show an increasing trend (particularly in the necking region) when compared to the engineering stress–strain response (see Figure 4.2a).

> Since a metal always strain hardens under stress, the true stress–strain response therefore describes a "true" deformation response of a metal experiencing dynamic changes in the stress state.

Also, engineering stress–strain response in the post-UTS region shows a decrease in stress, although the cross-section decreases faster than an increase in length (see Figure 4.2a). Such behavior is certainly not the true intrinsic response.

When considering the engineering stress–strain response in Figure 4.2a, plastic deformation of metals which occurs above YS consists of uniform deformation as well as non-uniform deformation. At this juncture, ultimate tensile strength (UTS) is defined, as the transition point from uniform deformation to non-uniform deformation in the stress–strain curve. UTS is an important design parameter, which signifies the maximum load that can be withstood by the specimen. In the uniform deformation regime, spanning between the YS and UTS, a power-law relationship between stress (σ) and strain (ε), is mathematically represented as,

$$\sigma = K\varepsilon^n \qquad (4.13)$$

where K denotes the strength coefficient and n denotes the strain hardening coefficient.

One fundamental concept is that the volume remains constant during plastic deformation and using simple mathematical calculation, we can arrive at the following relationship between the strain (ε_u) at the point of maximum load, that is, at UTS and strain hardening coefficient (n) as,

$$\varepsilon_u = n \qquad (4.14)$$

In the description of the mechanical response in reference to engineering stress–strain, the phenomenon of necking is important, which occurs after UTS. Necking is starkly different from other regimes of the σ–ε plot, as it is characterized by a rapid reduction in the cross-sectional area as compared to the elongation of the specimen. This also leads to a decrease in the stress–strain plot, since engineering stress is defined on the basis of the original cross-sectional area. In reality, the metal is strain hardened due to extensive deformation. Therefore, more stress is needed to be applied so that the metal can be continually deformed till the point of fracture. This physical concept is much better captured in the true stress–strain plot (see Figure 4.2). Also, as per the definition of true stress and engineering stress, one would wonder why the true stress, corresponding to the point of maximum load, is at the higher level and also at the right of the UTS co-ordinate. Similarly, the tensile failure occurs at larger value of true strain, as compared to that when defined in terms of engineering strain. Nevertheless, the extensive necking, after UTS in the non-uniform deformation region, leads to the fracture in metals.

The mechanical performance of a metallic material would be determined by elastic modulus, yield strength, UTS, and maximum strain at the point of maximum load.

4.3 Tensile Stress–Strain Response

Here, the difference in the mechanical response of metals/ceramics/polymers under tensile loading will be discussed. It is important to reiterate that an earlier chapter in this book describes the nature of bonding in these materials. The metals are characterized by metallic bonds with atoms floating in a sea of electrons. This attributes to good electrical and thermal transport properties. Also, the metallic bonding is isotropic in nature, implying the mechanical properties to be also isotropic. In contrast, ceramics predominantly either have ionic (oxides) or covalent (non-oxides) bonding. The ionic bonding demands electro-neutrality conditions to be maintained during deformation. The covalent bonding is directional in nature, and this would impart anisotropic mechanical properties. Polymers with macromolecular structure have mixed bonding. Along the backbone chains, the "mer" units are covalently bonded, while two neighboring chains are weakly bonded by either H-bond or Van der Waal's bonds.

Under mechanically loaded conditions, the three different classes of materials (metals/ceramics/polymers) show distinguishable responses, arising from the differences in bonding characteristics.

The major differences in tensile response of metals, ceramics and polymers, in terms of engineering stress versus strain plot, is shown in Figure 4.3. Clearly, ceramics have largely linear response followed by sudden fracture, while metals exhibit an initial linear response and then non-linear response up to the fracture point. In both the cases, the elastic modulus can be derived by measuring the slope of the curve in the linear regime. In contrast to metals and ceramics, a thermoplastic polymer characteristically shows a constant deformation at a lower stress level, which is preceded by a linear stress–strain response, like metals/ceramics. The slope of the initial linear response, a measure of elastic modulus, is markedly different in metals/ceramics/polymers.

When comparing the tensile deformation response of three classes of materials, it is obvious that all of them exhibit an initial linear response, which is a common factor. The major differences amongst these material classes are in terms of both elastic modulus and total strain to failure or UTS. On close observation of Figure 4.3, it can be seen that ceramic materials have high elastic modulus as compared to metallic materials, and much higher than polymeric materials. Moreover, the tensile failure strain exhibited by thermoplastic polymers exceeds that of the ductile metal by nearly one order of magnitude. While considering the transition of deformation from the linear to non-linear regime, that is, yielding, polymeric materials yield at stress levels which are much lower than metallic materials. From these discussions, and considering Figure 4.3, it is important for the material designers to know that metallic implants are to be used in applications involving large load-bearing capacity under tensile mode. However, for low load-bearing applications demanding large deformation without drastic failure, polymeric materials are preferred.

> Although tensile strength is low for ceramics, the compressive strength is typically eight times larger than the tensile strength.

When compared among different material classes, better compression strength is advantageous for ceramics in engineering applications. The elastic modulus is the highest for ceramics.

4.4 Deformation and Strengthening of Metals

Here, we shall explain the origin of the plastic deformation of metals. The non-linear deformation of metals is largely attributed to the motion of linear defects, that is, dislocations. These defects have the ability to glide on specific crystallographic planes (slip planes) along specific crystallographic directions (slip directions). This results in displacements among atomic planes. Slip planes are the most closely packed planes with the highest planar density of atoms, while slip directions are the closest packed directions with the shortest interatomic distances between two adjacent atoms on a slip plane. Most ductile metals have more than five slip systems. The movement of dislocations under shear stress requires localized bonds to be ruptured or remade. This is not possible in ionic ceramics so easily due to the restriction of electroneutrality conditions. The restriction of directional or non-deformable covalent bonds similarly restricts the dislocation movement in covalent ceramics. In polymers, dislocations are not available and a different defect type "crazes" evolve during deformation, as will be discussed later. The gliding of a dislocation can be described with reference to Figure 4.4.[24] In the ball and stick model, the dislocation is shown as the extra half-plane of atoms. In general, a crystal defect is usually described as a departure from the long-range ordering of atoms. The dislocation line lies perpendicular to the plane of the paper. In the presence of shear force acting to the parallel faces of the crystal, the extra half-plane of atoms moves from left to right and finally creates a step at the surface. The step formation is a measure of plastic deformation and in a real solid, a large number of dislocations glide simultaneously under shear stress, leading to plastic deformation. More discussion on the dislocation induced plasticity in metals is beyond the scope of this book.

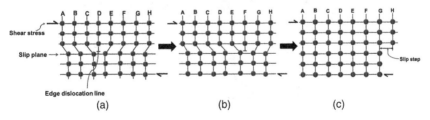

FIGURE 4.4 Schematic illustration of the gliding of edge dislocation on a slip plane: (a and b) dislocation movement and formation of slip step (c), a measure of plastic deformation in metals.

One of the important aspects of the mechanical behavior of metals is the strengthening mechanisms, which are based on physical mechanisms activated as dislocations find it difficult to glide on slip planes. These mechanisms are known to increase the strength often at the expense of ductility. Different strengthening mechanisms are briefly discussed below.

> The fundamental basis for strengthening mechanisms lies in restricting the dislocation motion through design of microstructures.

4.4.1 Solid Solution Strengthening

One classic example is the steels (e.g., Fe−C alloys), wherein the small interstitial solutes (here, C-atoms) can diffuse to the dislocation core, thereby restricting the dislocation motion in the matrix (here, Fe). This can be explained in reference to Figure 4.5. The solute segregation at the dislocation core makes them immobile, as dislocation has to glide with the solute atmosphere. In case of solutes of larger/smaller size (within 15% of the atomic size of the parent metal) the parent atom can be substituted by the foreign atom. This causes lattice strain, which can impede the dislocation motion. As a result, larger stress would be needed to move dislocations.

> The solid solution strengthening linearly scales with solute addition up to a critical alloying addition, beyond which intermetallic formation is likely to take place.

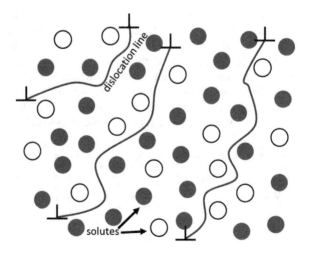

FIGURE 4.5   Solute atmosphere around a dislocation causing solid solution strengthening. (Adapted from Ref. 14).

4.4.2 Precipitation Hardening

This mechanism can be best described by taking the example of Duralumin (Al-4% Cu alloy). The heat treatment of this alloy results in finescale dispersion of precipitates in the matrix. Depending on the size, shape, volume fraction of precipitates as well as nature of matrix/precipitate interface; the strengthening can be realized. For better strengthening, a coherent/semi-coherent interface (minimal lattice plane bending) would be desired. If the precipitates grow in size or lose coherency, the magnitude of strengthening decreases.

Therefore, the precipitation hardening is most effective, when precipitates are non-deformable with finite inter-precipitate spacing. In a microstructure with such precipitates, a dislocation can overcome the barrier by forming dislocation loops. This is possible as dislocation has elastic properties, i.e., it can change its configuration, as and when required.

4.4.3 Dispersion Strengthening

When a second phase is added intentionally during processing stage, a dislocation can easily cut through the dispersoids, while gliding through the matrix (see Figure 4.6). During such interactions, the surface area increases, and more work is to be done to allow dislocation glide. Clearly, the dispersion strengthening increases with the second

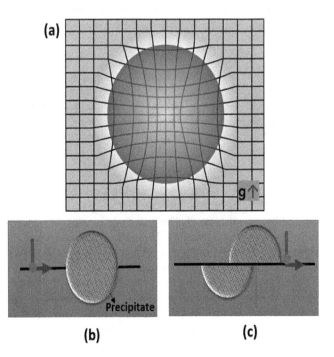

FIGURE 4.6 Schematic illustration of the coherent interface (a) and the dislocation precipitate interaction (b and c). (Adapted from Ref. 14).

phase amount. It must be mentioned that the last two mechanisms are sensitive to the heat-treatment conditions. Both the temperature and time are to be carefully tailored to optimize the strength increments.

4.4.4 Work Hardening/Strain Hardening

This kind of strengthening occurs when a metal undergoes repeated straining during fabrication. As a result, a large number of dislocations are generated, leading to "dislocation forest". The formation of the deformation substructure results in a decrease in the mobility of dislocations. For sustained deformation, a mobile dislocation has to cut through dislocation forest, and consequently, strength increases.

Any mechanism, which can potentially make the dislocation motion difficult, can make the metals stronger.

4.4.5 Grain-Size Strengthening

Prior to discussing this specific mechanism, we define the concept of a grain. During the solidification of a metal, the atoms are arranged in crystal lattices to form a solid. In crystalline materials, atoms are arranged along lattice planes in three-dimensional (3D) space with both long- and short-range orders. In amorphous materials, such as glass, only short-range order exists. In crystalline materials, the atomic arrangement in regular and periodic manner at a microscopic length scale varies from region to region.

A grain is defined with a characteristic microstructural length scale, wherein the atoms are arranged with identical lattice orientation, that is, all the lattice planes having identical orientation. On moving from one grain to another grain, the lattice orientation changes and the region of misorientation or the physical interface between two neighboring grains is defined as the grain boundary.

In polycrystalline materials, the grains having identical orientation on chemically etched surfaces, can be imaged using optical microscope (chemical reagent etches grain boundary and grain differently). The difference in crystal orientation between two neighboring grains is represented by the angle of misorientation. Based on the angle of misorientation, the low- and high-angle grain boundaries can be defined and the transition from one to another typically takes place at 5–10°.

The grain-size-induced strengthening can be explained by Hall–Petch relationship, which is based on the dislocation pile up against a low-angle grain boundary (see Figure 4.7). Clearly, more the dislocation piles up, more would be the stress required to continue deformation. The increment in the stress to ensure continued deformation can be expressed as,

$$\sigma = \sigma_0 + Kd^{-0.5} \tag{4.15}$$

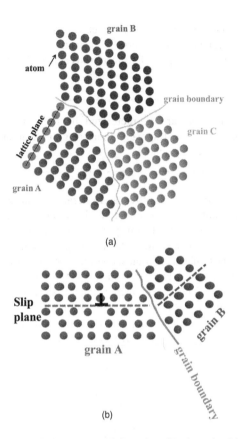

FIGURE 4.7 (a) Atomic model of grain and grain boundary, (b) schematic of dislocation propagation on a grain and possibility of getting stuck at a grain boundary. (Adapted from Ref. 14).

where d = grain size, and σ_0 = single crystal strength, that is, the strength of a crystal without any grain boundary. From Equation 4.15, it is clear that the strength increases non-linearly with the grain size. This also explains that finer the grain size, larger will be the strength for metals. However, the dislocation pile-up model is valid only for cases where the pile up contains at least 40–50 dislocations. Considering the length scale of the dislocations (~0.5 nm), it therefore appears that this strengthening mechanism fails, when the grain size is reduced to 20 nm or below.

Summarizing, this section briefly introduces various strengthening mechanisms, which can be thoughtfully considered while manufacturing metals for various engineering applications. It should be perceived that larger is the strength, more will be the ability of a metal to sustain more loads in service. Also, the elastic modulus should not be compromised, while designing the microstructure.

4.5 Brittle Fracture of Ceramics

It can be reiterated here that ceramics, under tension, exhibit a linear elastic response up to fracture. The absence of non-linear deformation is a consequence of brittleness.

This can be attributed to the ceramic microstructure, unless otherwise designed with toughening elements, can not resist the crack propagation. This means that the cracks will not find any resistance in its path. This results in catastrophic or ungraceful failure of ceramics.

> While ceramic materials are characterized by their high hardness, and compression strength, the widespread engineering applications are primarily restricted due to inherent brittleness.

We now explain the origin of brittleness in ceramic materials. In reference to Figure 4.2, it is important to recognize that non-linear deformation in metals is due to the dislocation movement. Analogously, the propagation of cracks determines the mechanical response of ceramics. Even though ceramics have outstanding combinations of hardness, elastic modulus, compressive strength; ceramics are more known for their brittleness (e.g., hydroxyapatite, Al_2O_3, etc.).

The brittleness can be better explained from the atomistic description of the fracture process. As shown in Figure 4.8,[25] the interatomic bonds behave more like "spring-like" structures. The elastic response of the stretching can be described as, $F = -K x$, where F is the external load, K is the spring constant, and x is the spring

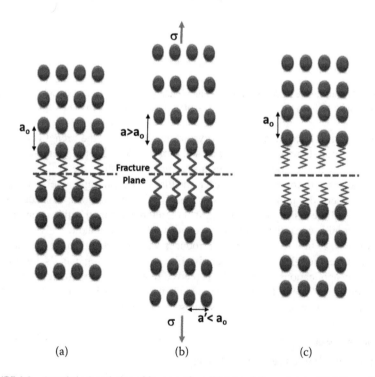

FIGURE 4.8 Atomistic description of fracture of a solid: (a) undeformed state, (b) deformed under tensile load, and (c) fracture due to the rupture of interatomic bonds. The interatomic bonding across the fracture plane is represented by spring-like behavior. (Adapted from Ref. 14).

displacement. As shown in Figure 4.8b, the bonds stretch from the equilibrium spacing of a_0 to a $(a > a_0)$ and as explained below, the stress concentration at the crack tip will subsequently rupture the spring (bond), leading to continued crack propagation. This sequence of physical events, as shown in Figure 4.8, takes place ahead of any advancing crack under external tensile loading.

The microscopic nature of the crack size distribution, their orientation and their influence on failure or damage progression under tensile and compression loading are shown in Figure 4.9. The formation of cracks in ceramics commonly occurs during their processing (i.e., sintering). Thus, it is important to tailor the sintering process to obtain ceramics with smaller/finer cracks. These cracks often vary in size and their orientation. The largest crack with most favorable orientation (close to 90° with respect to tensile axis) in all likelihood, would extend in a direction perpendicular to tensile axis. Once such crack grows to criticality, it leads to failure. The scenario is distinctly different in case of compression loading. Under compression, the cracks, oriented closer to compression axis, are more likely to grow (see Figure 4.9b).

In case pre-existing flaws are oriented at acute angle to compression axis, those would favorably grow in a manner to orient more towards the compression axis before further growth. This leads to the formation of secondary cracks (see Figure 4.9d).

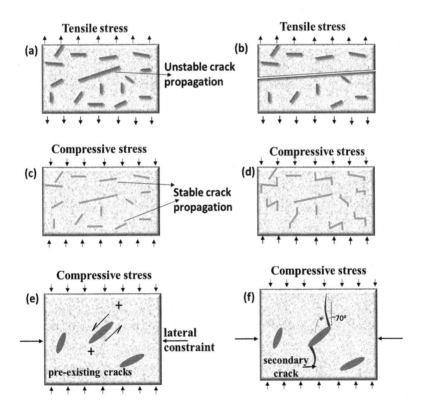

FIGURE 4.9 Illustration of crack propagation in ceramic implants under various loading modes, with pre-existing cracks. (a and b) Under tension; (c and d) under compression; (e and f) primary cracks leading to formation and growth of inclined secondary cracks under compression. (Adapted from Ref. 14).

The difficulty in crack propagation under compression also explains why ceramics have exceptional compression strength, that is, eight times that under tension. In case of metals, tensile and compression strength is nearly equal. For example, the tensile strength of stainless steel is 500–550 MPa, meaning their compression strength is also around 500–550 MPa. In contrast, the tensile strength of Al_2O_3 ceramic can be 150–200 MPa, but the compressive strength of Al_2O_3 can be as high 1200–1600 MPa. Therefore, ceramics are much better than metals in terms of compression strength.

In the following section, we explain more the brittle failure in ceramics/glass materials, by reviewing some of the important theories, in particular reference to quantitatively explain criticality of fracture.

> The initial theory of brittle fracture is based on interatomic bond breakage/ rupture, which gives rise to two additional surfaces.

Charles Inglis formulated a theory based on this aspect, by considering the stress concentration at the crack tip[17] (see Figure 4.10). In Inglis theory, a crack is physically considered as an elliptical hole or cavity in an otherwise solid continuum and a crack essentially represents the localized area of bond rupture or discontinuity in interatomic bonding.

The stress concentration, i.e., the modulation of stress at the crack tip can be expressed by the following equation, wherein σ is the externally applied stress, and c and ρ are, respectively, the edge crack length and crack tip radius of curvature,

$$\sigma_{max} = \sigma\left(1 + 2\sqrt{\frac{c}{\rho}}\right) \qquad (4.16)$$

As the crack tip radius is much smaller than the crack length (i.e., $\rho \lll c$), the above equation can be simplified and expression for maximum stress at crack tip edge is,

$$\sigma_{max} = 2\sigma\sqrt{\frac{c}{\rho}} \qquad (4.17)$$

The functional relationship between the maximum stress at crack tip and the crack tip radius as shown in the above equation illustrates that the stress experienced by the crack tip will be much larger than the externally applied tensile stress ($\sigma_{max} \ggg \sigma$). Inglis proposed that fracture will occur when this maximum stress exceeds a critical value, which is determined based on the breaking of interatomic bonds ahead of the crack tip.

$$\sigma_{max} = \sigma_{th} \qquad (4.18)$$

$$2\sigma\sqrt{\frac{c}{\rho}} = \left(\frac{E\gamma}{a_o}\right)^{1/2} \qquad (4.19)$$

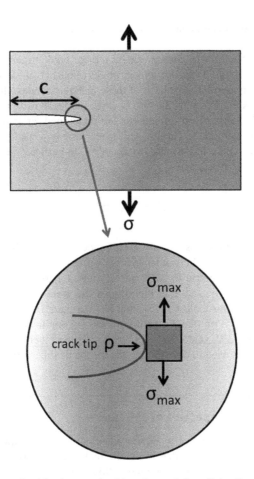

FIGURE 4.10 Schematic of the sharp crack subjected to mode I tensile loading, depicting the stress concentration at the crack tip. (Adapted from Ref. 14).

$$\sigma_c = \left(\frac{E\gamma\rho}{4a_oc} \right)^{1/2} \tag{4.20}$$

The brittle fracture of materials is dependent on both critical stress (σ_c) and critical crack length (c^*).

Thus, the combination of crack size and stress together determines the fracture criteria in brittle materials like, ceramics. At this juncture, the stress intensity factor is defined as,

$$K = Y\sigma\sqrt{\pi c} \tag{4.21}$$

where, Y is a factor dependent on the location/orientation of crack and loading conditions.

In the classical theory of fracture mechanics, three predominant modes of loading of the crack faces exist. They are tensile mode (mode I), shear mode (mode II), and tearing mode (mode III). For each of these different modes, the stress intensity factor is defined as K_I, K_{II}, and K_{III}. Of the three modes, tensile mode is critical as far as ceramics and other brittle materials are concerned, as they undergo failure predominantly due to mode I. Thus, K_I values are largely considered by researchers for determining the stress intensity of brittle solids. In particular, the K_I value under critical conditions is defined as,

$$K_{Ic} = Y\sigma_c\sqrt{\pi c^*} \tag{4.22}$$

where K_{Ic} is the crack tip stress intensity factor under mode I loading and c^* is half of the through-thickness critical crack length.

The value of the geometric factor Y varies around 1 to 1.1. The K_{Ic}, defined in Equation 4.22 is widely used as a measure of the fracture toughness. From the preceding discussion, it should be clear that K_{Ic} is a material property, that is, each material should have a unique K_{Ic} value. In various engineering applications, the external force induced K_I value should not exceed K_{Ic} and therefore, K_{Ic} is one of the important parameters used in determining the mechanical performance.

4.6 Mechanical Properties of Polymers

It can be reiterated in reference to Figure 4.11 that the deformation characteristics of polymers are different from that of ductile metals and brittle ceramics. In view of the wider use of thermoplastic polymers, this section discusses the deformation mechanisms of thermoplastic polymers. In Figure 4.11a, the typical stress–strain response of a thermoplastic polymer is shown, wherein the deformation of polyethylene (PE) can be described as "pseudo-ductile" behavior. In Materials Science literature, the deformation mediated by dislocation motion is described as "ductile" behavior. In case of polymers, the dislocation as a defect is not present and instead the deformation is primarily dominated by necking and crystallization, as a consequence of the chain realignments in response to stress. Therefore, the deformation of polymers is referred to as "pseudo-ductile" or more strictly, as viscoelastic behavior.

> The characteristic deformation of polymers is widely referred to as "viscoelastic" or "pseudo-ductile" deformation. Such deformation can be described as an integrated response of an elastic solid and shearing of parallel layers of polymeric chains.

In Figure 4.11b, the deformation mechanisms of a semi-crystalline polymer are schematically illustrated. Considering the fact that a polymer consists of randomly

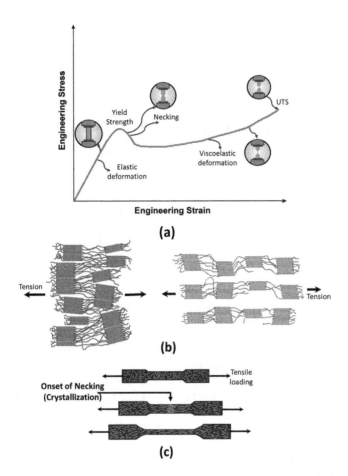

FIGURE 4.11 (a) Stress–strain response of typical thermoplastic polymers under uniaxial tensile loading. (b) Chain alignment in the amorphous regions of the semi-crystalline polymer along the loading (stress) direction. (c) The illustration of necking induced crystallization in polymers during tensile testing of the dog-bone geometry sample. (Adapted from Ref. 14).

oriented chains of different lengths, the short chains can align themselves along the tensile axis during the loading. Consequently, the parallel alignment of short chains is accomplished at specific location, leading to a localized orderly arrangement of polymeric chains. This phenomenon, known as crystallization, leads to necking in polymers. Once necking starts, the thermoplastic polymer maintains steady deformation almost at constant load, which allows it to undergo a large deformation, as shown in Figure 4.11c. The alignment of short chains across the entire gauge length (length of the tensile sample with a constant cross-section) is a time-dependent phenomenon, as it involves a steady rotation and alignment of the polymeric chains along tensile axis. This requires considerable time, and this is the origin of the viscoelastic deformation and anelasticity. While the crystalline region with more ordered arrangement of chains readily aligns along the tensile axis, the alignment of chains in the disordered or amorphous region is a rather slow time-dependent process (see Figure 4.11c). With continuous loading, a situation arises when most of the disordered regions with random chains is aligned along the tensile direction. The above phenomenon explains the viscoelastic response of polymers with large deformation and lower strength.

4.7 Numerical Approaches in Predicting Material Behavior

A number of interdisciplinary research problems demand the use of both experimental measurements of material properties together with computational or numerical analysis to predict such properties. This is particularly more relevant when experimental measurements pose significant challenge or simply impossible in reality. This may be because of the length scale of measurement or the faster kinetics of a process or the extreme physico-chemical or aerothermodynamic conditions. This is accomplished in two ways, (a) use of existing models or relevant equations from the literature to obtain predicted values of material properties depending on room temperature measurement or available data, or (b) the use of computational approaches or tools to capture physico-chemical properties. To this end, this section defines and describes the salient features of finite-element analysis (FEA)-based methodology, which are useful in solid mechanics or computational aerodynamics research.

Therefore, numerical analyses play a vital role in understanding the behavior of materials and evaluating new design concepts. With the computing advancements, numerical analysis enables the interdisciplinary researchers to understand certain phenomenon/physics, which cannot be monitored in real-time experiments. Though numerical analyses are based on the assumptions and simplifications, increasing the complexities of the problem together with modeling of required problem physics can significantly reduce the time taken from drawing table to device fabrication. With systematic methodology and proper validation, the numerical analyses can do a wide variety of parametric studies of critical independent process parameters, which may not be feasible to be tested in ground.

In general, there exists three major ways in which a typical engineering problem can be solved and they are described below.

4.7.1 Analytical Method

Analytical methods are based on closed-form solutions, which are obtained using a classical approach. This method can produce better results but are typically restricted to simple problems with primitive shapes/geometry. Due to its simplicity, the time involved for solving the given problem is very minimal with little cost involved. However, for complex shapes and loading conditions, the solution procedure becomes cumbersome and cannot be solved without simplifications. However, some approximations are used even in analytical methods in case the solution is not a closed form.

4.7.2 Numerical Method

This involves mathematical representation of the problem with suitable approximations and assumptions. This method can be used for real-life complex problems containing

3D geometries and varying loading conditions. The resulting solution is obtained by solving the governing equations with suitable approximations. As it uses approximations, the results need to be validated using experimental methods or analytical calculations. This method can solve the complex problem in considerable time and involves a minimal cost. Compared to analytical methods, the results obtained from numerical methods are approximate. A few numerical techniques are as follows:

  i. Finite-element method (FEM)
  ii. Finite difference method (FDM)
  iii. Finite volume method (FVM)
  iv. Boundary element method (BEM)

It is to be mentioned that FEA and FEM are synonymous and therefore refer to the same technique.

4.7.3 Experimental Method

This involves the measurement of desired parameters by designing the experiment. Fundamental measurement quantities such as strain, displacement, temperature, pressure, vibration, and so on can be measured to understand the material behavior. This method is applicable only if a physical prototype is available. The results here too cannot be trusted blindly and requires repeatability to ensure the accuracy of the obtained results under the desired experimental conditions. This method consumes a lot of time and needs expensive setup and sensors.

Depending on the type of problem, time, cost, and facility, either one of the above methods is chosen. Owing to the reduced design cycle time and faster visualization of optimal design, FEM is used for scientific research problems as it is proven to provide accurate results, if the assumptions and approximations are appropriately modeled into software module. As mentioned earlier, the results obtained from FEM can always be verified using analytical as well as experimental methods.

Some application areas of FEA include mechanical, aerospace, defense, civil, energy, electronics, structural/stress analysis (static/dynamic/thermal/fatigue/optimization), acoustics, electromagnetic field interactions, and biomechanics.

In recent research, the concept of integrated computational—experimental approach has been extensively utilized, in particular, by interdisciplinary researchers. In this section, the readers are introduced to one of the most used numerical techniques, that is, FEA, which is used to predict the behavior of the material/components in the simulated environment.

4.8 Finite-Element Method

The FEM is the most popular numerical method, which involves discretization of the physical domain into "finite elements" and determination of system response properties at these finite number of elements with desired initial and boundary conditions. The obtained result is interpolated to the entire domain using interpolation functions, depending on the type of element chosen for the analysis. The calculations

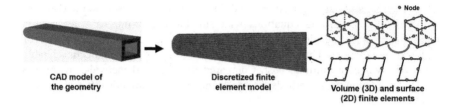

CAD model of the geometry   Discretized finite element model   Volume (3D) and surface (2D) finite elements

FIGURE 4.12 Schematic illustration to show discretization involved in finite-element modeling.

are performed, based on the governing equation in the matrix form, which depends on the type of analysis.

Fundamentally, the FEM is a numerical computational technique mostly used for solving engineering problems. A multitude of domain-specific problems can be handled by FEM, such as thermal analysis, structural analysis, modal analysis, fluid flow, electromagnetics, biomechanics, and so on. The governing equations for these problems are partial differential equations (PDEs) and can be solved analytically with required initial and boundary conditions. Though such process gives exact solution, the solution procedure can be cumbersome. Finite-element formulation, involves solving a system of algebraic equations, rather than PDEs. Though this technique yields approximate solutions as compared to analytical methods, the errors in FEM can be minimized by adopting best practices in discretizing and solution strategies, with recent trends in high-performance computing.[1]

As all real-life systems are continuous, their direct modeling becomes challenging. Thus, FEM cuts any given engineering structure into several elements and those are reconnected at "nodes" as if nodes were pins or drops of glue that hold elements together (see Figure 4.12). This process results in a set of simultaneous algebraic equations at the node. A field quantity is interpolated by a polynomial over an element and the adjacent elements share the degree of freedom (DOF) at connecting nodes. The connection of all the quantities provides the behavior of the specified region.

4.8.1 Defining Terms

In the following, some relevant terms related to FEA are defined.

Discretization: This process decomposes an irregular-shaped engineering structure into an equivalent system of many smaller bodies or units (finite elements) interconnected at points common to two or more elements (nodes or nodal points) and/or boundary lines and/or surfaces.

Node: It is a specific point in the finite element at which the value of the field variable is to be explicitly calculated.

Element: It is the specific geometric shape, which connects the nodes. The element can be one-dimensional (1D) (line/rod/bar), two-dimensional (2D) (tri/quad), and 3D (tet/hex/pyramid/prismatic).

DOF: It is the minimum number of independent coordinates required to completely define the position of an object in 3D space. A continuous system has an infinite DOF and hence not solvable. FEA reduces the DOF from infinite to finite through discretization.

Mesh: This is the geometric structure, which consists of nodes connected via elements with specified DOF at each node and when repeated in a regular manner forms the irregular structure under consideration.

Mesh convergence/Grid independence: It is defined as the minimal number of nodes/elements for which there are no significant changes in the field variable. This criterion ensures minimal computational cost and acceptable result approximation.

Element quality check: This is a routine which is used to determine how good the selected elements represent the physical models with minimal distortion. This is ensured by checking the skewness, aspect ratio, warpage, and so on of a given element.

Stiffness matrix: This is the characteristic property of the element, which is used for determining the unknown field variable with usually known load. The total DOF of the problem dictates the order of the stiffness matrix. For example, Equation 4.7 or 4.8 can be written in a matrix format for each element and subsequently, that matrix can be solved under the imposed boundary conditions.

Loads and boundary conditions: Loads define the external forces, which act on the object of interest. The loads can be point load, line load, surface load or body loads. Boundary conditions are defined as the value of a given field variable or its derivative, which is constrained at a specific node/free surface to replicate the physical problem under consideration. Though represent approximate, the combination of loads and boundary conditions are used to determine the equilibrium state of the body.

Convergence criteria: This specifies the allowable change in the conserved variable during iterative solving to attain equilibrium of conserved variables. Force and heat flow are considered as convergence criterion parameters in structural and thermal FEA, respectively, which is minimized.

Shape functions or interpolating functions, or basis functions: These are the functions, which find the results at any other point within an element, based on the computed nodal value. The shape functions may be linear or higher order, which are driven by the selection of elements.

Field variable: This is the main output, which is obtained as a result of solving the FE model. For structural problems, displacement is the primary field variable, while for heat-transfer problems, temperature is the primary field variable.

Derived variables: These are computed based on the field variables along with the shape function. In many cases, the derived variables will be of engineering interests. In case of structural FEA, strain and stress become derived variables. For thermal FEA, heat flux is the derived variable.

4.8.2 General Procedure for FEA

Various sequential steps involved in the three stages of FEA are mentioned below and also illustrated in Figure 4.13.

a. *Pre-processing*: This stage involves modeling the geometry (1D/2D/3D) and subsequently assigning finite-element type (beam/shell/solid) and associated geometric properties (length, area, etc.). Then, the material properties (elastic modulus, Poisson's ratio, thermal conductivity, etc.) are defined before meshing the model and checking the element connectivities. Subsequently, the loads and boundary conditions, are virtually imposed.

b. *Solution*: Once the problem is set in the preprocessing stage, the solution stage follows it, wherein the unknown values of the primary field variable(s) are computed through matrix algebra.

   Subsequently, computed values are used to compute additional derived variables, such as reaction forces, element stresses, and heat flow, by using the back-substitution process.

c. *Post-processing*: The elemental and nodal results computed in the solution stage, is visualized in the post-processing stage using contour plot, line plots, charts, animations, and so on. Typically, a dedicated post-processing software/inbuilt module contains sophisticated routines which enables this functionality.

In the following, some representative applications of FEA together with governing equations are described. Semi-discrete governing equations for thermal and structural analysis are given below.

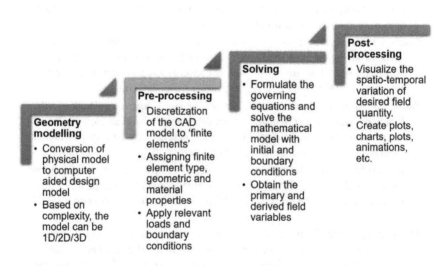

FIGURE 4.13 Flowchart to illustrate the conceptual workflow of Finite Element Analysis (FEA). (Adapted from Prakash, S., Sajin, J.B. and Sandeep, H., 2018. Finite Element Analysis for Deformation Studies in Natural Fiber (Human Hair) Composites. *International Journal of Applied Engineering Research*, 13(9), pp.6894-6897).

4.8.2.1 Thermal Analysis

This is relevant to capture temperature fields in 3D space and the fundamental equation to solve is,

$$[C]\{\dot{T}\}+[K]\{T\}=\{q\} \tag{4.23}$$

where C and K are the capacitance and conductance matrices, respectively, $q = q(t)$ is the applied load, and $T = T(t)$ is the temperature vector. For steady-state analysis, the above equations take the form $KT = q$.

4.8.2.2 Structural Analysis

This analysis is useful to obtain displacement fields at different nodes in response to load stimulus. The governing equation is as follows,

$$[M]\{\ddot{u}\}+[C]\{\dot{u}\}+[K]\{u\}=\{f\} \tag{4.24}$$

where M, C, and K are the mass, damping, and stiffness matrices, respectively, $f = f(t)$ is the applied force and $u = u(t)$ is the displacement vector. For steady-state analysis, the above equations take the form $Ku = f$.

4.8.3 Types of Coupled Field Analysis

A thermo-structural analysis using FEM packages can be performed in two ways, namely, sequentially coupled and fully coupled. A brief discussion of these methods is presented in this section.

4.8.3.1 Sequentially Coupled Analysis

The sequential (also load vector or weak) coupling method consists of two or more analyses, each involving a different physics. These physics are coupled by applying results from the first analysis as loads for the second analysis. A weak coupling involves coupling in element load vector as given below (for a thermo-structural problem)

$$\begin{bmatrix} M & 0 \\ 0 & 0 \end{bmatrix}\begin{Bmatrix} \ddot{u} \\ \ddot{T} \end{Bmatrix} + \begin{bmatrix} C & 0 \\ 0 & C^t \end{bmatrix}\begin{Bmatrix} \dot{u} \\ \dot{T} \end{Bmatrix} + \begin{bmatrix} K & 0 \\ 0 & K^t \end{bmatrix}\begin{Bmatrix} u \\ T \end{Bmatrix} = \begin{Bmatrix} f+f^{th} \\ q+q^{th} \end{Bmatrix} \tag{4.25}$$

where

$[M]$ = element mass matrix
$[C]$ = element structural damping matrix
$[C^t]$ = element specific heat matrix
$[K]$ = element stiffness matrix
$[K^t]$ = element diffusion conductivity matrix = $[K^{tb}] + [K^{tc}]$
$\{u\}$ = displacement vector
$\{T\}$ = temperature vector
$\{f\}$ = sum of the element nodal force and element pressure vectors = $\{f^{nd}\} + \{f^{pr}\}$
$\{f^{nd}\}$ = applied nodal force vector
$\{f^{pr}\}$ = pressure load vector
$\{f^{th}\}$ = thermal strain force vector

$\{q\}$ = sum of the element heat generation load and element convection surface
heat flow vectors = $\{q^{nd}\} + \{q^g\} + \{q^c\}$
$\{q^{nd}\}$ = applied nodal heat flow vector
$\{q^g\}$ = heat generation rate vector due to causes other than Joule heating
$\{q^c\}$ = heat convection surface vector

An example of this is a sequential thermal-stress analysis where the nodal temperature from thermal analysis is applied as body force loads in the subsequent stress analysis. The flowchart for sequential analysis is shown in Figure 4.14.

4.8.3.2 Direct Coupled Analysis

The direct (also matrix or strong or full or simultaneous) coupling method usually involves only one analysis in which a coupled-field element containing all necessary degrees of freedom is chosen. Coupling is handled by calculating element matrices or element load vectors, that contain all necessary terms. A strong coupling for a transient thermo–structural analysis involves coupling in element matrices as given below

$$\begin{bmatrix} M & 0 \\ 0 & 0 \end{bmatrix}\begin{Bmatrix} \ddot{u} \\ \ddot{T} \end{Bmatrix} + \begin{bmatrix} C & 0 \\ C^{tu} & C^t \end{bmatrix}\begin{Bmatrix} \dot{u} \\ \dot{T} \end{Bmatrix} + \begin{bmatrix} K & K^{ut} \\ 0 & K^t \end{bmatrix}\begin{Bmatrix} u \\ T \end{Bmatrix} = \begin{Bmatrix} f \\ q \end{Bmatrix} \tag{4.26}$$

$[C^{tu}]$ = element thermo-elastic damping matrix = $-T_0 [K^{ut}] \, T$
T_0 = absolute reference temperature = $T_{ref} + T_{off}$
T_{ref} = reference temperature
T_{off} = offset temperature from absolute zero to zero
T = current temperature

$[K^{ut}]$ = Element thermo-elastic stiffness matrix = $-\displaystyle\int_{vol} [B]^T \{\beta\}(\{N\}^T)d(vol)$

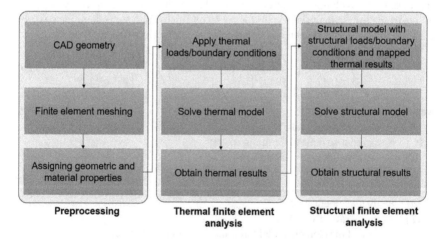

FIGURE 4.14 Flowchart to illustrate the conceptual framework of sequential thermo-structural modelling. (Adapted from Silva, R., 2013. Direct Coupled Thermal-Structural Analysis in ANSYS WorkBench. In *ESSS Conference & Ansys Users Meeting*, Atibaia, Brasil).

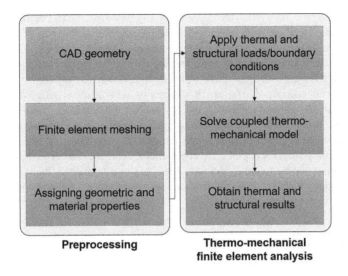

FIGURE 4.15 Schematic to illustrate the concept of coupled thermostructural modeling. (Adapted from Silva, R., 2013. Direct Coupled Thermal-Structural Analysis in ANSYS WorkBench. In *ESSS Conference & Ansys Users Meeting*, Atibaia, Brasil).

$[B]$ = strain−displacement matrix
$\{\beta\}$ = vector of thermo-elastic coefficients = $[D] \{\alpha\}$
$[D]$ = elastic stiffness matrix
$\{\alpha\}$ = vector of coefficients of thermal expansion = $[\alpha_x \ \alpha_y \ \alpha_z \ 0 \ 0 \ 0]^T$
$\{N\}$ = element shape functions

Coupled effects are accounted for by off-diagonal coefficient terms C^{tu} and K^{ut}. Using this coupling, the response in solution is obtained after one iteration. An example for this includes piezoelectric analysis, conjugate heat transfer with fluid flow, and so on, which will use a coupled field element. The flowchart for direct coupled analysis is shown in Figure 4.15.

The type of analysis is selected on the basis of degree of non-linear interactions between the different physics involved in the problem. For a thermal-stress analysis, it is better to choose a sequential approach as the user can perform transient thermal analysis separately and import the load at any desired time step to static structural analysis. Though this will miss out the complete dynamic behavior of the system, the results are not too far from a direct coupled analysis. The computational time and effort, processing memory, and resultant memory for the sequential analysis are much lesser than that of direct coupled analysis.

4.9 Closure

This chapter provides a conceptual understanding of stress and strain as tensorial quantities. The mechanical behavior of metals, ceramics, and polymers, in relation to the stress–strain response is also reviewed. Both the strengthening mechanisms and the theory of the brittle fracture are equally emphasized. In comparison with polymeric implants, metallic implants have clear advantages in terms of tensile and compressive strength behavior. However, when total strain to failure is concerned, polymer materials have superior functionality over ceramic materials, wherein the latter undergoes unpredictable failure during application. Despite having excellent hardness, and compression strength properties (even as compared to metals), poor fracture toughness remains a major concern for ceramics. While the large deformation to failure for thermoplastic polymers can be an advantageous property, the enhancement of elastic modulus and strength clearly demands the development of polymeric composites for engineering applications. More discussion on the major part of this chapter can be found in Ref. [14, 15, 16]. Towards the end, the concepts and methodologies adopted in finite element modeling to computationally, predict material behavior are briefly summarized.

REFERENCES

1. Barsoum M. W., *Fundamentals of Ceramics*, Taylor & Francis, Boca Raton, FL, 2003.
2. Carter C. B., M. G. Norton, *Ceramic Materials: Science and Engineering*, Springer, New York, 2007.
3. Chiang Y. M., D. P. Birnie, W. D. Kingery, *Physical Ceramics*, John Wiley & Sons, New York, 1997.
4. Richerson D. W., *Modern Ceramic Engineering: Properties, Processing, and Use in Design*, CRC Press, Salt Lake City, UT, 1992.
5. De Mestral F., F. Thevenot, Ceramic composites: TiB_2-TiC-SiC, Part I. Properties and microstructures in the ternary system, *J. Mater. Sci.*, 1991, 26, 5547–5560.
6. Graziani T., E. Landi, A. Bellosi, Oxidation of TiB_2-20 vol. % B_4C composite, *J. Mater. Sci. Lett.*, 1993, 12, 691–694.
7. Tampieri A., E. Landi, A. Bellosi, On the oxidation behaviour of monolithic TiB_2 and Al_2O_3–TiB_2 and Si_3N_4–TiB_2 Composites, *J. Therm. Anal.*, 1992, 38, 2657–2662.
8. Tampieri A., A. Bellosi, Oxidation of monolithic TiB_2 and Al_2O_3-TiB_2 composite, *J. Mater. Sci.*, 1993, 28, 649–653.
9. Kulpa A., T. Troczynski, Oxidation of TiB_2 powders below 900°C, *J. Am. Ceram. Soc.*, 1996, 79(2), 518–520.
10. Senda T., Oxidation behaviour of titanium boride at elevated temperatures, *J. Ceram. Soc. Jpn.*, 1996, 104(8), 785–787.
11. Yamada K., M. Matsumoto, M. Matsubara, Oxidation mechanism of TiB_2 particle dispersed SiC and oxidation protection by SiC-CVD coating, *J. Ceram. Soc. Jpn.*, 1997, 105(8), 695–699.
12. Brodkin D., A. Zavaliangos, S. R. Kalidindi, M. W. Barsoum, Ambient- and high temperature properties of titanium carbide–titanium boride composites fabricated by transient plastic phase processing, *J. Am. Ceram. Soc.*, 1999, 82(3), 665–672.
13. Barandika M. G., J. J. Echeberria, F. Castro, Oxidation resistance of two TiB_2-based cermets, *Mater. Res. Bull.*, 1999, 34(7), 1001–1011.
14. Bikramjit Basu, Biomaterials Science and Tissue Engineering: Principles and Methods, Cambridge University Press, ISBN: 9781108415156, August 2017.
15. Basu B., K. Balani, *Advanced Structural Ceramics*, John Wiley & Sons, Inc., USA and American Ceramic Society [ISBN: 9780470497111 (Print); 9781118037300 (online)], September, 2011.

16. Basu B., M. Kalin, *Tribology of Ceramics and Composites: Materials Science Perspective*, John Wiley & Sons, Inc., USA and American Ceramic Society (ISBN-10: 0470522631), September, 2011.

17. Inglis C. E., Stresses in a plate due to the presence of cracks and sharp corners, *Trans. Inst. Nav. Archit.*, 1913, 55, 219–241.

18. Du X., M. Qin, A. Rauf, Z. Yuan, B. Yangand, X. Qu, Structure and properties of AlN ceramics prepared with spark plasma sintering of ultra-fine powders, *Mat. Sci. Eng. A.*, 2008, 496, 269–272.

19. Duan R. G., G. D. Zhan, J. D. Kuntz, B. H. Kear, A. K. Mukherjee, Spark plasma sintering (SPS) consolidated ceramic composites from plasma-sprayed metastable Al_2TiO_5 powder and nano-Al_2O_3, TiO_2, and MgO powders, *Mat. Sci. Eng. A.*, 2004, 373, 180–186.

20. Bai L., M. Xiaodong, S. Weiping, G. Changchun, Comparative study of β-Si_3N_4, powders prepared by SHS sintered by spark plasma sintering and hot pressing, *J. Uni. Sci. Tech. Beijing*, 2007, 14(3), 271–275.

21. Venkateswaran T., B. Basu, G. B. Raju, D.-Y. Kim, Densification and properties of transition metal borides-based cermets via spark plasma sintering, *J. Eur. Ceram. Soc.*, 2006, 26, 2431–2440.

22. Munro R. G., Material properties of titanium diboride, *J. Res. Natl. Inst. Stand. Technol.*, 2000, 105(5), 709–720.

23. Basu B., G. B. Raju, A. K. Suri, Processing and properties of monolithic TiB_2 based materials, *Int. Mater. Rev.*, 2006, 51(6), 354–374.

24. (a) Khvorova I. A., *Materials Science,* Tomsk Polytechnic University, TPU Publishing House, Tomsk, 2012, p. 137. (b) https://www.nde-ed.org/EducationResources/Community College/Materials/Structure/linear_defects.htm

25. Lawn B., *Fracture of Brittle Solids*, 2nd edn, Cambridge Solid State Science Series, Cambridge University Press, 1993.

5

Conventional and Advanced
Manufacturing of Materials

Manufacturing of materials has a central place in the overall landscape of materials science. Manufacturing does not only refer to as how one can process a material from the "molten" or "powder" state, but also encompasses all post-processing techniques to obtain materials with a desired shape or size or architecture, using machining techniques. This chapter introduces to the reader, who may not have any formal knowledge on materials science, the commonly used manufacturing techniques for metals, ceramics, and polymers, particularly focusing on the principles. In addition, rapid prototyping (RP), particularly three-dimensional (3D) printing of materials, will be described together with 3D bioprinting for manufacturing biological cells or tissues. More discussion on the conventional and advanced implant fabrication processes can be found elsewhere.[1]

5.1 Conventional Manufacturing of Metallic Materials

The manufacturing techniques of metals are well established. A combination of these approaches is routinely used in industry to produce metals of various shapes and sizes.

In Figure 5.1, various manufacturing methods are summarized for shaping or making of metallic products. For ceramics, powder-based processing routes are employed. For polymers also, powders or granules are the starting materials. For consolidation of ceramics, high-temperature sintering is commonly used and for polymers, the injection/compression molding or extrusion is adopted.

For metallic materials, the liquid metal is first cast into a shape using a conventional casting route or in some cases by a continuous casting route. The cast products then are deformed using a set of manufacturing processes, including forging, rolling, extrusion, and wire drawing. In the following, we shall briefly discuss each of them.

In case of rolling, thick metal plates can be thinned upto the desired thickness by forcing through the gap between two oppositely rotating rolls (see Figure 5.2a).

Rolling is defined as a bulk deformation process, in which a workpiece will be deformed between two rotating rolls, while being subjected to compressive stresses from the rolls and surface shear stresses from the friction between the roll surface and workpiece surface.

Interdisciplinary Engineering Sciences

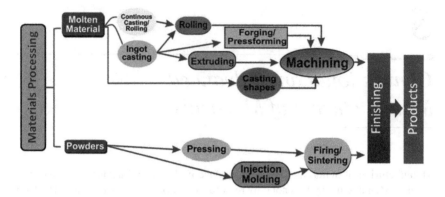

FIGURE 5.1 Flow chart showing the sequence of processing or post-processing approaches to obtain a metallic/ceramic product. (Adapted from Ref. 1).

The thickness reduction is largely influenced by the roll speed and the ability to undergo such deformation depends on the flow strength of workpiece metals. In the rolling process, the thickness of the plate is reduced and since the volume of the metal remains constant during plastic deformation, the velocity (v_o) of the rolled plate exiting from the roll gap can be written as,

$$v_o = v_i(h_i/h_o) \tag{5.1}$$

In the above, v_i is the entry speed of the plate entering the roll gap and h_i and h_o are the thickness of the initial plate (prior to rolling) and final thickness of the post-rolled plate.

For a large reduction, the rolled plates are heat treated and then passed to the next set of rolling stand. The ability to undergo plastic deformation increases with the temperature and this principle is applied in hot rolling.

Like rolling, forging is another bulk deformation process involving uniaxial compressive forces to reduce the thickness of the workpiece (see Figure 5.2b).

> Forging can be defined as the bulk metal deformation process, wherein the large compressive stress is applied with metal being vertically constrained between two platens, with or without lateral constraints.

The forging can be primarily classified as open die forging and closed die forging. Two alternative forging techniques are impression die and flashless forging. As shown in Figure 5.2b, the die cavity has a more regular geometric shape, allowing the forged part to undergo constrained deformation with the designed die cavity in flashless forging. Like in rolling, the extent of deformation during forging depends on the flow stress (i.e., the stress at which metal undergoes uniform plastic deformation at a given metal deformation temperature) as well as thickness or height change to be pursued.

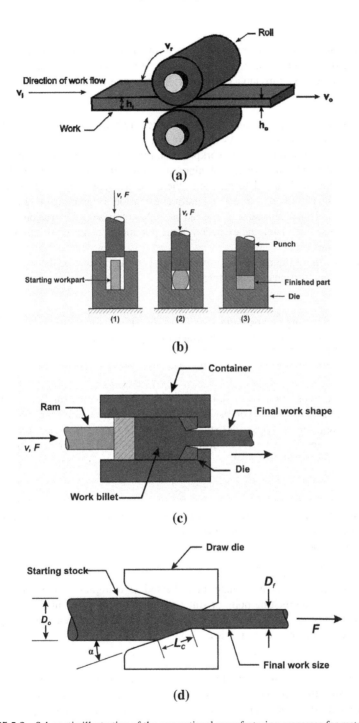

FIGURE 5.2 Schematic illustration of the conventional manufacturing processes for metals with an aim to obtain the final workplace with different sizes or shapes: (a) rolling; (b) flashless forging, (c) extrusion, and (d) wire drawing. (Reproduced with permission from Ref. 1).

Extrusion is a bulk deformation process to obtain rods or similar components with a specific cross-sectional profile, when a material is being pushed with compressive force through a die of the respective cross-section.

In contrast to rolling and forging, the lateral forces are being applied during extrusion and wire drawing processes. As shown in Figure 5.2c, a ram is used to transfer compressive forces to deform a workpiece and the extruded rod with an identical diameter of that of the die orifice is obtained at the die exit. During extrusion, the metal undergoes the maximum deformation at the die constriction region. The extrusion process can be classified as direct and indirect extrusion. Depending on ram pressure application and the die placement, one can use either of these two extrusion processes. As will be discussed later, the extrusion of the polymer depends on different process science aspects.

In contrast to all the above-described metal working processes, the wire drawing involves the application of tensile force.

The wire drawing can be defined as a metalworking process to obtain a wire with a reduced cross-section by pulling it through a single, or series of, drawing die(s) with a tensile force.

Clearly, the cross-sectional area of the drawn wire will be much smaller than that of the extruded rod. Depending on the die angle (α), the deformation zone length (L_c) in the wire drawing process extends to the die orifice (see Figure 5.2d). In both extrusion and wire drawing, the extent of deformation is quantified using % area reduction,

$$r = \frac{A_i - A_o}{A_i} \tag{5.2}$$

where A_i and A_o are the cross-sectional area prior to deformation and after deformation.

Heat treatment is one of the intermediate stages in the integrated cycle of materials processing of metals or ceramics.

During heat treatment, a material is heated to a higher temperature, but much below the melting point or sintering temperature of the metal/ceramic, respectively, before it is cooled to room temperature. The heat treatment changes the microstructure and properties.

On the basis of the nature of cooling, the heat-treatment process can be generically classified into three different classes, (a) annealing (the metal being cooled in the furnace itself), (b) normalizing (the metal taken out of the furnace and kept in ambient air atmosphere for normal cooling), and (c) quenching (the metal taken out of the furnace at the heat-treatment temperature and placed directly into a water bath/salt bath/oil bath for rapid cooling).

Depending on the targeted microstructure and properties, a combination of heat-treatment temperature, time, and nature of cooling is employed. Among the different heat treatment techniques, annealing provides poor hardness due to the coarse grain structure, while quenching can potentially provide better properties. The normalizing treatment provides intermediate properties. In the case of metals, strength scales with the microstructure. Therefore, as far as the microstructural length scale (e.g., grain size) is concerned, the microstructure of the annealed material is characterized by coarser, normalized by intermediate and quenched by a finer grain size. However, due to rapid cooling, surface distortion takes place as well as residual stresses are generated. In order to relieve the quenching associated residual stress, tempering of metals is carried out by heating the quenched metal to a temperature of 100–150°C above room temperature.

The distinction between annealing and other heat treatments (tempering/ aging) is that the former always involves the heating and holding a metal above the recrystallization temperature, whereas the temperature is much lower than the recrystallization temperature in the case of latter.

The annealing heat-treatment techniques are industrially used as an intermediate or an intermittent step between successive deformation processes. For example, a continuously cast or conventionally cast product is annealed before being forged or rolled or between the multiple passes of rolling or forging. The intermediate annealing softens the metal so that hot/cold working can be further performed.

5.2 Conventional and Advanced Manufacturing of Ceramics

In contrast to metals, the manufacturing of ceramic or polymeric materials start from fine powders or granules, due to the difficulty in adapting the melting–solidification route together with the bulk deformation process. In conventional processing techniques, the powder samples are generally pressed into a green body and subsequently, optimally sintered to obtain a solid ceramic compact.

Fundamentally, sintering refers to the process of consolidation of powders into a compact at temperature, $T > 0.5T_m$ which is accomplished by diffusional mass transport.

As shown in Figure 5.3, sintering involves neck formation between the powder particles and during heating, the necks can grow due to man transport from solid particles to the porous regions. The neck growth therefore is an important stage in powder consolidation.

Sintering necessarily involves heating of a powder compact which can be assisted with pressure (e.g., hot pressing [HP], hot isostatic pressing [HIP]) or without pressure (e.g., pressureless sintering in air/protective environment). Thermodynamically, the solid/vapor (s/v) interfaces are replaced by the solid/solid (s/s) interface in a dense

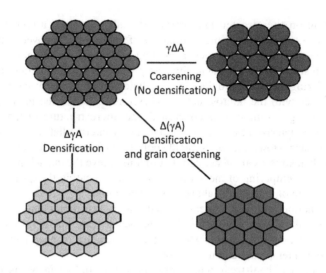

$\gamma\Delta A$
_____
Coarsening
(No densification)

$\Delta\gamma A$
Densification

$\Delta(\gamma A)$
Densification
and grain coarsening

FIGURE 5.3 Schematic demonstration of the sintering of a porous powder compact, either to a dense morphology of equiaxed grains, or to particle coarsening. (Adapted from Ref. 1).

compact, during the course of the sintering interface (see Figure 5.3). The sintering is the thermodynamically favored process, which occurs by reduction in total energy as

$$d(\gamma A) = A d\gamma + \gamma dA < 0 \qquad (5.3)$$

where γ and A are the interface/surface energy and surface area, respectively.

In general, solid/solid or solid/vapor is replaced by only solid/solid interface, therefore, the $A\,d\gamma$ term dominates during the initial/second stage of sintering (see Figure 5.3). The final stage of sintering often involves the competition between the closure of the residual pores and the grain coarsening. Therefore, the (γdA) term can dominate at the last stage.

Two sintering processes are widely used in lab-scale research on ceramics and those are HP and SPS. HP involves simultaneous application of pressure (30–50 MPa) and high temperatures (1200–1900°C) to the powder compact (see Figure 5.4). The HP can provide quite high density (>95%) compacts. The conventional sintering of ceramics is carried out without any pressure application. In view of a slow furnace heating rate (2–5°C/min) and holding at high temperature for a few hours, the entire consolidation process takes place over 18–20 hours for typical oxide ceramics, before it is taken out of the sintering furnace. In contrast, a higher heating rate and shorter holding time at lower sintering temperature are used in case of advanced consolidation approaches, for example, HP or spark plasma sintering (SPS). In general, a much shorter processing time is required for the consolidation of ceramics in the SPS route than the HP route.

SPS has been widely used as a faster consolidation route in the last few decades. It is primarily based on the electric spark discharge phenomenon. The SPS process uses the pulsed field along with the high current, leading to sparks at interparticle neck regions.[2] In SPS, the powder samples can be consolidated upto almost full densification even at sintering temperature of 200–500°C lower than that of conventional sintering.[2] Due to the high heating rates, the entire sintering cycle usually completes over a timespan of less than 1 h.

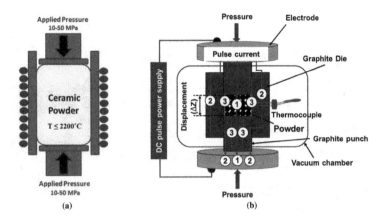

FIGURE 5.4 Schematic illustrating the mode of heat and pressure application in hot press (a) and SPS (b) of a powder compact.

From the process science perspective, the total current flowing from top punch is partitioned between graphite mold wall and the powder compact, depending on the electrical transport properties of the material to be consolidated, as shown in Figure 5.4. For many materials, the SPS process offers specific advantages vis-a-vis conventional HP and HIP,[4,5] such as relatively lower temperature sintering, short sintering time, and so on. All these lead to grain growth minimization, thereby allowing one to realize improved mechanical,[6,7] physical,[8] or optical[9] properties in spark plasma sintered materials.

5.3 Consolidation and Shaping of Polymers

The polymeric materials are usually consolidated to a given shape at much lower temperature (often below 300°C) as compared to those of ceramics. Like sintering of ceramics, both, the processing temperature and applied pressure are important process variables. These parameters play an important role in deciding the shear rate and viscosity that determine the flowability at the molding temperature. Depending upon the composition of raw materials as well as the application of end product, a number of processing techniques, such as compression/injection molding, extrusion, and so on are being used.

Extrusion, one of the widely used polymer processing techniques, has the capability to blend multiple polymers or to fuse different additives (antioxidants, UV stabilizers, etc.)/inorganic fillers with the base polymer (see Figure 5.5). During extrusion, the polymeric material in the molten form is allowed to shear between two screws and is collected through a die. The extrusion process is typically accomplished using a single-screw or twin-screw extruder (TSE). TSE is two parallely configured Archimedean screws. The pressure of dual screws enable a degree of positive conveyance as a means for material transport, which enables more uniform distribution of shear in the material. Two distinct design approaches are characterized by the respective direction of rotation of screws, namely co-rotating, in which the screws rotate in similar direction as well as counter-rotating, where the rotation of screws in opposite direction takes place.

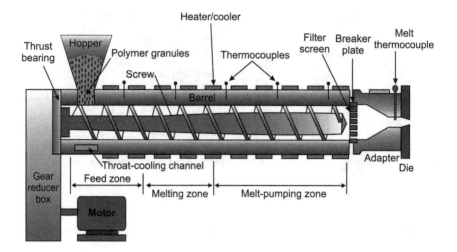

FIGURE 5.5   Schematic illustration of the extrusion process to extrude polymeric rods. (Reproduced with permission from Ref. 1).

The compression molding can be used to consolidate dense polymer compact by simultaneous application of compressive load and temperature.

Another polymer shaping process is injection molding. The injection molding is often attached to melt compounding facility, which is used either to mix co-polymers or mix polymers with inorganic fillers. Alternatively, injection molding is also used as a stand-alone or independent shaping facility for polymers. In this process, the polymeric powders are pre-mixed with the appropriate binders. Subsequently, the powder–binder mixture is fed into the injection molding chamber. The molded sample then undergoes a debinding process to remove the binder.

5.4 Additive Manufacturing of Materials

Phenomenologically, additive manufacturing is described as a layer-by-layer fabrication technology of 3D physical structures of a material, based on machine-readable design file, for example, stereolithography (STL) file.

RP is a broad term for several manufacturing techniques, where 3D structures can be developed in a layer-by-layer manner based on the computer-aided design (CAD) data. The main advantages of RP methods are the fabrication of complex 3D geometries with high reproducibility, waste minimization, extensive choice of materials domain (metals, polymers, ceramics, and composite), and customized product design. Furthermore, the RP processes can create micro- to nanometer details and but are not capable of precisely controlling the pore size, shape, interconnectivity, and special distributions within the fabricated construct. There are several RP technologies available, such as STL, selective laser sintering (SLS), 3D binder printing, electron beam melting (EBM), and laser-engineered net shaping (LENS™).

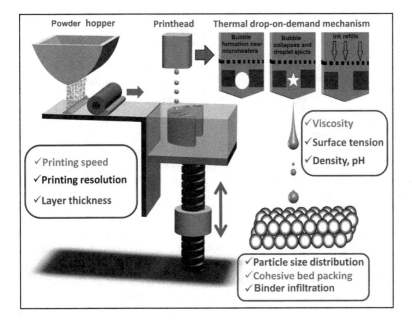

FIGURE 5.6 Schematic demonstration of the binderjet 3D printing, showing the stages in binder ejection and the relevant process/material related parameters of relevance to binder properties, and binder-powder interaction.

Although 3D printing can be accomplished in a number of different approaches, the generic process involves the following sequential steps. In the first step, a CAD model is created digitally for example using a CT scanner. Afterwards, from the CT scan data, the region of interest can be converted into a 3D design file, for example, STL file, by using MIMICS or CATIA (Figure 5.6).[19]

In the second step, the STL file is sliced (cross-sectioned) digitally into a number of 2D layers. In the slicing process, the layer thickness influences the model accuracy and the build time. Therefore, the layer thickness is an important parameter in this process. The model accuracy increases with a decrease in the layer thickness and an increase in the build time.

In the third step, the design data for the sliced layers are sequentially passed to the 3D printing machine through a computer interface. The prototype is made as a layer at a time over the top of earlier layers. Such a process is reported until the entire model prints completely.

Depending on the prototype composition, the post-processing treatment is followed as a last step to obtain 3D-printed prototypes with desired physical architecture and mechanical properties. Infiltration, high-temperature sintering, chemical conversion or combinations of them are among the common post-treatment processes.

The STL-based manufacturing process was first invented by Charles W. Hull in 1986.[10] The process described the fabrication of a 3D solid geometry from the physical state of a fluid through the selective photo-crosslinking in a layer-by-layer manner using UV radiation. The liquid phase used in STL is essentially a photosensitive resin or monomer solution (mainly acrylic or epoxy-based). The UV light (or electron beam) exposed onto

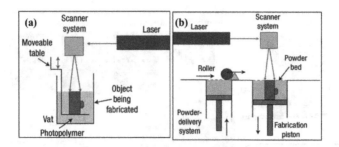

FIGURE 5.7 Schematic representation of different rapid prototyping process: (a) Stereolithography. (Reproduced with permission from, Hollister, S.J., Porous scaffold design for tissue engineering. Nature materials, 2005. 4(7): p. 518). (b) Selective laser sintering. (Reproduced with permission from, Hollister, S.J., Porous scaffold design for tissue engineering. Nature materials, 2005. 4(7): p. 518).

the liquid resin layer is used to initiate the polymerization chain reaction within the monomers. The activated monomers are instantly converted to polymers and then they solidify. The scanner system cured the liquid monomer by irradiating the UV laser beam, which is guided through the CAD modeling data (Figure 5.7a).[18] The STL machine consists of a build platform, which is connected with the moveable table and mounted in a vat of photopolymers. Initially, the UV beam solidifies the topmost layer of the exposed resin surface. Once the layer has been scanned, the moveable table moves downwards by exactly one layer inside the resin and the next layer is exposed to the radiated beam. The energy of the photo-radiation and the exposure time are the parameters, which control the layer thickness and ultimately determine the mechanical properties of the final product. After fabrication of the entire structure, the unreacted resins are removed from the final product through heating or photo-curing to get the desired mechanical strength. Not only polymer-based materials, but also ceramic materials can be constructed using this technique. Ceramic particles are suspended within the photocurable monomers or resins and using the UV laser beam, the monomers are polymerized in a layer-by-layer manner similar to the above-described method. After construction, the polymer binder is removed by pyrolysis and sintering provides the final strength.

SLS is an RP process, which allows us to generate a complex 3D structure using powders as raw materials. Figure 5.7b shows a schematic example of an SLS process.[20] The machine consists of three major operative parts, such as the powder delivery system, powder spreading roller, and the powder bed or build bed where the object is fabricated. The powder feed chamber is raised to supply the powder to the spreading roller (or scraper), which spreads the powder over the powder bed. The powder layers are selectively fused or sintered using the thermal energy of the laser beam. After sintering of each layer, the fabrication piston is lowered and the powder delivery system is raised. The subsequent powder layers are spread over the top of the previous layer and sintered by using the laser beam. During fabrication, the entire powder bed is heated to provide thermal energy to the printed layers and facilitate fusion to the previous layer. In the commercial SLS system, either a CO_2 or Nd:YAG laser is used, of power in the range of 25–50 W,[11] and the laser power depends on the type of materials under construction.

The fabrication principle of 3D binder printing is similar to that of SLS. The binder printer consists of a moveable inkjet printhead, which selectively expels a stream of adhesive droplets to the powder layers, as in Figure 5.6. The binder binds the powder particles to form a solid shape. The main challenge of this process is to develop a suitable powder–binder system. After fabrication, the construct is heat treated for

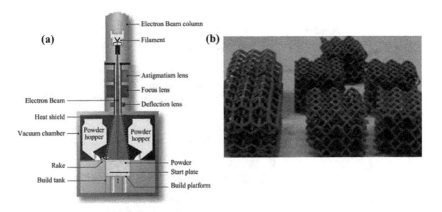

FIGURE 5.8 (a) Components of an EBM machine. (Reproduced with permission from, Galati, M. and L. Iuliano, A literature review of powder-based electron beam melting focusing on numerical simulations. *Additive Manufacturing*, 2018. **19**: p. 1–20). (b) Structures fabricated using EBM process. (Reproduced with permission from, Harrysson, O.L., et al., Direct metal fabrication of titanium implants with tailored materials and mechanical properties using electron beam melting technology. *Materials Science and Engineering*: C, 2008. 28(3): p. 366–373).

binder evaporation or chemically infiltrated, to get the final structure with desired mechanical strength.

EBM is a RP technology to fabricate fully dense 3D metal constructs directly from metal powder. An electron beam selectively scans the thin metal powder layer which is spread in a layer-by-layer manner in a high vacuum (Figure 5.8a).[21] The electron beam column is made up of an electron gun and magnetic lenses. A thermionic or field emission electron gun is used as a source of electron beam, which is focused by using the electromagnetic lenses. The specimen compartment consists of powder hoppers to deliver metal powder on the build table and the powder distributor spreads the powder to form a thin layer. The electron beam sinters the specific areas of each powder layer as directed by the CAD file. The mechanical properties of the fabricated construct depend on the sintering strength, which is controlled by the accelerating beam voltage, beam diameter, and exposure time. The vacuum environment is created to reduce oxidation and contamination of the fabricating materials. Generally, the electron beam gun compartment is kept in a higher vacuum ($\sim 10^{-7}$ Torr) compared to the specimen fabrication compartment ($\sim 10^{-5}$ Torr). Figure 5.8b shows EBM fabricated 3D structures made with Ti–6Al–4V alloys.[22]

Laser Engineered Net Shaping (LENS™) is another RP technology that was developed by Sandia National Laboratories and Stanford University, and was commercialized by Optomec for metal 3D geometries. Using this technology, fully dense metallic structures can be developed. The powder delivery nozzles transmit the powder particles through a gas flow jet and a laser beam is used to melt the powder particles (Figure 5.9)[23,24]. The mounted substrate moves in a raster manner on an X–Y stage according to the CAD model. The important parameter of these processes are the powder flow rate, laser power, and layer thickness. The main advantage of LENS™ is the multilateral deposition for *in-situ* alloying, composite material formation, wear and corrosion resistance surface coating, asymmetrical welding, and functionally gradient structure development.

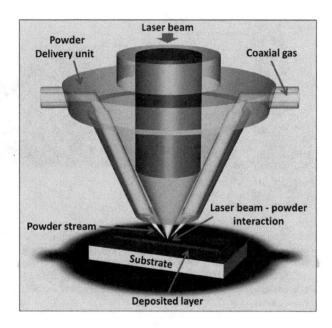

FIGURE 5.9   Schematic representation of the LENS based additive manufacturing process.

Most recently, 3D bioprinting has been investigated widely as a novel biofabrication method in regenerative tissue engineering to mimic the native-like tissue structure with specific dimensions and high reproducibility.[12,13] Bioprinting technologies get enormous attention for the precise placement of living cells within biomaterials into preprogrammed geometries using a 3D bioprinter. The raw material of 3D bioprinting is called bioink, which is a combination of living cells, growth factors, nutrients, and printing media. Biodegradable biopolymers are the raw materials of 3D bioprinting. The biopolymers loaded with living cells are deposited in a layer-by-layer manner, similar to that done using a conventional additive manufacturing technique. Several types of bioprinting strategies, such as extrusion-based, inkjet/droplet-based, and laser/light-assisted bioprinting strategies have been developed in the past few years. Each of the printing technologies has some characteristic features which control the cell viability within the bioprinted construct. Extrusion-based bioprinting is considered as the most promising method because the process is simple; there is a large window of choice of materials and chemically relevant tissues and organs can be fabricated with high accuracy. This printing technology involves mechanical pressure to extrude the bioinks through the nozzle of a syringe in a controlled manner (Figure 5.10a).[3] Living cells, growth factors, and microcarriers are loaded in the biopolymer and are inserted to the syringe barrel and printed as a continuous filament. Inkjet bioprinting is based on the drop-on-demand method, where hydrogel-encapsulated cells are placed within the inkjet cartridge, and bioink droplets are deposited in a controlled manner (Figure 5.10b).[3] The two different approaches developed to generate the ink droplets from the inkjet printhead are thermal and piezo-electric. For droplet bioprinting, thermal energy is applied to the bioink chamber to generate small air bubbles. The bubbles create a pressure pulse within the ink and inks are ejected as droplets with different diameters. For piezoelectric inkjet bioprinting, an electric current is applied

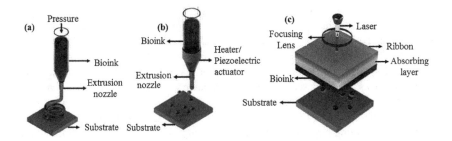

FIGURE 5.10 Schematic representation of (a) extrusion bioprinting, (b) inkjet bioprinting, and (c) laser-assisted bioprinting. (Reproduced with permission from, Heinrich, M.A., et al., 3D Bioprinting: from Benches to Translational Applications. *Small*, 2019. 15(23): p. 1805510).

to a polycrystalline piezoelectric ceramic, which creates a transient pressure to expel ink drops to the building substrate. The main advantage of the piezoelectric printhead is that it can generate a more precise droplet size compared to the thermal printhead. However, many researchers prefer thermal inkjet printing over piezoelectric, since the piezoelectric printing frequency within the range of 15–25 kHz can damage the cell membrane.[14] Laser-assisted bioprinting (LAB) is a nozzle-free approach, which allows bioprinting with a wide range of bioink viscosity (1–300 mPa/s) and high cell concentration in the order of 1×10^8 cells/mL without nozzle clogging, such as conventional extrusion or inkjet bioprinting.[15] The printhead setup comprises a ribbon that is typically a transparent glass slide or quartz. The donor side of the ribbon is coated with a laser-absorbing media (such as Ag, Au, Ti, and TiO_2), and the cell-encapsulated hydrogels are sprayed on it (Figure 5.10c).[3] The absorbing media also protects the hydrogel encapsulated living cells from the high power laser pulse. The laser impulse focuses the absorbing media through the ribbon and the absorbing media evaporated with the creation of high local pressure on the bioink film. The vapor pressure of the absorbing media generates cavitation-like bubbles toward the bioink film. The expansion and collapse of the bubbles creates jets within the bioink layer which leads to the creation of the bioink droplets, which are transferred to the printing substrate. The advantages of LAB are high printing resolution (≥ 20 μm) with no limitation on hydrogel viscosity and cell concentration. The process parameters of the LAB are the wavelength of the laser, pulse duration, beam focus diameter, viscosity and surface tension of hydrogels, and substrate properties.

5.5 Machining Processes

In the manufacturing of engineering components, the finishing and assembly are also equally important.

For metals, the desired shapes can be obtained by both, the conventional as well as non-conventional techniques. The conventional machining processes include turning, drilling, milling, grinding, and so on. In conventional processing, the removal of excess materials is done in the form of chips. In a cylindrical workpiece, the material removal results in the decrease of the diameter.

As far as the non-conventional machining is concerned, the electrical discharge machining (EDM) is among the key machining processes which can be used to fabricate the dies and molds of conducting materials.[16–18] EDM is a thermoelectric process, where multiple electrical discharge (sparks) is applied repeatedly in order to erode the conductive workpiece and electrode materials. In addition, EDM can potentially be used to machine brittle, hard, and difficult to machine materials with reasonable electrical properties.

EDM provides major advantages in terms of faster machining of components of desired shape as well as closer dimensional tolerance vis-à-vis conventional machining.

Die sinking and wire EDM are the two major non-conventional machining processes. In case of die-sinking EDM, die cavity is directly impressed on the workpiece, while accomplishing large material removal (Figure 5.11a). In order to minimize the

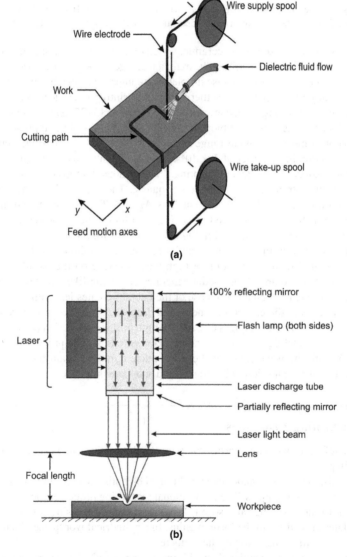

FIGURE 5.11 Schematic illustration of non-conventional machining processes: (a) electrodischarge machining using wire and (b) laser beam machining. (Reproduced with permission from Ref. 1).

electrode wear, materials with good thermal and electrical properties as well as a high melting point are preferred. For a given material, the efficacy of EDM is generally assessed in terms of materials removal and tool wear rates. For wire EDM, oxygen-free high conductivity copper (OFHC) is used as wire to cut or complex shaped products (see Figure 5.11b).

Laser beam machining is another non-conventional process which uses a high power laser source. The laser–material interaction allows the metal to locally melt and vaporize (Figure 5.11). Based on the path traversed by the laser beam, the metal can be machined. The major process variable is the laser power. In order to avoid oxidation, the laser beam machining of metals are performed in an inert gas atmosphere.

5.6 Closure

At closure, it should be emphasized that more than one single processing approach is often used to fabricate polymers with various shapes and composition. For example, a melt-mixer together with injection molding is used to process shaped polymers with homogeneous bulk chemistry and properties. This chapter introduces to any non-specialist reader the conventional manufacturing of metals, ceramics, and polymers with a particular focus on process methodology in a qualitative manner. For ceramics, advanced processing routes are briefly described. More importantly, the additive manufacturing routes together with 3D bioprinting was also emphasized.

REFERENCES

1. Bikramjit Basu, Biomaterials Science and Tissue Engineering: Principles and Methods, Cambridge University Press, ISBN: 9781108415156, August 2017.
2. Tokita M., Trends in advanced SPS spark plasma sintering systems and technology, *J. Soc. Powd. Technol. Japn.*, 1993, 30(11), 790–804.
3. Heinrich M. A. et al., 3D bioprinting: From benches to translational applications, *Small*, 2019, 15(23), 1805510.
4. Langer J., M. J. Hoffmann, O. Guillon, Direct comparison between hot pressing and electric field-assisted sintering of submicron alumina, *Acta. Mater.*, 2009, 57, 5454–5465.
5. Bernard-Granger G., A. Addad, G. Fantozzi, G. Bonnefont, C. Guizard, D. Vernat, Spark plasma sintering of a commercially available granulated zirconia powder: Comparison with hot-pressing, *Acta. Mater.*, 2010, 58, 3390–3399.
6. Shen Z., Z. Zhao, H. Peng, M. Nygren, Formation of tough interlocking microstructures in silicon nitride ceramics by dynamic ripening, *Nature*, 2002, 417, 266–268.
7. Kumar R., K. H. Prakash, P. Cheang, K. A. Khor, Microstructure and mechanical properties of spark plasma sintered zirconia-hydroxyapatite nano-composite powders, *Acta. Mater.*, 2005, 53, 2327–2335.
8. Orru R., R. Licheri, A. M. Locci, A. Cincotti, G. Cao, Consolidation/synthesis of materials by electric current activated/assisted sintering, *Mater. Sci. Eng. R.*, 2009, 63, 127–287.
9. Aman Y., V. Garnier, E. Djurado, Influence of green state processes on the sintering behavior and the subsequent optical properties of spark plasma sintered alumina, *J. Eur. Ceram. Soc.*, 2009, 29(16), 3363–3370.

10. Hull C. W., Apparatus for production of three-dimensional objects by stereolithography. 1986, Google Patents.
11. Pham D. T., R. S. Gault, A comparison of rapid prototyping technologies, *Int. J. Mach. Tools Manuf.*, 1998, 38(10–11), 1257–1287.
12. Murphy S. V., A. Atala, 3D bioprinting of tissues and organs, *Nat. Biotechnol.*, 2014, 32(8), 773.
13. Ozbolat I. T., Y. Yu, Bioprinting toward organ fabrication: Challenges and future trends, *IEEE Trans. Biomed. Eng.*, 2013, 60(3), 691–699.
14. Cui X. et al., Thermal inkjet printing in tissue engineering and regenerative medicine, *Recent Pat. Drug Deliv. Form.*, 2012, 6(2), 149–155.
15. Guillotin B., F. Guillemot, Cell patterning technologies for organotypic tissue fabrication, *Trends Biotechnol.*, 2011, 29(4), 183–190.
16. Petrofes N. F., A. M. Gadalla, Processing aspects of shaping advanced materials by electrical discharge machining, *Adv. Mater. Manuf. Process*, 1988, 127–153.
17. Lauwers B., J. P. Kruth, W. Liu, W. Eeraerts, B. Schcht, P. Bleys, Investigation of material removal mechanisms in EDM of ceramic composite materials, *J. Mater. Process. Tech.*, 2004, 149, 347–352.
18. Sanchez J. A., I. Cabanes, L. N. Lopez de Lacalle, A. Lamikiz, Development of optimum electro discharge machining technology for advanced ceramics, *Int. J. Adv. Manuf. Tech.*, 2001, 18(12), 897–905.
19. Kumar A., S. Mandal, S. Barui, R. Vasireddi, U. Gbureck, M. Gelinsky, B. Basu, Low temperature additive manufacturing of three dimensional scaffolds for bone-tissue engineering applications: Processing related challenges and property assessment, *Mater. Sci. Eng. R.*, 2016, 103, 1–39.
20. Hollister S. J., Porous scaffold design for tissue engineering, *Nat. Mater.*, 2005, 4(7), 518.
21. Galati M., L. Iuliano, A literature review of powder-based electron beam melting focusing on numerical simulations, *Addit. Manuf.*, 2018, 19, 1–20.
22. Harrysson O. L. et al., Direct metal fabrication of titanium implants with tailored materials and mechanical properties using electron beam melting technology, *Mater. Sci. Eng., C.*, 2008, 28(3), 366–373.
23. Obielodan J., B. Stucker, Characterization of LENS-fabricated Ti6Al4V and Ti6Al4V/TiC dual-material transition joints, *Int. J. Adv. Manuf. Technol.*, 2013, 66(9–12), 2053–2061.
24. Hofmeister W. et al., Investigating solidification with the laser-engineered net shaping (LENSTM) process, *JOM*, 1999, 51(7), 1–6.

6

Electrochemistry and
Electroanalytical Techniques

The chemical reactions involving either the usage or generation of electrical current (i.e., flow of electrons in different media) lead to the foundation of a special branch of chemistry, widely known as "electrochemistry." Unlike homogeneous reactions, electrochemical reactions are heterogeneous in nature, as they occur primarily on a surface. The possibility of spontaneous or even non-spontaneous reactions to occur on surfaces of interest, at desired rates and to desired extents, leads to a wide spectrum of technologically important applications, ranging from structural, functional, biomedical, to energy generation and storage. In this context, after a brief discussion on some of the basic concepts in chemical thermodynamics, this chapter will introduce the basic principles governing electrochemistry, connect the same with the associated concepts in thermodynamics, and also describe some of the techniques related to electrochemical measurements/analysis.

6.1 The Laws of Thermodynamics

There are three basic laws that govern the concepts of thermodynamics. The *first law* is a manifestation of the law of conservation of energy; which states that in an isolated or closed system, energy can be transformed across various forms, but can be neither created nor destroyed. In more specific terms, the change in the internal energy (ΔU) of a closed system is equal to the amount of heat energy supplied to the system (Δq) less the amount of work done by the system on its surroundings (Δw). This can be expressed via the following relationship:

$$\Delta U = \Delta q - \Delta w \qquad (6.1)$$

Furthermore, an extensive variable, i.e., a variable that depends on the size of the system, viz., entropy (S), needs to be defined before proceeding to the second law of thermodynamics. Entropy is a manifestation of the degree of randomness or disorder in a system. In more specific terms, it is a function of (or proportional to) the number of states or configurations (microstates or Ω) that are possible to exist in a system (viz., $S \propto \ln \Omega$). Now, the *second law* of thermodynamics states that the overall entropy of a closed system cannot decrease with time. Rather, when considering the net entropy of the system and its surroundings, it can either increase in the case of spontaneous or irreversible process or remain constant when the system is in thermodynamic equilibrium and/or undergoing a reversible process. In a way, the increase in entropy is associated with the irreversibility of various spontaneously occurring natural processes. In mathematical terms, an infinitesimal increment in entropy (δS) of a closed system due to an infinitesimal transfer of heat (δq) at a given temperature (T) from the surroundings, in equilibrium with the system, can be expressed as

$$\delta S = \delta q / T \tag{6.2}$$

Combining the first and second law, that is, Equations (6.1) and (6.2), for a work done just involving a change in the volume (ΔV) at a constant pressure (P) (as is more applicable to condensed matters), the change in the internal energy (ΔU) may be expressed as

$$\Delta U = T\Delta S - P\Delta V \tag{6.3}$$

In very generic terms, the *third law* of thermodynamics states that the entropy of a closed system approaches a constant value as the temperature approaches absolute zero. As mentioned in the preceding paragraph, entropy is proportional to the number of microstates, and, in case there is only a unique ground state with minimum energy at absolute zero, then the entropy will not only be constant, but also will be equal to zero. Nevertheless, if the concerned system lacks a well-defined order and the minimum energy state is not well-defined (as is typically for glassy systems), then there is a possibility of some remnant finite entropy at temperatures close to absolute zero, which is then called the system's residual entropy.

6.2 Auxiliary Functions and Gibb's Free Energy

For a closed system undergoing a change in the volume (ΔV) at a constant pressure (P) while moving from state (i), having an internal energy of U_i to state (ii), having an internal energy of U_{ii}, Equations (6.1) and (6.3) can be rearranged to write an expression for the associated change in heat energy of the system under constant pressure (q_p);

$$q_p = (U_{ii} - U_i) + P(V_{ii} - V_i) \tag{6.4}$$

$$q_p = (U_{ii} + PV_{ii}) - (U_i + PV_i) \tag{6.5}$$

Now, the summation of $U + PV$ is known as the enthalpy (H) of the system. Hence, $q_p = \Delta H$, which implies that the heat admitted to or withdrawn from a system at constant pressure is equal to the change in enthalpy of the system.

At constant pressure, the parameters H and entropy (S) can be combined to lead to a concept, popularly known as Gibbs free energy (G), such that

$$\Delta G = \Delta H - T\Delta S \tag{6.6}$$

Gibbs free energy is a manifestation of the internal energy that is available for a system to perform certain work under the given conditions of constant pressure (and preferably, temperature also). When a spontaneous process is performed by the system, its free energy decreases. Hence, for a system involving chemical reactions, the change in Gibbs free energy signifies whether the reaction is thermodynamically favorable (for $-ve$ ΔG) or not (for $+ve$ ΔG). Now, if a system performs some other work (w_{ext}), in addition to the basic $P\Delta V$, then Equation (6.3) can be re-written as

$$\Delta U = T\Delta S - P\Delta V - w_{ext} \tag{6.7}$$

Since, Equation (6.6) can also be re-written as

$$\Delta G = \Delta U + P\Delta V - T\Delta S \tag{6.8}$$

combining Equations (6.7) and (6.8), ΔG may be expressed as

$$\Delta G = T\Delta S - P\Delta V - w_{ext} + P\Delta V - T\Delta S \tag{6.9}$$

$$\Delta G = -w_{ext} \tag{6.10}$$

This expression has immense practical significance, as in the case of an electrochemical cell, which will be discussed later in Section 6.5.

6.3 The Chemical Potential

If for a system at constant pressure and temperature, the addition or removal of a species (i) leads to the change in chemical free energy (G) of the system, then the net change in the free energy (i.e., ΔG) of the system, normalized by the net change in the quantity of the concerned species (i.e., ΔN_i), forms an important parameter known as the chemical potential (μ_i) of the species, as per the following expression:

$$\mu_i = \Delta G / \Delta N_i \tag{6.11}$$

N_i may either be the number of particles or the number of moles of the species (i), with the resultant values just differing with respect to the Avogadro's constant ($\sim 6.023 \times 1023$; i.e., the number of constituent particles in one mole of the species or NA). Since a basic assumption involved in the expression for chemical potential is that the composition of the system does not change appreciably upon addition/removal of the concerned species by an amount equivalent to ΔN_i, the normalization with respect to the number of moles of the species is usually valid for systems that are large enough such that the composition remains fairly similar despite the variation of N_A number of particles of species (i). Furthermore, a basic chemical potential assumption is that the concerned species (i) is not electrically charged and hence do not interact with the inner potential of the system.

If not so, then μ_i is further modified to lead to electrochemical potential, as defined in the following section.

6.4 Redox Reactions, Free Energy Change, and Electrochemical Potential

Let us consider the reduction of an oxide (XO) in the presence of another element, y, which involves the reduction of the oxidation state of X from +2 to 0 (i.e., reduction of X), with a concurrent increase in the oxidation state of Y from 0 to +2 (i.e., oxidation of Y).

$$XO + Y \rightarrow YO + X \qquad (6.12)$$

This is a very simple example for the simultaneous occurrence of both oxidation and reduction reactions and electron transfer reactions; more commonly known as the "redox" (or e^- transfer) reaction. Like any other reaction, this will be associated with an overall change in the chemical free energy (or ΔG), irrespective of the way in which it is conducted.

Any electrochemical reaction is associated with the overall change in energy (viz., ΔG). In this case, ΔG will be determined by electrochemical parameters with the driving force being the change in electric voltage arising from the difference in the electrical potential. Electrochemical potential (μ_e) of a charged species (having charge z) participating in an electrochemical reaction is related to the chemical potential (μ_0; see Section 6.3), via the electric voltage (Φ), as per the following equation:

$$\mu_e = \mu_0 + z\mathcal{F}\Phi \qquad (6.13)$$

where $\mathcal{F}$ is the Faraday's constant ($\sim$96,500 Coulombs per mole of electron; or the net charge present in Avogadro's number of electrons). μ_0 can be conceived as the work required to add a particle to the system at constant pressure and temperature.

> In very fundamental terms, it is the electrochemical potential difference of the electrons between the two "reactants" which set-up the electrical potential (or electrochemical cell voltage) that drives electrochemical reactions.

The electrochemical potential in turn depends on the Fermi energy level (or equivalent) of the concerned material, with electron transfer taking place from the material having a higher Fermi level to the lower for equilibrating the same; eventually leading to a potential difference at the contact.

6.5 Cell Voltage, Nernst Equation, and Effects of Concentration

Continuing from the preceding section, the difference in energy level of the electrons participating in the concerned redox reaction determines the overall voltage ($\Delta\Phi$) of the cell with electrodes in an electrolytic bath. In the extreme situation, when no current is being withdrawn from the cell, $\Delta\Phi$ is also called the open-circuit voltage (OCV) of the cell and is the electromotive force (or EMF) for driving the cell reaction at that point of time. As will be discussed in the following, the OCV depends on the state of the electrodes and the electrolyte. The OCV dynamically changes with the occurrence of the cell reactions (or current being withdrawn). The same is also true during recharging of a cell in the case of a rechargeable system or secondary cell. It is important to note that the feasibility of a cell to be recharged or not depends on the reversibility of the concerned redox reactions.

Now let us try to establish a relationship between the electrochemical parameters and very basic thermodynamic parameters. As a recap, according to the *first law* of thermodynamics (see Section 6.1) and the auxiliary functions (see Section 6.2), when a system undergoes a reversible process (at constant temperature and pressure), the decrease in Gibbs free energy changes (ΔG) should equal the work done (Δw) by the system (in this case, the cell), which can be expressed by the following relationship (as also expressed in Equation 6.10 earlier)

$$\Delta w = -\Delta G \tag{6.14}$$

An electrochemical cell (more specifically, a galvanic cell) performs electrical work by transporting the electric charge (z) across a voltage difference ($\Delta\Phi$); such that Δw_{elect} is equal to $z\mathcal{F}\Delta\Phi$. This voltage difference under equilibrium conditions is called the EMF of the cell. Since, $\Delta w = -\Delta G$ (as per Equation 6.14);

$$\Delta G = -\Delta w_{elect} = -z\mathcal{F}\Delta\Phi \tag{6.15}$$

Equation (6.15) is the very basic Nernst equation, relating the chemical free energy change to the electrochemical energy. For any general reaction, as given in Equation (6.12), ΔG will depend on the activity of the reactants. Therefore, an electrochemical system will be governed by the concentrations of redox active species, as per the following equations

$$\Delta G = \Delta G^0 + RT \ln[\text{YO}]/[\text{XO}] \tag{6.16}$$

where [YO] and [XO] denote the concentrations of the product and reactant, respectively, as in Equation (6.12). The concentrations of the elemental species, namely, X and Y, have not been included in the above equation, since they are considered to be in their standard states. Additionally, in Equation (6.16), R is the universal gas constant ($\sim$8.314 J/mol/K) and T is the temperature (in K) under consideration. Accordingly, by combining Equations (6.15) and (6.16), it can be seen that under the non-standard states, the EMF of the cell becomes dependent on the concentration of the active species. This is known as the modified Nernst equation, showing a dependence on the concentration or activity of the redox active species in the electrolytic solution, as shown below.

$$\Delta\Phi = \Delta\Phi^0 - (RT/z\mathcal{F})(\ln[\text{YO}]/[\text{XO}]) \tag{6.17}$$

In the above context, it can be understood that with the progress of the electrochemical reaction, the overall cell voltage changes due to the changes in the concentrations of the active species in the electrochemical cell.

As will be seen in Chapter 17, the changes in the composition of electrode active materials (via insertion/removal of foreign species or oxidation states of redox-active species in the electrode) during electrochemical cycling will also lead to a change in the overall cell voltage, in addition to changes in the electrolyte composition.

6.6 Polarization and Overpotential

As discussed in the previous section, the maximum (electrical) energy that can be drawn from an electrochemical cell (as delivered by the chemicals stored within or supplied) depends on the ΔG of the cell reaction or the electrochemical couple. However, the actual energy that can be harnessed in the form of the desired electrical energy depends on other factors, including the magnitude of the current withdrawn or the rate of discharge and the kinetics of the concerned redox reactions.

In more specific terms, losses due to polarization occur when the current (I) passes through the electrodes. These losses primarily include (1) *activation polarization*; which drives the redox reaction at the electrode surface, and (2) *concentration polarization*, which arises from the concentration differences of the reactants and products at the electrode surface and in the bulk as a result of (sluggish) mass transfer. Furthermore, the internal resistance of the cell (even though not truly a polarization) also causes a voltage drop during operation (or when current is drawn), which also results in dissipation of a part of the energy as waste heat. The total internal resistance (R_{int}) of an electrochemical cell is contributed by the ionic resistance of the electrolyte, the electronic resistances of the active mass, the current collectors and electrical tabs of both electrodes, and the contact resistance between the active mass and the current collector. It is to be noted that in a typical electrochemical cell, the current collectors and external circuit carry the electronic current, while the electrolyte (an electronic insulator) carries the ionic current (i.e., the ionic species that participate in the redox reactions at the electrode surfaces). The voltage drop due to internal impedance is usually referred to as "ohmic polarization" or "IR" drop; and is proportional to the current drawn from the system. It may be noted here that "IR" drop, in true sense, is not a "polarization" (or overpotential) per se; but leads to a similar effect.

When connected to an external load (R_{ext}) and a current (I) is passed, the cell voltage (Φ) can be expressed as

$$\Phi = \Delta\Phi^0 - [IR_{int} + |(\eta_{ct})_c| + |(\eta_{ct})_a| + |(\eta_{conc})_c| + |(\eta_{conc})_a|] \qquad (6.18)$$

where Φ^0 is the electromotive force or open-circuit voltage of the cell, $(\eta_{ct})_c$ and $(\eta_{ct})_a$ are the activation polarizations at the cathode and anode, respectively, $(\eta_{conc})_c$ and $(\eta_{conc})_a$ are the concentration polarization at cathode and anode, respectively, I is the operating current of the cell, and R_{int} is the internal resistance of the cell.

"Overpotential" (ΔV) is typically associated with "polarization". In very basic terms, it is the excess energy required to drive an electrochemical reaction (at a desired rate) in order to compensate for the polarization losses and is the difference in the actual potential (V_{act}) at which the concerned reaction is taking place at the electrode (at the associated current/rate) and the equilibrium potential (V_{eqm}) for the redox reaction (i.e., $\Delta V = V_{act} - V_{eqm}$). Let us take a slightly different view here. Consider that there are no polarization losses. Even in that case, at the equilibrium potential both the anodic and cathodic processes are expected to occur and at the same rate, resulting in the net current being zero (or no apparent net reaction in a particular direction; namely, either reduction or oxidation). Accordingly, a slightly negative potential with respect to the equilibrium potential will tilt the favor toward (or make it faster) the reduction reaction, as compared to the reverse oxidation reaction, leading to a net negative or reduction current and vice versa. This is the basic necessity for the overpotential in the case of a reversible redox reaction and will be detailed more in Chapter 7 on electrochemical kinetics.

> The polarization losses increase the magnitude of the overpotential needed to support particular current or the rate of reaction in one direction (either reduction for negative overpotential or oxidation for positive overpotential).

6.7 Electroanalytical Techniques: Concepts and Applications

Various steady-state and impulse electroanalytical techniques are used to determine different mechanistic aspects of a cell, and to evaluate/predict the performances that are expected when the cells eventually get translated to technologies. These techniques are described below.

6.7.1 Cyclic Voltammetry

In cyclic voltammetry, the current is monitored as a function of cell voltage, when the cell is subjected to reversible linearly changing voltage (at a constant voltage sweep rate) within a pre-set voltage window (viz., the lower and upper cut-off voltages). The voltage window is usually chosen to facilitate the occurrences of the redox reactions, while the electrolyte remains stable. Voltage scan from the lower to the upper cut-off voltage leads to net oxidation of the electro-active species in the electrolyte, primarily at the electrode−electrolyte interface, as commonly referred to as oxidation or anodic scan. During the subsequent scan back from the upper to the lower cut-off voltage, the oxidized species get reduced in the case of a reversible system, as commonly referred to as the reduction or cathodic scan.

During the voltage ramp, as the applied voltage approaches that of the reversible potential for the concerned redox reaction (or equivalent electrochemically driven phenomena), current starts flowing beyond the background current. The magnitude of this current, associated with the redox reaction/process, increases rapidly as the voltage crosses the equilibrium potential due to the associated overpotential (see Section 6.6). With further scanning, the current becomes limited at a potential slightly beyond the standard potential by the subsequent depletion of reactants, despite the fact that the overpotential keeps increasing. The depletion of reactants establishes concentration profiles which spread out into the solution and becomes flatter as time progresses (see Figure 6.1a). As the concentration profiles extend into solution, the concentration gradient driving the mass transport goes down and, accordingly, the rate of diffusive transport at the electrode surface causes the current to decrease. "Diffusion layer" (as indicated in Figure 6.1a) is defined as the region ahead of the electrode surface (and

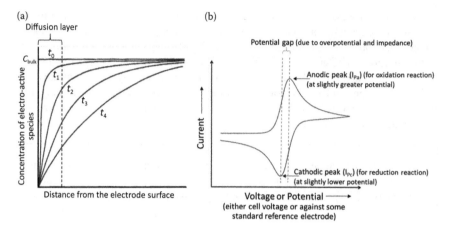

FIGURE 6.1 (a) Concentration profiles at different times ($t_4 > t_3 > t_2 > t_1 > t_0$) of the electroactive species in the electrolyte away from the electrode surface (and toward the bulk of the electrolyte); (b) typical current versus potential profile obtained during cyclic voltammetry (CV), showing the cathodic (I_c) and anodic (I_a) currents, with the currents peaking at the potentials where the concerned reduction and oxidation reactions take place. (Partly adapted from *Handbook of Batteries*, D. Linden, T. B. Reddy, Copyright 2002, 1999, 1994, 1972 by The McGraw-Hill Companies, Inc., NY, USA.)

toward the electrolyte bulk) within which the concentration gradient of a given species in the analyte primarily develops at the beginning. This occurs concurrently with the concentration just beyond this diffusion layer being assumed to be constant and equal to the bulk concentration of the electrolyte. Based on this definition, it can be said in simpler terms that with further scanning of voltage beyond the equilibrium potential (i.e., with passage of time), for a fixed or non-rotating electrode, the thickness of the "diffusion layer" keeps increasing (with the electrode surface concentration close to 0). This leads to flattening of the concentration profile from the electrode surface toward the bulk of the electrolyte. Accordingly, the current passes through a well-defined maximum (leading to the formation of a current peak), as illustrated in Figure 6.1b. This is the characteristic feature of cyclic voltammograms recorded with systems having well-defined reversible redox reactions.

The voltage gap between the oxidation and reduction peaks (i.e., oxidation peak being at a higher voltage than the corresponding reduction peak) is primarily associated with the overpotential(s) needed beyond the equilibrium potential for the corresponding redox reaction(s) to take place. The overall impedance of the cell also contributes toward the voltage difference between the oxidation and reduction peaks. The greater the overpotential needed and greater the impedance, the larger is the gap between the oxidation and reduction peaks.

6.7.2 Chronopotentiometry

Chronopotentiometry involves the study of voltage transients at an electrode upon imposition of constant current. It is sometimes also known as galvanostatic voltammetry. The voltage responses under constant current conditions indicate the changes in electrode processes occurring at the interface.

In the case of practical electrochemical devices, like batteries, reversible chronopotentiometry is used to discharge and charge the devices, when it is referred to as galvanostatic cycling.

Galvanostatic cycling is also used to determine the charge storage capacity of the cell, which is typically given by the product of the current and the time taken to fully discharge or charge.

As a very basic example, we consider the reduction of a species O to R, as expressed by $O + ne \leftrightarrow R$. As the constant current is applied, the concentration of O in the vicinity of the electrode surface begins to decrease. As a result, O diffuses from the bulk solution into the depleted layer, and a concentration gradient grows out from the electrode surface into the solution. This is very similar to the picture at the electrode surface in the case of CV, but the current is maintained constant here and voltage is an independent variable. As the cycling continues, the concentration profile extends further into the bulk solution. When the surface concentration of O falls to zero, the electrode process can no longer be supported by reduction of O. Usually a somewhat flat potential profile is observed till the time the reaction gets "completed" at the surface (due to depletion of the concentration of the reactant) (see Figure 6.2). Once that happens, an abrupt change in potential occurs till another cathodic reaction gets initiated. Similar behavior is observed during the oxidation half cycle in the case of a reversible system. The time interval between the commencement of the concerned electrochemical redox reaction and the sudden change in potential upon "completion" of the same is known as the "transition time" (as indicated in Figure 6.2). As depicted in Figure 6.2, the profile obtained with chronoamperometry usually displays segments due to the initial "IR" drop, the subsequent charging of the electrochemical double layer, and the onset-cum-occurrence of the faradaic redox reaction.

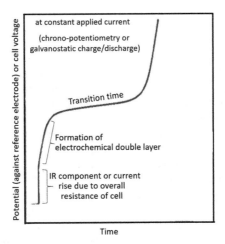

FIGURE 6.2 Typical chronopotentiogram or potential (i.e., cell voltage) versus time profile obtained with galvanostatic cycling (i.e., at constant current). (Partly adapted from *Handbook of Batteries*, D. Linden, T. B. Reddy, Copyright 2002, 1999, 1994, 1972 by The McGraw-Hill Companies, Inc., NY, USA.)

6.7.3 Chronoamperometry

The principle of this technique can be better understood phenomenologically from the following example. Consider that in an electrochemical cell at a given instant, the electrode is at a potential (E_i) above that for the concerned R–O couple (say, E_0). Also, assume that only O is present in the solution, with the flux to the electrode being diffusion controlled. If in the next instant, say t_0, the potential is stepped to a value (say E_s, which is negative compared to E_0), nearly all the species O in the electrolyte solution at the surface of the electrode will get reduced to R. This can potentially result in a very high instantaneous current (depending on the activation overpotential). This leads to the development of a steep concentration gradient for species O across a "diffusion layer" from a value of nearly 0 at the surface of the electrode to a value close to the electrolyte bulk concentration just beyond the "diffusion layer" (refer back to Figure 6.1a). This steep concentration gradient leads to the rapid transport of the reactant species O to the electrode surface, thus able to support the instantaneous high current. With progress in time (say, from t_0 to t_3) at the "stepped" potential, the "diffusion layer" thickens away from the electrode surface to the electrolyte bulk, resulting in the concentration gradient becoming progressively less steeper and concomitantly the magnitude of the current. This phenomenon can be supported by the continually lower transport rate of the concerned O species to the electrode surface (as understood based on Figure 6.1a). This basic form of the current response during chronoamperometry measurements, with the assumptions of diffusion limited current being valid and capacitive current (which varies as t^{-1}) being neglected, allows the determination of the diffusion co-efficient (D) of the redox species in the electrolyte solution. Accordingly, the current (I) typically varies with time (t) as per the Cottrell equation (i.e., $I \propto t^{-1/2}$), as shown below

$$I = n\mathcal{F}AC_0D^{1/2}\pi^{-1/2}t^{-1/2} \tag{6.19}$$

where n is the stoichiometric number of electrons involved, F is the Faraday's constant, A is the planar electrode surface area, and C_0 is the concentration of electroactive species, in this case, O.

In addition to the determination of diffusivity of the redox species, chronoamperometry can also be used to detect the occurrence/non-occurrence of nucleation-growth-induced phase transformations during the electrochemical reaction and determine the kinetics of the same. This is so because, unlike the basic case (i.e., in the absence of phase transformation) where the current monotonically decreases with time upon imposition of the potential jump (as described above and presented in Figure 6.3a), when nucleation and growth govern the electrode kinetics, the current response shows a momentary increase or hump after the initial drop, followed by a further continual drop in current (as shown in Figure 6.3b). After the initial usual decrease, the current begins to increase (say, at point *i* in Figure 6.3b) upon initiation of nucleation, which leads to an increase in the number of two-phase boundaries (i.e., regions where the new phase grows). The current then reaches a maximum (say, at point *ii* in Figure 6.3b), which corresponds to the maximum number of phase boundaries, followed by again the usual monotonic decrease from point *ii* to point *iii*. This indicates that the new phases merge with each other to form large domains.

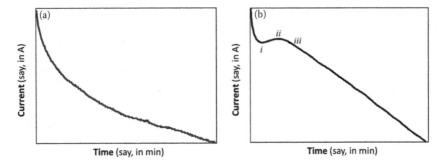

FIGURE 6.3 Typical chronoamperograms involving (a) no phase transformation and (b) first-order nucleation and growth induced phase transformation, as indicated by the hump. (Partly adapted from Satam et al., *Scripta Mater.* 2016, 124, 1.)

An approximate fit of the resultant chronoamperogram (i.e., the *I-t* response) to the classical Kolmogorov–Johnson–Mehl–Avrami (KJMA) model predicts the fraction of newly formed phase (*f*) as a function of time (*t*), as per the following equation:

$$f = 1 - \exp(-kt^n) \tag{6.20}$$

where k and n are fitting parameters. n (Avrami exponent) can be further expressed in terms of nucleation index (a), dimensionality of growth (b), and growth index (c), as $n = a + bc$. Here, the nucleation index or (a) indicates the time dependence of the nucleation rate. The scenario characterized by $a = 1$ implies the constant nucleation rate. In general, a varies between 0 and 1 and this indicates that the nucleation rate decreases with time. The parameter (b) indicates the dimension of the nuclear growth, with $b = 1$, 2 and 3 corresponding to one-, two-, and three-dimensional nuclear growth, respectively. The growth index or c indicates the rate-limiting step of the phase transformation, with $c = 0.5$ and 1 corresponding to diffusion and phase-boundary movement controlled growth, respectively.

6.7.4 Electrochemical "Titrations"

Moving ahead from basic electrochemical testing, galvanostatic intermittent titration (GITT) and potentiostatic intermittent titration (PITT) are two of the most commonly used techniques to evaluate the diffusivity of the redox-active species within the solid electrode lattice during electrochemical reduction/oxidation (especially in alkali metal ion battery systems). The methodologies used for determining the diffusivity of the species in the electrolyte solution, in addition to the chronoamperometry technique discussed in Section 6.7.3, will be discussed in Chapter 7 (Section 7.4). While introducing the basic electrochemical titration techniques, this section discusses the use of the same for determination of diffusivity of species within an insertion electrode.

The principles involved in the two titration techniques are not too different from the basic aspects discussed earlier, with the time taken for the relaxation (or achieving uniform distribution of the inserted species) within the electrode after current (for GITT) or potential (for PITT) pulses being primarily used for estimating the diffusion co-efficient. These basic GITT and PITT measurements and the equations, as below,

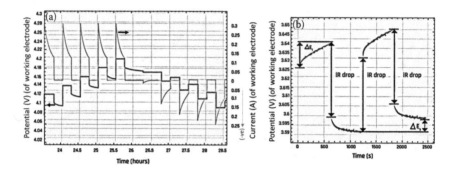

FIGURE 6.4 Excerpts of representative (a) PITT scans and (b) GITT scans (with ΔE_t and ΔE_s representing steady-state voltage change due to the current pulse and the actual voltage change during the constant current pulse, respectively). (Partly adopted from *Metrohm Autolab Application Notes BAT03 and BAT04*.)

are strictly valid only in the case of solid–solution electrodes (i.e., in the absence of phase transformations upon insertion/removal of the foreign species).

In more specific terms, in the case of PITT, a small voltage step is applied and the resulting current is measured as a function of time. Excerpts from a typical current–potential profile recorded during the PITT experiment is presented in Figure 6.4a. Using Fick's law, the diffusion coefficient of the inserted species in the solid electrode is estimated as per the following equation:

$$D = (4L^2 / \pi^2)(-d \ln I(t)/dt) \tag{6.21}$$

where $I(t)$ is the current recorded during the constant voltage step and L is the characteristic thickness of electrode materials (i.e., the length over which the diffusion is expected to occur from the point of insertion), with all the other symbols having the usual meaning.

In the case of GITT, the electrode system is subjected to small constant current pulses, followed by relaxations (or ~ 0 current). The concomitant potential changes are measured as a function of time. Excerpts from a typical potential–time profile obtained during GITT experiments are presented in Figure 6.4b. Assuming one-dimensional diffusion of the inserted species in the solid electrode, the diffusion coefficient (D) is estimated using Fick's law

$$D = (4/\pi)(IV_m/z_A \mathcal{F}A)(dE(x)/dx)^2(dE(t)/dt^{0.5})^{-2} \tag{6.22}$$

where z_A is the charge number of electroactive species, S is the surface area of the electrode (in contact with the electrolyte), I is the applied constant current during the pulses, and V_m is the molar volume of the electrode material. In deriving Equation (6.21), ohmic drop, double-layer charging, charge-transfer kinetics, and phase transformation are neglected. The value of $dE(t)/dt^{0.5}$ is obtained from the plot of voltage versus the square root of time during the constant current pulse, and $dE(x)/dx$ is measured by plotting the equilibrium electrode voltage (i.e., at the end of the relaxation period) against the electroactive material composition (or state of charge) after each full current pulse. One of the advantages of GITT over PITT is that the ohmic effects arising from the voltage drop due to the uncompensated resistances in the electrode and electrolyte do not affect the results obtained with GITT, whereas such effects are not usually controllable in the case of PITT (which does not fix the current).

6.8 Closure

In summary, the fundamental concepts of electrochemistry, starting with understanding of the associated basic thermodynamic principles and electrochemical redox reactions, are discussed in this chapter. Such discussion is primarily based on the well-known Nernst equation. Keeping an application, namely, electrochemical energy storage, in mind, slightly advanced concepts concerning overpotential and polarization and their relation to the equilibrium reaction potential have been discussed. This is followed by description of some of the basic electroanalytical techniques, which are not only applicable to research, but also form the foundation of electrochemical operations/ testing of actual electrochemical devices. In fact, it is galvanostatic cycling, which is used to discharge and charge a battery. Additionally, some of the techniques discussed, namely, cyclic voltammetry and electrochemical titrations, can have several applications toward further evaluation and understanding of various aspects, such as diffusion and kinetics, of electrochemical processes. This will be further discussed in a latter chapter of this book.

FURTHER READING

Gaskell D. R., *Introduction to the Thermodynamics of Materials*, 5th edn, CRC Press, 2003.

Bard A. J., L. R. Faulkner, *Electrochemical Methods: Fundamentals and Applications*, 2nd edn, John Wiley & Sons, Inc.

Linden D., T. B. Reddy, *Handbook of Batteries*, Copyright @ 2002, 1999, 1994, 1972 by The McGraw-Hill Companies, Inc., NY, USA.

7

Chemical and Electrochemical Kinetics

In Chapter 6, it has been emphasized that the feasibility of any reaction can be predicted by thermodynamics. However, most engineering applications are typically influenced by the rate at which various reactions (or phenomena) take place. For instance, the current drawn from an electrochemical energy storage device depends on the rate of the performance-determining electrochemical reactions and the associated transport processes. These, in turn, influence the efficiency and life of a device. Similarly, the rate of homogeneous reactions in an industrial reactor ultimately determines the productivity and performance. Against this backdrop, this chapter will discuss the basic aspects concerning general reaction kinetics, followed by application of such concepts toward electrochemical reactions and the associated performances of electrochemical systems.

7.1 Reaction Kinetics, Arrhenius Relation, and Activated Complex

In this section, we shall reiterate the salient features concerning the theory of chemical reaction kinetics in very general terms, without specific reference toward heterogeneous reaction kinetics (like electrochemical reactions at electrode surfaces). For a reversible reaction, say $R \leftrightarrow P$, the rate of forward reaction (v_f in moles/s) can be written as $k_f C_R$ where k_f is the rate constant for the forward reaction (per s) and C_R is the concentration of the reactant (in moles). Similarly, the rate of backward reaction (v_b) can be written as $k_b C_P$. Accordingly, at any given instant, the net rate of the reaction (v_{net}) in one of the directions, say forward direction, is given by

$$v_{net} = v_f - v_b = k_f C_R - k_b C_P \tag{7.1}$$

which becomes zero at equilibrium. It may be recalled here that, in case of electrochemical redox reactions (Sections 6.4 and 6.6), the net current at equilibrium potential is zero, even though both the oxidation and reduction reactions take place simultaneously (at the same rate). The rate of both the forward and backward reactions at equilibrium (which are equal) is known as "exchange velocity" (v_0) of the reaction [i.e., $v_0 = k_f(C_R)_{eqm} = k_f(C_P)_{eqm}$].

The rate constant or k of a reaction varies directly with temperature (T). In more precise terms, $\ln k$ varies almost linearly with $1/T$, as per Arrhenius equation

$$k = F \exp(-E_a/RT) \tag{7.2}$$

where E_a is the activation energy (or barrier height; see Figure 7.1), F is the frequency factor, and R is the universal gas constant. The exponential term on the right-hand side is associated with the probability of using thermal energy to surmount the energy

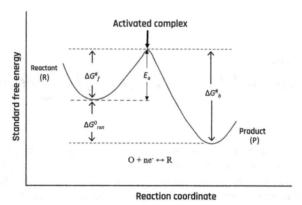

FIGURE 7.1 Free energy changes ($\Delta G^\#$) being manifestations of the activation energies (i.e., E_a) required for forward and backward reactions. (Adapted from Bard A. J., L. R. Faulkner, *Electrochemical Methods: Fundamentals and Applications*, 2nd edn, John Wiley & Sons, Inc.)

barrier of height (E_a). Accordingly, F is related to the frequency of attempts made to surmount the barrier. Equation 7.2 or the basic Arrhenius equation can be rearranged to yield a form, as below

$$k = F' \exp(-\Delta G^\#/RT) \tag{7.3}$$

where $\Delta G^\#$ is the standard free energy of activation (which is a manifestation of E_a).

In a bid to satisfactorily predict the values of F and E_a in terms of quantitative molecular properties, the most popular theory is the transition state theory. This is also known as the activated complex theory. According to this theory, the reactions proceed through a fairly well-defined transition state or activated complex (AC), as indicated in Figure 7.1, with the standard free energy of activation being associated with the standard free energy change in going from the reactant state to the state of the activated complex in both the directions, that is, $\Delta G^\#_f$ for the forward reaction (from R → AC) and $\Delta G^\#_b$ for the backward reaction (from P → AC).

7.2 Electrochemical Reaction Kinetics

Continuing from the previous section, which talks about the kinetics of homogeneous chemical reactions, a typical unit electrochemical redox reaction may be written in a simplified form as

$$O + ne \leftrightarrow R \tag{7.4}$$

On similar lines, in this case, the rate of forward or reduction/"cathodic" reactions (v_f; now in moles/cm²/s) can be written as $k_f C_O$. Similarly, the rate of backward or oxidation/"anodic" reactions (v_b) can be written as $k_b C_R$. However, since electrochemical reactions at the electrode surface are heterogeneous in nature, some modification in the parameters is needed. The concentrations in this case have to be defined more precisely, since that would vary from the electrode surface (say, $x = 0$) to the bulk of the electrolyte solution (i.e., different at any given x). This assumes a one-dimensional situation for simplicity and the typical concentration profiles can be seen in Figure 6.1. Accordingly, for the heterogeneous electrochemical reactions occurring at the electrode surface, the concentrations at any given time t may be denoted as $C_O(0,t)$ and $C_R(0,t)$. Hence, $v_f = k_f C_O(0,t)$ and $v_b = k_b C_R(0,t)$.

The current (i) is related to the corresponding rate of reaction as $v = i/n\mathcal{F}A$, where n is the number of moles of electrons participating in the unit redox reaction (as depicted above), $\mathcal{F}$ is Faraday's constant (connecting number of moles to charge $\sim$96,500 C/mole) and A is the area of the electrode surface. Accordingly, the net electrochemical reaction velocity (v_{net}) and concomitantly, the net current (i_{net}) at the concerned electrode are given by the following relations:

$$v_{net} = v_{cathodic} - v_{anodic} = k_f C_O(0,t) - k_b C_R(0,t) = i_{net}/n\mathcal{F}A \qquad (7.5)$$

$$i_{net} = n\mathcal{F}A[k_f C_O(0,t) - k_b C_R(0,t)] \qquad (7.6)$$

Since the concentrations of the redox active species are still expressed in mol/cm³, the heterogeneous rate constants have units of cm/s. The next section will shed some light on the formulation of the rate constants for electrochemical reactions, eventually leading to the formulation of the *Butler–Volmer* and *Tafel* relations.

It is to be noted here that the entire formulation of the electrochemical reaction kinetics deals with the forward and reverse reactions occurring simultaneously at one particular electrode.

7.3 *Butler–Volmer* and *Tafel* Relations for Electrochemical Reaction Kinetics and Their Applications

In the case of electrochemical reactions, the interfacial potential difference controls the reactivity. It is therefore essential to predict, in very precise terms, how the rate constants depend on the potential. Again considering the basic unit redox reaction, now with $n = 1$ for simplicity, the standard free energy profiles along the reaction coordinate can be represented as in Figure 7.2; which, in a way, is adopted from Figure 7.1 for the homogeneous reaction kinetics.

At the equilibrium potential, E_{eqm}, the cathodic ($\Delta G^{\#0}_c$) and anodic ($\Delta G^{\#0}_a$) activation energies (or barrier heights) are equal. If the potential is changed to E, the relative energy of the electron resident on the electrode changes by $\mathcal{F}(E - E_{eqm})$ and concomitantly the O + e curve in Figure 7.2 moves up or down by that amount. As can be seen from Figure 7.2, the barrier for oxidation, $\Delta G^{\#}_a$, gets reduced by a fraction of the total energy change [$\mathcal{F}(E - E_{eqm})$], if $E > E_{eqm}$. Conventionally, this fraction is denoted as $(1 - \alpha)$, where α is called the transfer co-efficient, which varies between 0 and 1 depending on the shape (symmetry) of the intersection of the free energy curves. Accordingly, the $\Delta G^{\#}_a$ and similarly $\Delta G^{\#}_c$ can be written as

$$\Delta G^{\#}_a = \Delta G^{\#0}_a - (1 - \alpha)\mathcal{F}(E - E_{eqm}) \qquad (7.7)$$

$$\Delta G^{\#}_c = \Delta G^{\#0}_c + \alpha\mathcal{F}(E - E_{eqm}) \qquad (7.8)$$

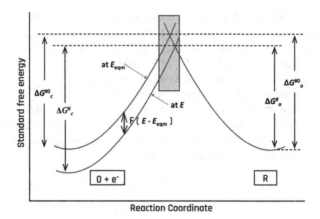

Reaction Coordinate

FIGURE 7.2 Effects of change in electrochemical potential (w.r.t. the equilibrium potential) on the standard free energies of activation for the elementary electrochemical oxidation and reduction reactions taking place on an electrode. (Adapted from Bard A. J., L. R. Faulkner, *Electrochemical Methods: Fundamentals and Applications*, 2nd edn, John Wiley & Sons, Inc.)

Also, the rate constants (k_f and k_b) in their Arrhenius forms (as in Equation 7.3) can be written as

$$k_f = F_f \exp(-\Delta G^{\#0}_c / RT)\exp[-\alpha f(E - E_{eqm})] \tag{7.9}$$

$$k_b = F_b \exp(-\Delta G^{\#0}_a / RT)\exp[(1 - \alpha)f(E - E_{eqm})] \tag{7.10}$$

where $f = \mathcal{F}/RT$. It can be noted that the first two factors in the above expressions are independent of the potential and their products can be interpreted as being the rate constants at $E = E_{eqm}$. These rate constants are equal to each other and may be denoted as the standard rate constant (k^0). Accordingly, replacing the first two terms by k^0 and combining Equations 7.6, 7.9, and 7.10 results in the basic form of the current–potential relationship or the *Butler–Volmer* equation (as below).

$$i_{net} = \mathcal{F}Ak^0[C_O(0,t)\exp\{-\alpha f(E - E_{eqm})\} - C_R(0,t)\exp\{(1 - \alpha)f(E - E_{eqm})\}] \tag{7.11}$$

In a general sense, the *Butler–Volmer* equation predicts the variation of the net current with the magnitude of the overpotential (i.e., $E - E_{eqm}$; usually referred to as η; as introduced in Section 6.6).

k^0 is a measure of the inherent kinetics of the concerned redox reaction, with a larger k^0 implying that the reaction is capable of achieving equilibrium in a shorter time period and vice versa.

At equilibrium, the net current (or i_{net}) is zero, and the electrode surface adopts a potential based on the bulk concentrations of O and R (i.e., C_O^* and C_R^*, respectively; which are also present at the surface) as per the Nernst equation. However, despite the zero net current, both the anodic and the cathodic processes keep occurring at finite rate and the current corresponding to that is expressed in terms of *exchange current*

(i_0), which is equal in magnitude for either component current, namely, i_c or i_a (i.e., $i_0 = i_c$ or i_a; at equilibrium).

Exchange current may also be re-defined as a measure of the rate of exchange of charge between oxidized and reduced species at equilibrium potential, but without net overall change.

Based on Equations 7.9, 7.10 and 7.11, i_0 can be written as

$$i_0 = \mathcal{F}Ak^0 C_O^{*(1-\alpha)} C_R^{*\alpha} \tag{7.12}$$

Using this form of i_0 and substituting for k^0 in Equation 7.11 lead to a more widely used form of the *Butler–Volmer* relation as mentioned below. For looking into the more detailed assumptions and formulations involved in obtaining this form of the Butler-Volmer relation, the reader is directed to the reference 1 of this chapter (i.e., Bard A. J., L. R. Faulkner, *Electrochemical Methods: Fundamentals and Applications*, 2nd edn, John Wiley & Sons, Inc.).

$$i_{net} = i_0[\{C_O(0,t)/C_O^*\}\exp\{-\alpha f(E - E_{eqm})\} - \{C_R(0,t)/C_R^*\}\exp\{(1 - \alpha)f(E - E_{eqm})\}] \tag{7.13}$$

The variations of the currents (i_{net}, i_c, and i_a) at the electrode with respect to the overpotential (η, which is $E - E_{eqm}$) are depicted in Figure 7.3. It can be seen that for large negative η, the anodic component is negligible; hence the curve for the net current (i_{net} or the solid curve) merges with that for i_c, with the reverse happening at large positive overpotentials. As per the theoretical predictions, in going either direction from E_{eq}, the magnitude of the current increases rapidly due to the domination of the exponential factors in the *Butler–Volmer* relation. However, at more extreme overpotentials, the current tends to level off since it becomes mass transfer limited, rather than limited by the actual redox (heterogeneous) kinetics at the electrode surface.

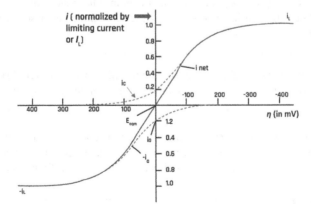

FIGURE 7.3 Representation of the basic *Butler–Volmer* relationship or current–overpotential curves for the system $O + e^{-1} \leftrightarrow R$, with $\alpha = 0.5$ and at constant temperature. Limiting current or I_L is the maximum current that is achieved under a given mass transfer limited condition, that is, when the electrode surface concentration of the concerned electro-active species falls to 0 and no further increment in the concentration gradient driving the species from the bulk to the electrode surface can be achieved (refer to Figure 6.1 and Sections 6.7.1 and 7.4). (Adapted from Bard A. J., L. R. Faulkner, *Electrochemical Methods: Fundamentals and Applications*, 2nd edn, John Wiley & Sons, Inc.)

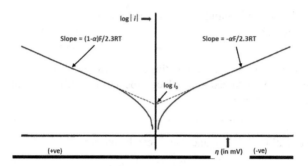

FIGURE 7.4 Typical *Tafel* plots for anodic and cathodic branches of the current–overpotential curve, again for the system O + e ↔ R (where "e" stands for electron), with $\alpha = 0.5$ and at constant temperature. (Adapted from Bard A. J., L. R. Faulkner, *Electrochemical Methods: Fundamentals and Applications*, 2nd edn, John Wiley & Sons, Inc.)

Considering no mass transfer effects, that is, if the solution is well stirred or currents kept low enough that surface concentrations do not differ notably from the bulk values, the *Butler–Volmer* equation can be further simplified as

$$i_{net} = i_0[\exp\{-\alpha f\eta\} - \exp\{(1-\alpha)f\eta\}] \tag{7.14}$$

Under conditions of very large overpotentials, the *Butler–Volmer* equation (i.e., Equation 7.14) can be further simplified since one of the terms within the brackets (viz., $e^{-\alpha f\eta}$ or $e^{(1-\alpha)f\eta}$) becomes negligible. Considering the case at very large negative overpotential (corresponding to net cathodic or reduction reaction), $e^{-\alpha f\eta} \gg e^{(1-\alpha)f\eta}$, Equation 7.14 can be re-written as

$$i_{net} = i_0[\exp\{-\alpha f\eta\}] \tag{7.15}$$

The above is the basic form for the *Tafel* equation. *Tafel* plots are typically plots of log|*i*| versus η (see Figure 7.4), where the anodic branch has a slope of $(1-\alpha)F/2.3RT$ and the cathodic branch has a slope of $-\alpha F/2.3RT$. Both linear segments can be extrapolated to an intercept of log i_0.

It is to be noted that the *Tafel* equation is more valid when the electrode kinetics are very sluggish and significant activation overpotentials are needed.

Accordingly, good fit to *Tafel* behavior is often considered an indication for irreversible kinetics. No significant current flow is seen, except at high overpotentials, where the faradaic process is nearly unidirectional and chemically irreversible.

7.4 Determination of Diffusivity of Species and Analytes in Electrolyte

The methodologies and the principles involved in determining the solid-state diffusion in insertion electrode types have already been discussed in Chapter 6, as an application for two of the electro-analytical techniques (in Section 6.7.4). This section primarily discusses the methodologies for the determination of diffusivity of species in the electrolyte solution (transported to/from the electrode surface before/after electrochemical reactions). Diffusion or mass transport under concentration gradients of ionic species in solution has an immense influence on the kinetics of various chemical and electrochemical processes. For instance, suppressed diffusion kinetics of the analytes or the active species in the case of electrochemical reactions may lead to mass transport limitations for otherwise kinetically favorable redox reactions; thus enhancing the overpotential. This renders it imperative to know precisely the diffusion co-efficient of the concerned species under the relevant electrochemical reaction conditions.

The more commonly used and reliable method to measure the same, particularly for a redox species, is the method involving the determination of "limiting current" (I_L) while sweeping the potential at a fixed scan rate (similar to a cyclic voltammetry experiment), but using varying revolutions per minute of a rotating disk electrode (see Figure 7.5). Based on the as-determined limiting currents (I_L), the diffusivity (or D) of the concerned ion (i.e., the species getting reduced) can be estimated based on the Levich equation[1,2]

$$I_L = 0.620 n F A V^{-1/6} C D^{2/3} \omega^{1/2} \tag{7.16}$$

where V is the kinematic viscosity of the electrolyte medium, A is the area of the electrode surface, C is the concentration of the concerned species, ω is the angular velocity of the rotating disk electrode (in rad/s). All the other symbols retain the same meanings, as defined earlier. The limiting current (i.e., the maximum current density, which remains constant) is achieved, because the rotating disk electrode helps in keeping the thickness of the "diffusion layer" (as defined in Section 6.7.1) constant during the potential sweeping, unlike in the case of a fixed (or non-rotating) electrode (which would lead to a current peak, as in cyclic voltammetry, due to thickening of the "diffusion layer" with continued scanning; see Section 6.7.1).

Even though the limiting current methodology, as described above, is possibly the most reliable method for estimating the diffusivity (of analytes in a solution), another more commonly used method is based on cyclic voltammetry. However, cyclic voltammetry is

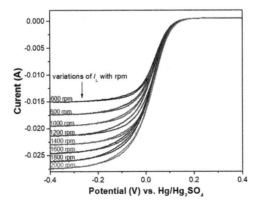

FIGURE 7.5 Current versus potential plots showing limiting currents (I_L) recorded during electrochemical reduction of Fe^{3+} to Fe^{2+} in $FeCl_3$ solution (using bulk Pt rotating disk electrode) at 10 mV/s using the bulk Pt disk electrode rotating at different rpms with electrolyte containing (a) 0.25 M $FeCl_3$ solution.

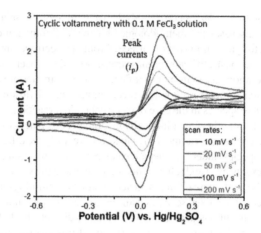

FIGURE 7.6 CVs recorded at different potential scan rates (during electrochemical reduction of Fe^{3+} to Fe^{2+} in $FeCl_3$ solution), showing the increase in intensity of the peak currents with scan rate.

usually used for determining the diffusion co-efficient of species within solid insertion electrodes before/after electrochemical reaction at the interface (the aspect discussed in Section 6.7.4). In this method, variations of intensities of the concerned current peak (i_p) in the cyclic voltammograms (CVs) recorded with varying current densities are noted (see Figure 7.6). If the current is assumed to obey the power law, the relationship between the peak current (i_p) and sweep rate (V) is expressed as in the following[3]:

$$i_p = av^b \tag{7.17}$$

where a and b are adjustable parameters, with $b = 1$ indicating purely surface controlled mechanisms and $b = 0.5$ indicating purely diffusion controlled mechanisms. If the current is diffusion limited, which is the condition for being able to estimate the corresponding diffusivity, the diffusivity can be estimated based on the following equation:

$$i_p = 0.4958 nFACD^{1/2} \left(\frac{\alpha n_a F}{RT} \right)^{1/2} v^{1/2} \tag{7.18}$$

where α is taken as 0.5 and n_a is taken as unity. All the other symbols retain the same meanings, as defined earlier.

7.5 Overview of Implications of Reaction Kinetics toward Technological Demands

Electrochemical reactions form important aspects of various technological applications and are increasingly becoming important, primarily led by the ever increasing societal needs for advance technologies and at the same time depletion of various natural resources, such as fossil fuels and green cover (i.e., natural forest) (more details to be provided in Chapters 11 and 13). For instance, the direct consequences of rapidly increasing demand for automobiles and reduction in trees are the excessive environmental pollution arising from the usage of fossil fuels for running automobiles, which (*i.e.*, the fossil fuels), in turn, are also getting depleted by the day. Accordingly, there is an enormous thrust toward the development of automobiles which can run without the usage of fossil fuels. The primary solution to this is to run all types of vehicles (including very heavy vehicles) on rechargeable batteries, which are electrochemical energy storage devices.

There are various aspects concerning the successful usage of the electrochemical energy storage devices in automobiles, such as the necessities for enhanced energy density, power density, cycle life, and safety aspects. Even though discussing about the principles of such batteries and about all the parameters is beyond the scope of this chapter (which will be done subsequently in Chapter 13), among the aforementioned parameters, power density is probably one of the most important ones. This is because power density governs how fast can a battery be charged and how fast can the energy be drained from it whenever required (for instance, during rapid acceleration).

If you consider yourself driving an electric vehicle on the highway and running out of "battery charge," ideally you would like to get it "recharged" at a "recharging station" in a few minutes to continue your journey further for at least another 100 miles.

For this to happen the battery in the car needs to have superior power density, or the ability to get charged very fast, even compared to most of the state-of-the art advanced batteries (like Li-ion batteries). In fact, it is desired that most car batteries in electric vehicles get recharged during breaking from the kinetic energy generated during the same; a process known as "regenerative braking," which would last for a few seconds. Charging and discharging of batteries are based on electrochemical redox reactions, primarily occurring at the surface of the concerned electrodes (say, the charge carrier ions, namely, Li-ions in Li-ion batteries, getting reduced at the cathode surface during discharge and vice versa), followed by the insertion of the reduced ions to within the solid electrodes (which store them reversibly). Accordingly, for achieving greater power density or the ability to get charged/discharged very fast, the basic electrochemical reactions (i.e., reduction of Li-ions in this case), which are partly governed by the diffusivity of the concerned ions (Li-ions) in the electrolyte and the transport of the reduced Li-ions to within the electrode bulk, that are in turn governed by the solid-state diffusivity of Li-ions in the concerned electrode material, need to be very fast (see Sections 6.7.4 and 7.4 for more fundamental aspects). This in itself highlights the importance of electrochemical reaction kinetics toward the relevant application in electric vehicles.

As mentioned toward the beginning of this section, the depletion of green cover is partly responsible for the increased environmental pollution. This is because trees have been the natural absorber of atmospheric CO_2. In this context, the efforts to reduce the environmental CO_2 content via artificial/technological means are important. The two most commonly investigated methods are the chemical and electrochemical reduction of CO_2, which can also convert CO_2 into useful fuels. In both the methods, reaction kinetics are important, which affect the efficiency to a considerable extent (since thermodynamic feasibility is always ensured first in all the cases). However, the kinetics assume a special importance in the case of the electrochemical reduction of CO_2, which in fact is presently being investigated with enormous thrust and is deemed to be the dominant CO_2 reduction process in the next few years. The reason kinetics is very important in the case of electrochemical reduction is because, not only does it affect the efficiency of the process, but also it determines the feasibility of the desired reactions taking place in a given set-up. For instance, usage of aqueous electrolyte is always preferred, which however has a limited potential window beyond which it breaks down. Accordingly, the concerned electrochemical reactions (for CO_2 reduction and associated desired product formation) need to take place within the concerned potential window. In other words, the overpotential (see Section 6.6) requirements for the concerned reactions need to be adjusted/reduced so that they take place within the potential window. As can be envisaged, this is then all about reducing the activation energy and concomitantly improving the electrochemical kinetics (see Sections 7.1 through 7.3 for more fundamental aspects).

One of the avenues is the use of electro-catalysts and accordingly, various materials have been investigated, starting from the more expensive noble metals like Pt and Au, followed by the relatively less expensive metals like Sn, Fe, Pd, Cu, etc. and presently metal-oxides/perovskites as superior electro-catalysts for CO_2 reduction (for reducing the overpotentials and improving the efficiencies of the as-desired reactions). It must be mentioned here that the research and developments on electro-catalysts for electrochemical CO_2 reduction have been somewhat similar to the more traditional search for superior electro-catalysts for room temperature fuel cells, which are electrochemical energy conversion devices (converting chemical energy to the desired electrical energy), where reactions at the anode and cathode, especially the oxygen reduction reaction at the cathode needs good electro-catalysts.

7.6 Closure

Moving ahead from Chapter 6, where the feasibility of occurrence of electrochemical reactions was primarily dealt with, in this chapter, the concepts related to the rate of electrochemical reactions have been discussed. It is the relative rates of the individual cathodic and anodic reactions associated with an overall reaction that determines whether the concerned electrochemical reduction or oxidation (as the net reaction) takes place at a given electrode and at a certain electrochemical potential. Accordingly, the rate of the overall reaction also depends on the electrochemical potential. In addition to the potential, the overall rates also get governed by the diffusivity of the species in the electrolyte; which are, in turn, possible to estimate via some basic electroanalytical methods. Subsequently, the implications of these basic understandings and determinations on the overall societal needs have been discussed, which will be further elaborated in more specific terms in Section II.

REFERENCES

1. Bard A. J., L. R. Faulkner, *Electrochemical Methods: Fundamentals and Applications*, 2nd edn, John Wiley & Sons, Inc.
2. Linden D., T. B. Reddy, *Handbook of Batteries*, Copyright 2002, 1999, 1994, 1972 by The McGraw-Hill Companies, Inc., NY, USA.
3. Sonia F. J., M. K. Jangid, B. Ananthoju, M. Aslam, P. Johari, A. Mukhopadhyay, Understanding the Li-storage in few layers graphene with respect to bulk graphite: Experimental, analytical and computational study. *J. Mater. Chem. A.*, 2017, 5, 8662.

8

Introduction to the Biological System

The last few decades have witnessed significant efforts by engineering scientists, to address research problems related to biological sciences without any formal training in biological sciences. Against this backdrop, this chapter describes the structural characteristics and important properties of the major elements of a biological system, that is, proteins, cells, bacteria, and tissue.

8.1 Introduction

For any living system, a biological cell is the basic building block. The different cell types and their functionality are summarized in Table 8.1. The most abundant macromolecule in a biological cell is protein. In structural terms, proteins can be defined as biological heteropolymers containing at least 20 different amino acids. Figure 8.1 shows the schematic reaction between two amino acids, leading to the formation of peptide bonds. Structurally, proteins represent polypeptide chains, which exist in different formats, like primary (helical shape), secondary (a helix, b sheets), tertiary (mixture of a and b helix), and quaternary (globular shaped). Proteins in the coiled state express appropriate biological functions. The aggregates of macromolecules, when bound by a double-layer membrane, are known as organelles. In multicellular organisms, a self-organized array of similar types of cells with identical functionalities is known as tissue. Similar types of tissues with common functions form organs and the functionally related organs constitute a higher level of organization known as an organ system. One of the unique characteristic of human physiology is, therefore, the hierarchy in structural complexity and organization of different elements at different length scales.

8.2 Protein: Structure and Characteristics

Each cell contains thousands of different proteins, each serving a variety of functions in the cell, both catalytic and structural. The typical protein concentration in the cytoplasm is 180 mg/mL and considering the volume of a 15 μm-sized fibroblast cell as 2×10^{-9} cm^3, the average number of proteins is 4×10^9 molecules per eukaryotic cell. Since the size of a prokaryotic cell is mall (considering 2 μm length and 0.8 μm diameter of *E.coli*, the volume 1×10^{-12} m^3), the number of proteins is 2×10^6 molecules per prokaryotic cell.

Proteins are biological polymers made up of smaller units (monomers). The monomers are arranged in a linear manner and this arrangement is widely known as a

TABLE 8.1

Various Cell Types and Their Functionality Can Be Potentially Differentiated from Different Types of Stem Cells under Various Biophysical or Biochemical Cues

Cell Type	Functionality
Adipocyte	Cells that make and store fat compounds
Astrocyte	A type of glia (glue) cell that provides structural and metabolic support to the neurons
Cardiomyocyte	Cells that form the heart; also called myocytes
Chondrocyte	Cells that make cartilage
Endothelial cell	Cells that form the inner lining (endothelium) of all blood vessels
Hematopoietic cell	Cells that differentiate into red and white blood cells
Keratinocyte	Cells that form hair and nails
Mast cell	Associated with connective tissue and blood vessels
Neuron	Cells that form the brain, spinal cord, peripheral nervous system
Osteoblast	Gives rise to osteocytes or bone-forming cells
Pancreatic islet cells	Endocrine cells that synthesize insulin
Smooth muscle	Muscle that lines blood vessels and the digestive track

polypeptide sequence. These monomers are known as amino acids. In general, amino acids consist of an acidic Carboxyl group (-COOH), a basic Amino group ($-NH_2$), a hydrogen atom (H), and a variable "R" group. Each group is bonded to a central carbon (C) atom (also called alpha carbon). Two amino acids are linked together by a peptide bond (shown in Figure 8.1). The "R" group is specific to each amino acid which gives different identities to these protein monomers (Figure 8.1).

For particular proteins, the polypeptide chain has a unique sequence of amino acids, which is verified with the help of the information provided in the cellular genetic code. In general, these sequences are written at left with the N-terminal and at right with the C-terminal.

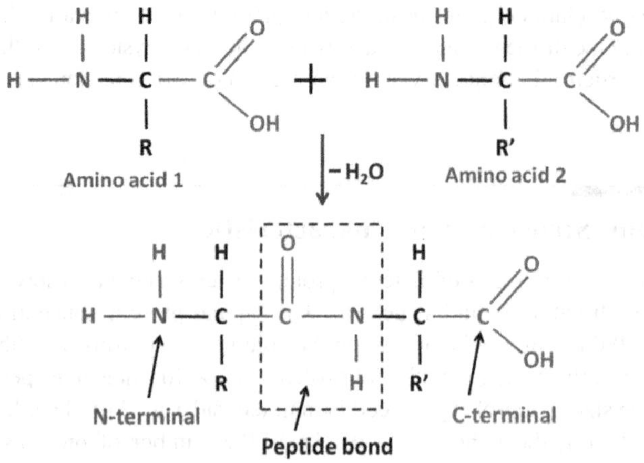

FIGURE 8.1 Proteins play an important role in diverse biophysical functions in living organisms. Schematic representation of the reaction between two amino acids leading to covalent polypeptide bond formation along the backbone chain of a protein molecule. (Adapted from Ref. 5).

8.3 Eukaryotic and Prokaryotic Cells

Eukaryotic cells are, in general, enclosed with a bilayer structure of lipid molecules and they possess the genetic material, required to channelize their replication as well as for continuity of life. Overall, cells are the smallest structural as well as functional units of life. In natural tissues, the cell density is about $1-3 \times 10^9$ cells/mL. For example, an adult of 50 kg weight will possess approximately 10^{14} cells in his body. All the living cells are made up of major elements: hydrogen (59%), oxygen (24%), carbon (11%), nitrogen (4%), and others (2%)—phosphorus (P), sulfur (S), and so on.

The bilayer structure of a cell membrane (made up of phospholipids) is semi-permeable in nature, which allows only some of the selective ions/molecules to pass in or out of the cell (see Figure 8.2). A number of protein molecules are also embedded across the bilayer membrane structure which helps in the transportation of various kinds of ions. As far as the replication of the cell is concerned, nucleic acid is a key genetic material that contains and helps in expressing a cell's genetic code. Deoxyribonucleic acid (DNA) and ribonucleic acid (RNA) are the major classes of nucleic acids (see Figure 8.2). DNA is the genetic molecule, which contains the necessary information, required to replicate and maintain the cell. RNA is associated with the expression of the information stored in DNA and protein synthesis. DNA is present in the nucleus of a cell and uses RNA as intermediary for communicating with other cells.

The living cells are categorized into two basic types; prokaryotes (*pro-* before; *karyon-* nucleus) and eukaryotes (*eu-* = true). The single-celled organisms such as bacteria and archaea are classified as prokaryotes, while all the eukaryotes are multi-cellular organisms. The prokaryotic cells do not contain nucleus and any membrane-bound organelle. However, eukaryotic cells possess membrane-bound organelles within the cell with the nucleus containing all the genetic information (DNA). The size of eukaryotes is much bigger (25–30 μm) than that of prokaryotes (2–3 μm). The doubling times of prokaryotes and eukaryotes are about 20–60 minutes, and 12 hours (or longer), respectively.

In prokaryotic cells, a single circular DNA strand, which lies at the center of the cytoplasm, is known as a central nucleoid that contains the genetic material, whereas in eukaryotic cells, there is a presence of membrane-bound distinct nucleus. A prokaryotic cell is defined by a cell membrane separating a single compartment (cytoplasm) from the surrounding environment. Outside the cell membrane, prokaryotic cells contain a relatively rigid cell wall that acts as an additional layer for protection. The cell membrane plays an important role in maintaining the shape as well as helps in dehydration. It is generally made up of peptidoglycan.

The sizes of eukaryotic and prokaryotic cells are about 10–100 μm and 0.1–5.0 μm, respectively.

8.4 Structural Details of Eukaryotic Cell

Eukaryotic cells possess a cell membrane or plasma membrane having a total thickness of 7–10 nm. It consists of a phospholipid bilayer along with embedded proteins separating the interior of cells from their external environment (Figure 8.2).

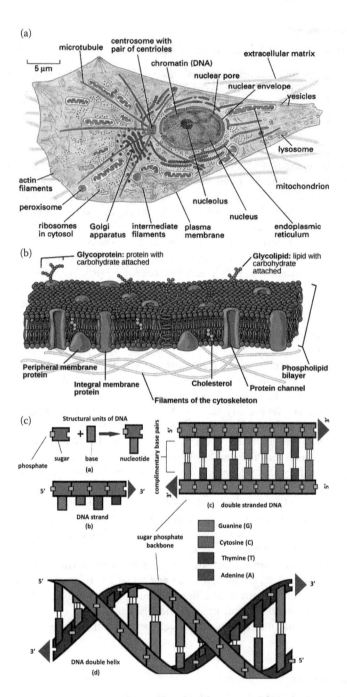

FIGURE 8.2 (a) Schematic structural details of an eukaryotic cell, showing the morphological features of various cellular organelles as well as their morphology. (b) The plasma membrane of eukaryotes comprises of a phospholipid bilayer with embedded proteins and cholesterol.[4] (c) Schematic representation of the DNA structure. (Adapted from Ref. 5).

Apart from phospholipids and proteins, other compounds such as carbohydrates and cholesterol are also observed in the membrane.

Biochemically, a cell membrane is composed of lipids (40%), proteins (55%), and carbohydrates (5%). The lipid bilayer of the plasma membrane makes it semi-permeable which regulates the passage of molecules, ions, and water (Figure 8.2). The bilayer structure also prevents these substances to enter the cytosol to maintain internal conditions. The cell surface receptor proteins play an important role in cell–cell interactions or cell–material interactions. The biological membranes can be best described as a phospholipid bilayer with several transmembrane proteins and voltage-gated ion channels. Such membrane physically separates intracellular organelles from the cytoplasm and a cell from the extracellular matrix.

The lipid bilayer of plasma membrane makes it semi-permeable, regulating the passage of some substances, such as organic molecules, ions, and water. Cytoplasm is an aqueous colloidal solution (cytosol) containing numerous functional and structural elements, which exist in the form of molecules/organelles, and can be broadly categorized into three groups: organelles, inclusion bodies, and cytoskeleton. A number of membrane-bound cellular organelles regulate the metabolic activities of the cells and these include Mitochondria (sites for adenosine-tri-phosphate (ATP) production through aerobic respiration), Endoplasmic Reticulum (sites for protein synthesis), Golgi Apparatus (sites for the synthesis and transport of proteins and other macromolecules), Ribosomes (sites where the RNA undergoes translation into proteins). Other cellular organelles include Lysosomes (sites for the breakdown of proteins, polysaccharides, lipids, nucleic acids, etc.); Peroxisomes and Centrosomes (active role in cell division). One of the important load-bearing structures in the cytoplasm is the Cytoskeleton, which consists of actin filaments, intermediate filaments, and microtubules. The major functions of cytoskeleton include, a) cell shape maintenance and b) facilitating cell migration/motility. Another important organelle is the nucleus, which accounts for most of (10 percent) the cell's volume as compared to other cellular organelles. The nucleus (plural = nuclei) houses the cell's DNA in the form of chromatin and directs the synthesis of ribosomes and proteins. Typically, the nucleus of a eukaryotic cell consists of a nuclear membrane (nuclear envelope), nucleoplasm or karyoplasm (matrix present inside the nucleus), nucleolus, and chromosomes. Nuclear membrane is punctuated with several openings, known as nuclear pores that control the exchange of large molecules (ions, molecules, proteins, and RNA) between the nucleoplasm and cytoplasm, permitting some to pass through the membrane, but not others. All these pores form a nuclear pore complex. The chromosomes are packed inside the nucleus of every cell and DNA is the hereditary material with proteins ultimately forming a complex structure called chromosome. DNA and proteins are packaged inside chromatin, which is further classified as heterochromatin and euchromatin, based on their functions.

8.5 Structure of Nucleic Acids

DNA and RNA along with proteins and carbohydrates constitute the essential biological macromolecules for all known forms of life. Some basic understanding of the way the DNA and RNA structures are built up is an essential precursor to probe into the cell differentiation and gene expression.

8.5.1 Structure of DNA

DNA is a biomacromolecule that contains the genetic codes, which are utilized in the propagation or replication of all known living organisms, including eukaryotic, prokaryotic cells, and some viruses. The functioning of DNA is also involved in cell growth, functioning, and organism development. In eukaryotic cells, the nucleus contains most of the DNA and some of the DNA are stored by other organelles, including mitochondria, chloroplasts, and so on. In contrast, in prokaryotic cells, the cytoplasm contains entire DNA.

As shown in Figure 8.2, most of the DNA molecules are structurally characterized by two long polypeptide strands uniquely coiled around each other which form a double helix structure. Each of these strands, known as polynucleotides, are composed of four types of nucleotides, that is, nitrogen containing nucleobase, either cytosine (C), guanine (G), adenine (A), or thymine (T) as well as a sugar called deoxyribose and a phosphate group linked covalently into the polynucleotide chain with a sugar-phosphate backbone. This unique structural arrangement results in a characteristic sugar–phosphate backbone. The DNA structure is determined by the famous base pairing rules (A with T, and C with G). The complementary base pairing enables packing in energetically most favorable arrangement.

As far as the structural length scale is concern, each helical coil has a pitch of 3.4 nm and a radius of 1 nm. A number of nucleotides can form a large molecule of DNA polymers. The largest chromosome in human contains around 220 million base pairs, which is absent 85 mm long on straightening. It is the ATGC sequence along the backbone that encodes the biological information, which is replicated following the separation of two strands.

By using the genetic information from DNA, RNA strands specify the amino acids sequence to form functional proteins by the process of translation. These RNA strands are formed on the DNA template via transcription. It is worthwhile to note that a significant portion (>98% for humans) does not have any ability for transcription and therefore do not code for any protein sequences.

8.5.2 Structure of RNA

Ribonucleic acid (RNA) is another important biomacromolecule involved in gene expression and regulation. Structurally, RNA also consists of nucleotides chain-like DNA. However, the important distinction is the single-stranded nature of RNA that often fold onto itself to form complex three-dimensional structures, rather than the double helical structure of DNA. While the chemical structure of RNA has a large similarity with the DNA structure, a fundamental structural difference exists in reference to the following two aspects (see Figure 8.3)

    a. Instead of deoxyribose as in DNA, RNA contains ribose sugar in sugar-phosphate backbone, which also makes RNA more labile than DNA.

    b. The complementary base pairing in case of RNA is adenine (A) to uracil (U), which is adenine (A) to thymine (T) in DNA.

The most biologically active RNAs can be classified as messenger RNA (mRNA), ribosomal RNA (rRNA), transfer RNA (tRNA), etc. One of the essential cellular processes influencing the cell functionality is the protein synthesis, wherein mRNA

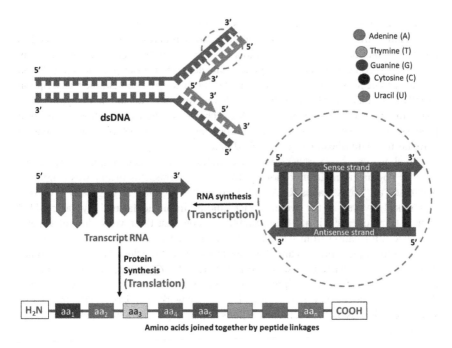

FIGURE 8.3   Transcription–translation process is universally regarded as the "central dogma" of molecular biology. Schematic representation of the transformation of DNA to RNA to protein synthesis in an eukaryotic cell.

molecules facilitate the assembly of polypeptides on ribosomes. In simple terms, mRNA carries the codon for the amino acid sequence, that synthesizes protein. Some RNA molecules also take part in stimulating some of the biological reactions, for example, gene expression and cell signaling. As the name suggests, rRNA comprises the basic constituent of ribosomes and also catalyzes the protein synthesis. Typical responsibility of tRNA is to encode the information carried by mRNA and transfer the specific amino acids to the growing polypeptide chain, while that of rRNA is to link amino acids to form proteins. Like DNA, different RNA types (mRNA, rRNA, and tRNA) have self-complementary sequences on the same strand. As a result, the RNA pair fold onto itself and forms loop like double helices. These double helices are not similar to that of the long helices of DNA, rather it is a group of many short helices, arranged together in a protein-like structure.

8.6   Transcription and Translation Process

It is important to mention how proteins are constantly synthesized inside a cell by a sequential process of transcription and translation. Transcription is a molecular biology process wherein the genetic information encoded in a part of DNA is transcribed into a newly formed mRNA by specific RNA polymerase enzyme and other transcription factors. In simple terms, one strand of DNA during transcription, is copied to produce a single stranded RNA, which is complementary to the template strand of DNA. In eukaryotic cells, such a process involves the splicing, which leads to the formation of a mature mRNA chain. Another complementary process, translation is defined as the transformation of mature mRNA to a ribosome. In prokaryotic cells, without any well-defined compartmentalized organelles/nucleus, the transcription and translation process can occur simultaneously in cytoplasmic space. In contrast, the transcription usually takes place inside the nucleus, while the translation occurs in the cytoplasmic space in eukaryotic cells (see Figure 8.3). Therefore, the transcription–translation process in eukaryotic cells requires the transportation of mRNA from nucleus to cytoplasm, where ribosomes get attached to it.

The above described transcription–translation process is commonly referred as "central dogma" in molecular biology. In simplest terms, this theory was originally postulated as "DNA makes RNA and RNA makes protein," as originally proposed by Crick. However, as the overall understanding in molecular biology has evolved over last few decades, it has been established now that the reverse flow of genetic sequence information from RNA to DNA is also possible, which is known as the reverse transcription process and this was originally not precluded by Crick. In other words, transcription and reverse transcription are indeed possible in living organisms. However, the transformation of protein to RNA does not commonly occur. Nevertheless, this important dogma is widely regarded as a framework to understand the transfer of sequence information between biopolymers containing genetic information in living organisms.

8.7 Cell Fate Processes

The cells coordinate and communicate through principal cellular fate processes. These cellular fate processes are described below, with a particular focus on cell division.

8.7.1 Cell Differentiation

It is a process derived from differential gene expression, leading to the phenotypic changes by cells to become a matured cell with different physiological functions. It starts with irreversible changes in a set of genes that are expressed in the cell. This process can be best described by a well-synchronized switching off and on of gene families. The specific family of genes, which is specific to mature cell types, are expressed. The cell differentiation process has been widely investigated experimentally using stem cells, an unspecialized cell type. Under a given set of signalling processes in the cellular microenvironment, stem cells can differentiate to bone cells (osteoblasts), cartilage cells (chondrocytes), muscle cells (myoblasts), neuron-like cells, etc. Accordingly, these processes are known as osteogenesis, chondrogenesis, myogenesis, neurogenesis, respectively etc.

8.7.2 Cell Migration

Like a baby crawls on a floor, similarly a cell also can physically crawl on a solid material substrate, albeit at a much slower speed than baby's crawl. Also, a cell can change its direction during crawling. This process is known as cell migration or cell motility. In adult organisms, cell migration plays an essential role in eliciting a proper immune response. Abnormal cell migration is seen in various pathological situations, like cancer metastasis. Indeed, the migratory process is also important in wound healing and angiogenesis. It is known that cells migrate in response to a variety of stimuli.

Biophysically, the cell migration process involves the depolymerization or repolymerization of cycloskeletal proteins, leading to dynamic changes in anchorage of cells on a given material substrate. Also, cells normally adhere to a material substrate by the formation of focal adhesion complexes (FACs), which are aggregates of cell surface receptors and protein adsorbed on a material surface. These FACs are either destroyed or remade during cell migration.

8.7.3 Cell Division

The cell division is a highly distinguishable characteristic of biological systems. An eukaryotic cell undergoes the repeating and ordered series of events, known as cell cycles comprising four phases (Figure 8.4). During the first gap phase (G_1), the cell enlarges physically, copies organelles, and prepares itself for DNA replication that occurs in the subsequent S phase. A cell can stay in the G_1 phase for variable time periods, while the S phase is continued for 8 hours. In the G_2 phase, the cell further grows in size, which is about 2–3 hours long. Then, finally in the M phase (mitotic phase), the genetic content of the cell is splitted into two copies followed by cytokinesis. During the whole cell cycle, the combined time period of the $S+G_2+M$ phase remains constant ($\sim$12 hours), while the G_1 phase has variable time duration.

The phases of the cell cycle are unidirectional and are controlled by checkpoints (Figure 8.4). The cell's size, certain environmental conditions, and DNA integrity are checked before the cell enters the S phase. If the cell fails to pass these checkpoints, it may initiate apoptosis and the cell will die. The G_2 checkpoint senses for unreplicated DNA, which can potentially decide the cell cycle arrest. The cell cycle progression can be stopped at the G_2 checkpoint due to DNA damage. DNA damage also restricts the cell cycle at G_1 checkpoint. Cells typically divide at a rate proportional to the cell population at a given point of time. For unconstrained growth, the rate of formation of new cells is proportional to the cell population,

$$dX/dt = \mu X = X(t) = X_o \exp(\mu t)$$

Based on the above equation, the growth rate can be defined as $\mu = \ln(2)/t_d$; where t_d is the doubling time. For human cells, $t_d \sim 12$ hours and therefore, $\mu_{max} \sim 0.06$/day. The ability of cells to undergo division varies with the cell type. For example, the doubling time of hematopoietic progenitors, dermal foreskin fibroblasts, and adult chondrocytes are 11–12 hours, 15 hours, and 24–48 hours, respectively.

8.7.4 Cell Death

Cell death may occur for many reasons. For example, cells, that die from tissue damage, undergo a cell death process, known as necrosis. Cell necrosis usually occurs due to external factors, like sudden mechanical shock, exposure to UV irradiation or

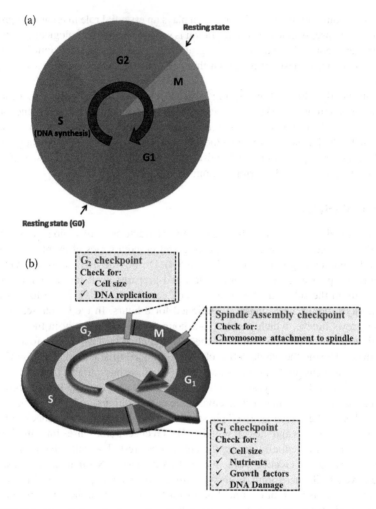

FIGURE 8.4   Description of eukaryotic cell cycle (a) and various checkpoints regulating the cell cycle (b). (Adapted from Ref. 5).

an electromagnetic field, resulting in irreversible injury. But a major part of tissue development experiences programmed cell death (apoptosis), which occurs as a result of a series of events leading to tissue morphogenesis. During apoptosis, first of all cell shrinks, followed by DNA fragmentation and chromosomal condensation, and finally cells split into small apoptotic bodies. These apoptotic bodies are membrane-bound structures and do not spill into the extracellular space to cause local peripheral damage. Thus, apoptosis is a highly ordered process, and a failure often leads to tumor formation. Due to unfavorable cellular microenvironments, if a cell is unable to repair DNA damage (see Figure 8.4), it will initiate a suicidal mechanism. Cell apoptosis can be characterized morphologically as (a) cell shrinkage, (b) membrane blabbing, and (c) DNA fragmentation and nuclear condensation.

8.8 Tissue

Tissue is a self-assembly of groups of cells having a similar structure and origin, which together perform a specific function. Various cell types and their functionality are summarized in Table 8.1. Based on the structure and function, tissues are classified into four major types, connective tissue, muscle tissue, nervous tissue, and epithelial tissue. Among these, the connective tissues consist of fibrous tissues embedded in ECM and provide structural framework to an organ. Connective tissue also contains spindle-shaped fibroblasts. Bone, adipose tissues, tendon, ligament, and blood are classical examples of connective tissues. As shown in Figure 8.5, the muscle tissue consists of muscle cells which have contractile nature that produce force and control the motion (movement or locomotion) in an organism. Among different types of muscle tissues, smooth muscle constitutes the inner linings of hollow organs like digestive tracts, blood vessels, etc., while the skeletal muscle are found to be attached to bone. Another type of muscle tissue is cardiac muscle, which is found only in the heart, has self-contracting nature that helps in rhythmic blood pumping throughout an organism. The neural tissue consists of neurons and neuroglia. In the central and peripheral nervous system, the nerve tissue constitutes the brain and spinal cord, and cranial nerves and spinal nerves, respectively. The epithelial tissues consist of closely packed epithelial cells, which cover the outer and inner surfaces of the organ.

Besides this, tissues can also be categorized as soft and hard tissues on the basis of their stiffness and mechanical strength. As per the nomenclature, hard tissues are mechanically stronger than the soft tissues. The hard tissue, also named as the mineralized tissues, is composed of minerals (hard) and collagenous matrices (soft). Bone, dentin, tooth enamel, tendon, and cartilage are some examples of mineralized tissues. Some of these mineralized tissues have the characteristic adaptability and multifunctional properties. Due to its inorganic and collagenous constituents, mineralized tissues exhibit diverse properties like low weight, toughness, strength, and stiffness. Due to the characteristic structural arrangement of collagen fibers and the calcium phosphate minerals, the mechanical stresses are transferred through several length scales, from macro to micro to nano and these results in energy dissipation and damage tolerance.

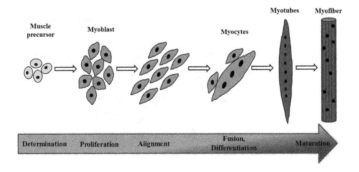

FIGURE 8.5 Schematic description of the formation of myotube in muscle tissue involving the fusion of the myoblast cells. (Adapted from Ref. 5).

8.9 Generic Description of Bacterial Cells

> Bacteria are unicellular microscopic prokaryotic organisms with a very simple structure and few micrometers (0.5–2 μm) in length. Cell division is usually accomplished by binary fission.

As shown in Figure 8.6, the bacterial cell is enclosed by a lipid membrane, which houses proteins, ribosomes, and other necessary components of the cytosol. As mentioned earlier in this chapter, membrane-bound organelles are absent in bacteria and thus have few intracellular structures. The nucleoid is an area of the bacterial cells, which is composed of chromosomes associated with small amounts of RNA and proteins. Like monkeys have tails, the tail-like structures of bacteria are known as flagella, which help in motility or migration on a material substrate. The flagella are proteinaceous structures of about 20 μm in length and up to 20 nm in diameter.

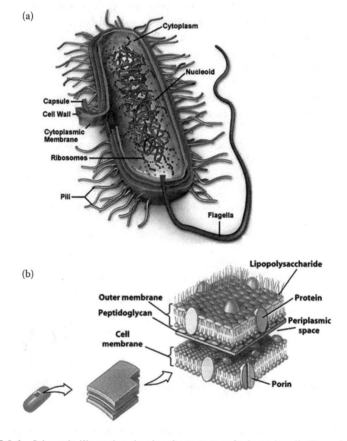

FIGURE 8.6 Schematic illustration showing the anatomy of a bacteria cell. (Reproduced with permission from Ref. 5).

Pili are essentially extracellular structures, which can transfer DNA/RNA from one bacterial cell to the other, in the process of bacterial conjugation.

The cell wall of Gram-positive bacteria is composed of a thick peptidoglycan layer (50%–90% of cell wall) and is positive on Gram's staining, while Gram-negative bacteria have a thin peptidoglycan layer (10% of the cell wall), sandwiched between the outer lipopolysaccharide layer and inner plasma membrane.

Bacteria can also be grouped as "Gram-positive" bacteria, which have a thick peptidoglycan layer and "Gram-negative" bacteria, which possess a thin peptidoglycan layer.

A comprehensive view of the 3D structure of peptidoglycan (responsible mainly for structural rigidity and osmotic balance) in the bacterial cell wall of Gram-negative bacteria has been shown in Figure 8.6.

8.10 Bacteria Growth

Like any eukaryotic cell, bacteria also grow or multiply in a given cellular microenvironment. A representative growth curve is shown in Figure 8.7.

Bacterial growth generally occurs by binary fission in which a single bacterial cell is divided into two daughter cells.

In a favorable *in vitro* environment, the bacterial population grows exponentially and doubles in 20–30 min time interval. Bacterial growth generally follows a geometric progression: 1, 2, 4, 8, etc. or $2^0, 2^1, 2^2, 2^3 \ldots \ldots 2^n$ (where n represents the number of divisions or generations). This is characteristic of the *exponential growth phase*.

In general, bacterial cells initially tune itself to the new surrounding and prepare for cell division (*lag phase*). Then they start dividing periodically via binary fission (*exponential/log phase*). After some time, the cells stop dividing due to the depletion of growth nutrients (*stationary phase*), and eventually, they show a sharp decline in viability (*death phase*). The growth is demonstrated as a change in the number of live cells over time. The growth rate can be calculated from the slope of the log phase in the growth curve. The culture duration timescale for bacteria is in hours while the generation/doubling times are much shorter, of the order of 20–30 min. Different phases of the bacterial growth curve are recognized and are discussed below (see Figure 8.7).

The exponential growth rate of bacteria in culture can be expressed in terms of *doubling time* or *generation time* of the bacterial population. Generation time (G) can be defined as the time (t) per generation. More precisely, it can be described as

$$G = \frac{t}{3.3 \log(b/B)} \tag{8.1}$$

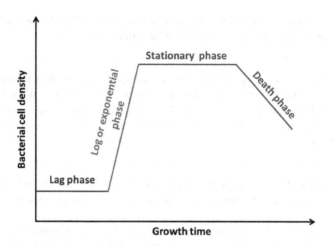

FIGURE 8.7 A typical bacterial growth curve. (Adapted from Ref. 5).

where *t*, *B*, and *b* represent the incubation time in hours or minutes, number of bacterial cells before incubation, and number of bacterial cells after incubation, respectively.[5]

8.11 Bacterial–Material Interaction and Biofilm Formation

Bacterial adhesion can be described in terms of physicochemical interactions, which enable bacteria to adhere firmly to a material substrate or biological cell/tissues. It is the outcome of the balanced attraction and repulsion between the bacteria and the material surface. Bacterial adhesion followed by cell growth on a material surface have a positive/negative impact on a variety of systems, including prosthetic infection.[1,2]

The precursor to bacterial adhesion is slime formation. Importantly, the consequence of a large number of bacteria adhering to a surface leads to biofilm formation.

> Slime is best described as an extracellular substance (mainly polysaccharides) synthesized by the bacteria. A biofilm is an accumulated biomass of bacteria and extracellular materials (basically slime) on a material surface.

Biofilm formation starts with the adhesion of microorganisms over the substrate (Figure 8.8). The next stage is the synthesis of an extracellular polysaccharide matrix (EPS), which promotes cell–cell and cell–surface interactions. After secure attachment, the biofilm grows into a highly complex environment.

> Biofilms are a complex dynamic environment, wherein cells from the growth medium attach to the biofilm and are spread into the surrounding.

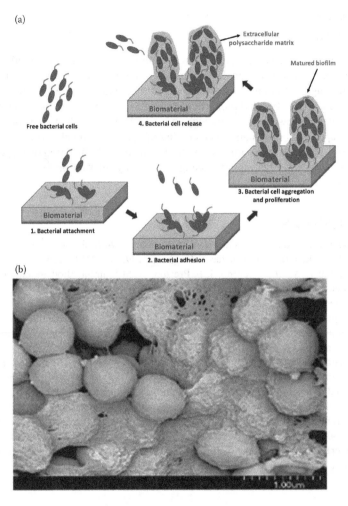

FIGURE 8.8 (a) Schematic representation of biofilm formation on the material substrate. (b) SEM image of a staphylococcal biofilm. It is noted that the multiple layer of bacteria is covered with a polysaccharide matrix. (Adapted from Ref. 5).

Microbial biofilms have several unique resistive features against antimicrobial agents. For example, bacterial biofilms deter the effect of antibacterial metal ions, such as copper, zinc, and silver, whereas these agents are highly effective against floating/planktonic bacteria.

A single bacterium has the capability to form a bacterial colony, which further grows on the implant surface and form biofilms.

The resistance mechanisms for the biofilm involve the suppression of bacterial growth, and structural modification in the cell envelope, following attachment to soft and hard tissues.[3]

8.12 Closure

This chapter provides non-biologists a foundation of the biological system. Recognizing various aspects in the context of bioengineering and general importance to engineering community, the discussion in this chapter largely focuses on the structural characteristics and properties of proteins, cell, bacteria, tissue, and so on. More detailed discussion on cell biology and microbiology can be found elsewhere.[5,6]

This chapter also elicits an understanding of the bacterial cell structure, as well as bacterial colonization, leading to biofilm formation. Sufficient knowledge of the biological system can help in addressing biologically relevant problems for human healthcare.

REFERENCES

1. Ong Y. L., A. Razatos, G. Georgiou, M. M. Sharma, Adhesion forces between *E. coli* bacteria and biomaterial surfaces, *Langmuir*, 1999, 15, 2719–2725.
2. Schmidt L. M., J. J. Delfino, J. F. Preston, G. St Laurent, Biodegradation of low aqueous concentration pentachlorophenol (PCP) contaminated groundwater, *Chemosphere*, 1999, 38, 2897–2912.
3. Cooper M., S. M. Batchelor, J. I. Prosser, Is cell density signalling applicable to biofilms? in *The Life and Death of Biofilm*, eds. J. Wimpenny, P. Handley, P. Gilbert, H. Lappin-Scott, Bioline Press, Cardiff, 1995, pp. 93–97.
4. Boundless. The Plasma Membrane and the Cytoplasm. *Boundless Biology*. Boundless, 19 September 2014. Retrieved 27 Mar. 2015 from https://www. boundless.com/biology/textbooks/ boundless-biology-textbook/cell- structure-4/ eukaryotic-cells-60/the-plasma-membrane-and-the- cytoplasm-314-11447/.
5. Bikramjit Basu, Biomaterials Science and Tissue Engineering: Principles and Methods, Cambridge University Press, ISBN: 9781108415156, August 2017.
6. Alberts B., A. Johnson, J. Lewis et al. Molecular Biology of the Cell, 4th edition. ISBN-10: 0-8153-3218-1, ISBN-10: 0-8153-4072-9.

9

Elements of Bioelectricity

Bioelectricity plays an important role in regulating various metabolic processes in living tissues. In view of inherent electromagnetic phenomenon of the living tissues, the external electrical stimulation can be used to regulate the various metabolic processes in cells/tissues. Therefore, the coupling between the internal and external electric field is critical in deciding its successful implication. In this perspective, understanding of the electrical phenomenon in living beings can be effectively utilized in mimicking these inherent phenomena using external electrical stimuli. Overall, understanding the science of bioelectricity will help in developing such technologies to simulate tissue physiologically as well as pathophysiologically. This becomes important as the inherent electricity of the body is actively used to monitor and regulate a number of physiological functions in the body. This chapter emphasizes the importance of interdisciplinary approach towards the development of comprehensive understanding for the interaction of living cell with external electrical stimulation along with its consequences.[35]

9.1 Introduction to Cell Membranes

The fundamental unit of matter, an atom, consists of charged particles such as protons, electrons, and neutrons. Together, electrons and protons make an atom neutral along with a neutron, however, when this balance shifts, atoms change to ions. These negative and positive ions of a particular element or compound play a key role in governing the various metabolisms of a living organism.

> The electrical phenomenon in living beings is concerned with the mechanism of flow of ions such as, Na^+, K^+, Ca^{2+}, Mg^{2+}, and various other charged species as well as their potential gradients.[1]

The cell, which is the fundamental unit of any living organism, is the prime center for all the physiological processes. A single cell of all the living organisms possesses all the properties, required to sustain life. The presence of various inorganic ions, inside and outside the cell such as Ca^{2+}, Na^+, K^+, Cl^-, and HCO_3^-, participate in various electrical activities across the cell membrane. These activities include nerve signaling, heart pumping, muscle contraction, wound healing, osmoregulation, etc.[2]

It is widely known that the cell membrane of a eukaryotic cell is composed of a lipid bilayer along with the embedded transmembrane proteins. The transmembrane protein forms various ionic paths such as voltage-gated Ca^{2+} channels, and Na^+/K^+

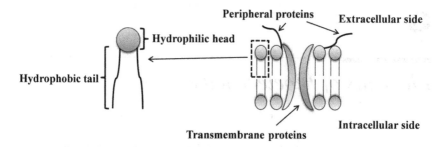

FIGURE 9.1 Schematic representation of the cell membrane consisting of a lipid bilayer and proteins.

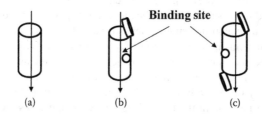

FIGURE 9.2 Schematic illustration of various channels present in the membrane. (a) Open channels, (b) gated channels, and (c) carriers.

channels. Figure 9.1 demonstrates a generalized cell membrane having a lipid bilayer and proteins. The bilayer arrangement of lipids resists the flow of ions across the membrane, while proteins provide the pathway to the ions to cross the membrane. These transmembrane proteins in the cell membrane constitute various ion specific channels, pumps, carriers, and transporters (see Figure 9.2). Their primary function is to regulate the flow of ions across the cell and to maintain ionic homeostasis. Various mechanisms of flow of ions across the cell membrane play an important role in the electrical phenomenon across it.

> The maintenance of the ionic homeostasis signifies that there is an inherent gradient in the concentration of various ions across the cellular membrane, which contributes to a small electric potential across the membrane. This electric potential is called the resting potential of the cell.

The significance of this potential is that whenever an external/internal stimulus is sensed by the cell, it responds accordingly toward that specific task, for example, contraction of a muscle. The motor neuron senses the stimuli and develops an action potential that travels through its synapse to various muscle cells, attached to the synapse of a particular motor neuron. This phenomenon in turn develops an action potential in muscle fibers, and contraction of the muscle proceeds.

In view of the importance of ionic transport through various channels extending across the cell membrane, a brief description is provided below.

9.2 Integral Membrane Proteins

9.2.1 Open Channels

These membrane protein channels are always open. The cell membrane of many cells contains channels that provide an open pathway to water molecules and these water channels are called aquaporins. These aquaporins are also found in membranes of bacteria and plant cells. They provide different mechanisms of transport of water as well as neutral molecules. Some ions also pass unimpeded through open channels along with water molecules and contribute significantly to maintain the resting potential of the cell.

The transport mechanism through these channels follows the phenomenon of electro-diffusion.

9.2.2 Gated Channels

These membrane proteins contain a movable barrier, which alternatively opens and closes depending upon the voltage across the membrane (voltage gating), mechanical stress across the membrane (mechanical gating), and synaptic signals across the membrane (ligand gating). Gating is the process of opening and closing of the channels in response to particular external or internal stimuli. The selectivity filter gives specificity to the ion channels, that is, a particular class of ions should pass through a particular channel. Mostly, Na^+ channels are voltage gated, that is, they open whenever the potential across the membrane changes and in this process Na^+ ions are allowed to enter the cells. This contributes to the generation of action potential. The mechanism of flow through Na^+ channels is the same as through the open channels. Apart from Na^+ channels, K^+, Ca^{2+}, proton, and anion (Cl^-, HCO_3^-) channels are also gated.

9.2.3 Carriers

These channels are gated from both sides and do not provide a continuous pathway to the ions or hydrophilic solutes. The mechanism of flow through these channels is known as facilitated diffusion, that is, it is passive uncoupled transport. For a small concentration gradient, it follows the Fick's law of diffusion and electro-diffusion phenomenon.[3] For a large concentration gradient, the flow of ions and hydrophilic solutes become constant.

The carriers are having one or two binding sites for the ions and solutes, which can be trapped during the transport. Therefore, the speed of transport becomes constant at some instant, based on trap and release mechanisms. Glucose transporter (GLUT1), is an example of membrane protein that mediates facilitated diffusion. Other examples are urea transporter and organic cation transporter. These carriers provide non-coupled mode of transport of ions across the membrane, that is, passive transport.[1] However, other carriers, for example, pumps, cotransporters, and exchangers provide coupled mode of transport, which requires energy.

The coupled mode of transport, called active transport, is a process, which transfers solutes and ions against their electrochemical potential energy difference.

The energy, which is provided for the transport, is provided by ATP hydrolysis as well as thermodynamic free energy of various ions and solutes which is available in their concentration gradient. The ATP hydrolysis is utilized by pumps via primary active transport. In the secondary active transport, thermodynamic free energy available in the concentration gradient of solutes and ions is utilized for the transport by cotransporters and exchangers. Pumps, which utilize the energy provided by ATP hydrolysis, are called as ATPases. Most prominent ATPase is the Na–K pump, which extrudes three Na^+ ions from a cell with a concurrent uptake of two K^+ ions into the cell for hydrolysis of one molecule of ATP. This pump is generally called as electrogenic, because there is a net transfer of charge across the membrane. The Na–K pump is the only membrane protein that transfers Na^+ ions through primary active transport.[1,2,4]

9.3 Transport Kinetics across Cell Membrane

All the viable cells maintain a biologically relevant fluidic environment inside and outside the cell through various transport processes across the cell membrane. The cytoplasm and extracellular cellular matrix (ECM) contain abundant inorganic ions that contribute to the highly conductive nature of the cell membrane and ECM. Therefore, the electrical activity of the cell membrane is fundamentally based on the transport of ions. The concentration gradient of various ions present across the membrane gives an electrically polarized membrane, which act as a miniature battery and handles many transport processes (Figure 9.3). The various transport processes are briefly discussed as below.

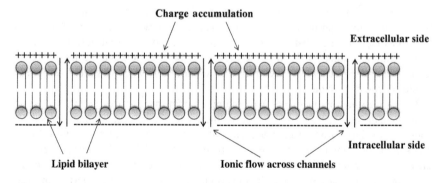

FIGURE 9.3 Schematic representing the accumulation of charge and flow of ions across the membrane.

9.3.1 Passive Transport

The concentration and electrical gradient of ions distributed across the cell membrane provide the driving force to the flow of ions across the membrane and this flow is facilitated by the transmembrane proteins. Under resting conditions, only few ion-specific channels are open, which do not allow the other ions to dissipate their electrochemical gradient, which consequently builds a charge gradient around the cell membrane. This phenomenon is responsible for the membrane potential. The accumulated charge is mainly concentrated on the lipid bilayer just inside and outside the cell membrane. Skeletal muscle cells, cardiac cells, and neurons have a resting membrane potential in the range of -60 to -90 mV. Smooth muscle cells have a membrane potential in the range of -55 mV.[1] However, for human erythrocytes, it is only about -9 mV.[2] Overall, the resting membrane potential of any cell type is negative with the reference to the outer side of the cell membrane, that is, the ground potential.

The passive non-coupled transport refers to the flow of ions across the membrane depends on their electrochemical potential energy difference.

However, a favorable pathway should exist for the ions to cross the membrane. The pathway in the form of channels is present in the membrane, but is ion-specific. Therefore, only few ions can pass through it. The net flux of the permeable ions across the membrane determines the equilibrium state of these ions. Let us now discuss this briefly.

When the membrane is permeable to a particular ion only, the potential of the membrane is said to be equivalent to the equilibrium potential for that ionic species, like in the case of glial cells whose membrane is only permeable to K^+ ions in the resting state.[2] Therefore, its resting membrane potential is the equilibrium potential of K^+, which is determined by the Nernst equation as,

$$E_X = \frac{RT}{zF} \ln \frac{[X]_o}{[X]_i} \tag{9.1}$$

$[X]_o$ and $[X]_i$ represent the concentration of ion X outside and inside the cell, respectively, and E_X is the equilibrium potential (in volts).

The equilibrium potential of the particular ionic species depicts that the net flux of that particular ion X across the membrane is zero.

This Nernst equation can be derived from the Goldman–Hogkin–Katz current equation, which is based on various assumptions about the ionic current through the cell membrane. These assumptions include, (a) the membrane potential is constant (constant field assumption), (b) the movement of a specific ion across the membrane is independent of other ions (independence principle), and (c) permeability coefficient is constant. Against these assumptions, the current carried by single ion X (I_X) through the membrane can be given as,

$$I_X = \frac{z^2 F^2 V_m P_X}{RT} \left(\frac{[X]_i - [X]_o e^{(-zFV_m/RT)}}{1 - e^{(-zFV_m/RT)}} \right) \tag{9.2}$$

The above equation is called as the Goldman–Hodgkin–Katz (GHK) current equation.[4] Here, V_m and P_X are the membrane potential and the permeability coefficient of the ion X, respectively. This equation is simply based on Ohm's law, because the current and voltage are directly proportional in this case. If the current of the single ionic species is set to zero, then the above equation reduces to Nernst equation, giving the equilibrium potential for that particular ionic species. Nernst equation is also called the reversal potential, for the cations and anions. For the cations, if the potential of the membrane is less positive as compared to the reversal potential, the current of the cation will be directed inwards. If it is more positive as compared to the reversal potential, the current of cations will be directed outwards. Therefore, the direction of the ionic current, changes at reversal potential. Similar phenomenon is true for the case of anions as well.

Thus, equilibrium potential of a particular ion gives a zero net flux. However, the membrane potential is far different from the equilibrium potential for any ionic species across the membrane, when it is permeable to multiple types of ions. Therefore, ions present across the membrane would experience a net driving force in volts ($V_m - E_X$), for that particular ion.

Here, V_m and E_X are the membrane potential and the equilibrium potential for the ion X across the membrane, respectively. Furthermore, assuming that the membrane is only permeable to the monovalent ions (K^+, Na^+, Cl^-, etc.), the total ionic current carried by these ions across the membrane is the sum of the individual ionic currents.

$$I_{Total} = I_K + I_{Na} + I_{Cl} \tag{9.3}$$

To find out the resting membrane potential, the individual ionic currents are added and the total current is set to zero. While solving Equation 9.2, we derive the GHK voltage equation or the constant field equation as,

$$V_m = \frac{RT}{F} \ln \left(\frac{P_K[K^+]_o + P_{Na}[Na^+]_o + P_{Cl}[Cl^-]_i}{P_K[K^+]_i + P_{Na}[Na^+]_i + P_{Cl}[Cl^-]_o} \right) \tag{9.4}$$

This gives the resting potential of the cell when the net current is zero, that is, the cell membrane is in steady state. Note that, here the pumps or transporters are not playing any role in this equation (not considered), however, pumps and transporters do maintain the original concentration gradient, which brings the cell back to its resting state.

9.3.2 Active Transport

Another transport phenomenon across the cell membrane is called as active transport, which is a mechanism by which ions and other hydrophilic solutes are transported against their electrochemical potential energy difference. Membrane proteins, called carriers, facilitate such a type of transport process. The energy required for active transportation comes from the ATP hydrolysis and thermodynamic free energy available due to the concentration gradient of ions. Carriers, which utilize energy from ATP hydrolysis, are pumps and are also called as ATPases. The transport through these ATPases is called as primary active transport. Some of the ATPases are Na–K pumps, H–K (hydrogen-potassium) pumps, and Ca^{2+} pumps.[1]

The H–K pump extrudes two H^+ ions from the cell with the uptake of two K^+ fueled by hydrolysis of one molecule of ATP. Ca^{2+} pumps extrude one Ca^{2+} ion with the intrusion of one H^+ ion for each molecule of hydrolyzed ATP. Ca^{2+} pumps also exist on the membrane of intracellular organelles, such as sarcoplasmic reticulum in muscle cells and endoplasmic reticulum in other cells for the active sequestration of Ca^{2+} ions into intracellular stores. There are other types of ATPases, present in the cell membranes and membranes of intracellular organelles, which are not discussed here.

Other carriers, which are present in the membrane, are cotransporters (symporters) and exchangers (antiporters). Cotransporters transport the driving solute (one whose electrochemical gradient provides the energy) and driven solute (which is pushed against its electrochemical gradient) in the same direction. Cotransporters such as Na^+/glucose (SGLT), Na^+-driven cotransporters for organic solutes, Na^+/HCO_3, Na^+-driven cotransporters for other inorganic anions, Na/K/Cl, Na/Cl, K/Cl, and H^+ generally utilize the energy from the electrochemical gradient of Na^+ ions except the K/Cl cotransporter, which is independent of Na^+.[1] Exchangers move the solute/ion, whose electrochemical gradient provides energy in the opposite direction with that of the ion/solute, which is pulled against its electrochemical gradient. In general, cotransporters exchange cations for cations and anions for anions. Some of the exchangers are Na–Ca, Na–H, Na^+-driven Cl-HCO_3, and Cl-HCO_3. These exchangers utilize the energy from the electrochemical gradient of Na^+, except the Cl-HCO_3 exchanger. These cotransporters and exchangers transport the ions or solutes through secondary active transport. A schematic diagram of primary active transport and secondary active transport is shown in Figure 9.4.

The most prominent ions involved in the transport process across the cell membrane are Na^+ and K^+. Na^+ ions are abundantly present in the extracellular matrix (ECM), while K^+ ions are present in intracellular space. Therefore, the electrochemical

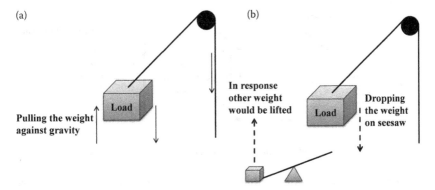

(a)

Pulling the weight against gravity

Load

(b)

In response other weight would be lifted

Load

Dropping the weight on seesaw

FIGURE 9.4 Schematic representation of (a) primary active transport and (b) secondary active transport. (Adapted from B. F. Walter and B. L. Emile, *A Medical Physiology: A Cellular and Molecular Approach*, 2012, ISBN: 978-1-4377-1753-2.)

gradient for each cation tends to dissipate, but this is maintained through the Na–K pump. This phenomenon also plays a major role in maintaining a negative membrane potential, because the pump itself is electrogenic, that is, it extrudes a net positive charge out of the cell. This causes a positive current outside the cell. Another factor is the intrusion of K^+ inside the cell, which develops an electrochemical gradient for K^+ directed outwards. Thus, K^+ tends to exit through K^+ channels, leaving a negative charge inside the cell.

9.4 Electrical Equivalent of the Cell Membrane

The current equation given by Golmann–Hodgkin–Katz for the individual ion describes the current–voltage relationship for that particular ion. The current varies with the voltage of the membrane for that individual ion, that is, the difference between the membrane voltage and the equilibrium potential for the particular ion provides the driving force to flow across the membrane. As a result, whenever this driving force varies, the current changes. The equilibrium potential for the single ionic species acts as a battery, because of the ionic gradient. Therefore, each ion channel can be considered as a resistor in series with a battery, representing its equilibrium potential. The lipid bilayer can be represented as a capacitor, because it has the tendency to block/hinder the flow of ions (Figures 9.5 and 9.6).

The individual current may vary because the resting potential of the cell membrane is not equal to the equilibrium potential of any ionic species. Therefore, all ions are in constant flow across the membrane. However, ionic pumps constantly maintain this potential across the membrane. Therefore, ionic pumps can be represented as a constant current source with a membrane lipid bilayer represented as a capacitor in parallel. The equivalent circuit for the whole membrane can be represented as shown in Figure 9.7.

If this circuit is solved by taking individual equilibrium potential values of each ionic species and their conductance values (neglecting the role of pumps), the equivalent circuit for the cell membrane appears as shown in Figure 9.8.

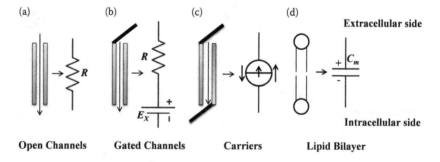

FIGURE 9.5 Schematic illustration of (a) open channel, represented by only a resistor, (b) gated channels, represented by resistor and battery in series, (c) carriers, represented by the constant current source, and (d) lipid bilayer, represented as a capacitor. E_x represents the equilibrium potential for that ion. (Adapted from Sterratt D. et al. *Principles of Computational Modelling in Neuroscience*, Cambridge University Press, ISBN-978-0-521-87795-4.)

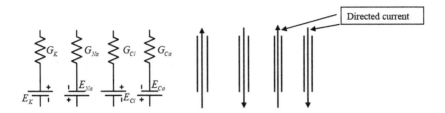

FIGURE 9.6 Schematic illustrating the individual circuit for each ion channels along with its equilibrium potential (arrow indicates the directed current for each ionic species). (Adapted from Kandel E. R. et al. *Principles of Neural Science*, 5th edn, McGraw-Hill Medical, 2013, ISBN: 978-0-07-181001-2; Sterratt D. et al. *Principles of Computational Modelling in Neuroscience*, Cambridge University Press, ISBN-978-0-521-87795-4.)

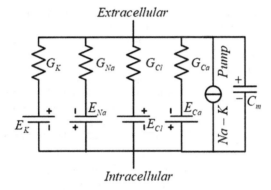

FIGURE 9.7 Electrical equivalent circuit of the cell membrane, representing individual ion channels, pumps, and the lipid bilayer. (Adapted from Kandel E. R. et al. *Principles of Neural Science*, 5th edn, McGraw-Hill Medical, 2013, ISBN: 978-0-07-181001-2; Sterratt D. et al. *Principles of Computational Modelling in Neuroscience*, Cambridge University Press, ISBN-978-0-521-87795-4.)

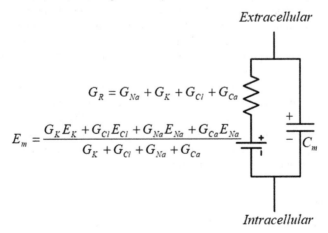

FIGURE 9.8 Electrical equivalent circuit for the cell membrane (neglecting the role of pumps). (Adapted from Kandel E. R. et al. *Principles of Neural Science*, 5th edn, McGraw-Hill Medical, 2013, ISBN: 978-0-07-181001-2; Sterratt D. et al. *Principles of Computational Modelling in Neuroscience*, Cambridge University Press, ISBN-978-0-521-87795-4.)

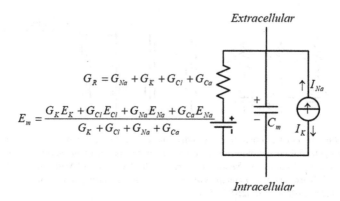

FIGURE 9.9   Complete electrical equivalent of the cell membrane. (Adapted from Kandel E. R. et al. *Principles of Neural Science*, 5th edn, McGraw-Hill Medical, 2013, ISBN: 978-0-07-181001-2; Sterratt D. et al. *Principles of Computational Modelling in Neuroscience*, Cambridge University Press, ISBN-978-0-521-87795-4.)

However, if pumps are included, the equivalent circuit can be represented with a constant current source in parallel with the above circuit. Here, only a single pump (Na–K pump) is included and this pump is electrogenic, because it extrudes 3 Na^+ from the cell in exchange for 2 K^+ during one cycle of ATP hydrolysis. However, other ATPases pumps (Ca^{2+} pumps and H–K pumps) can also be included. The complete circuit can be represented as in Figure 9.9.

This model can be used to construct an electrical equivalent of an entire living cell, which can include cytoplasmic resistance, nuclear membrane capacitance, and its resistance (if only nucleus is considered as single species). Because of the presence/movement of various ions inside/outside the cell membrane, the human tissues and organs possess a significant amount of conductivity and permittivity values. These values are summarized in Table 9.1. A significantly higher amount of permittivity is due to the presence of various ionizable residues/ions, as these values are directly imaged from the body.

According to third and fourth law of Maxwell in electrodynamics, that is, changing the electric field can create a magnetic field and vice versa. Therefore, both electricity and magnetism exist inside the body. Figure 9.10 summarizes the biomagnetic field, present in various tissues and organs of the human body. From Figure 9.10, it is clear that the human heart (100 pT) is the strongest source of biomagnetic field in the human body, which is followed by the muscle ($\sim$50 pT), eye ($\sim$10 pT), and the brain (50–500 fT).[8]

Now, because of the presence of inherent bioelectricity in living cells, their interaction with externally applied electrical stimulation is anticipated. Toward such understanding, the subsequent section elaborately discusses the theoretical aspects of coupling of a single living cell with external electrical stimulation.

9.5   Interaction of Living Cells with E-Field

9.5.1   E-Field Effects on Living Cells

In living cells, the cell membrane is an electrically active nano-sheet mediating the regulated flow of inorganic ions such as Na^+, K^+, Ca^{2+}, Cl^-, and HCO_3^- between the extracellular and intracellular fluids.[9] In addition, the cell membrane also plays

TABLE 9.1

Conductivity and Permittivity Values for Tissues and Organs
in the Human Body

Tissue	Conductivity σ (Ω/m)	Permittivity (ε_r)
Muscle	0.86	434,930
Fat	0.04	24,104
Bone	0.04	12,320
Cartilage	0.04	32,042
Skin	0.11	1136
Nerve	0.12	69,911
Intestine	0.11	69,911
Pancreas	0.18	56,558
Heart	0.50	352,850
Blood	0.60	5259
Gland	0.11	56,558
Liver	0.13	85,673
Kidney	0.27	212,900
Lung	0.04	145,100
Bladder	0.20	51,418
Eye	1.66	54,553
Stomach	0.11	56,064
White matter	0.12	69,811
Grey matter	0.12	164,060
Marrow	0.04	5632
Muscle	0.86	434,930
Lens	0.11	105,550
Mucous membrane	0.52	32,135
Spleen	0.18	106,840
Gall bladder	0.20	600
Intestine contents	0.11	69,911

Source: Reproduced with permission from Bencsik M. et al., *Phys. Med. Biol.*, 2007, 52, 2337–2353. IOP Publishing.

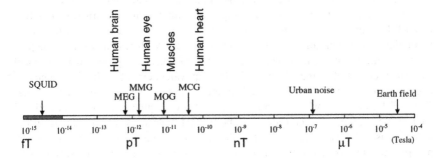

FIGURE 9.10   Schematic representing the biomagnetic field in various tissues and organs of the human body. (Reproduced from Zhou S.-A., M. Uesaka, Bioelectrodynamics in living organisms, *Int. J. Eng. Sci.*, 2006, 44(1–2), 67–92. With permission from Elsevier.)

a crucial role in the maintenance of the transmembrane potential (TMP) and also assists in neuronal signalling as well as muscular contraction through the development of chemical gradients of ions.[10] These gradients and the polarizable nature[11] of the cell membrane induce endogenous electric fields. These endogenous electric fields play a significant role in wound healing,[12] growth, and development of embryos,[13] modifying the cellular response[14] and modulating neuronal signalling.[15] Therefore, it is expected that the application of an external E-field can effectively manipulate/ guide the transport processes across cell membrane as well as can control various functionalities of living cells. In addition, living cells consist of various membrane-bound organelles and together they maintain homeostasis through synchronized intracellular transport processes. Thus, the application of optimally tuned external electrical stimulation can also control intracellular processes.

The E-field-induced membrane level effects include electroporation,[16] increased/ reduced cell proliferation,[17] necrosis, drug delivery, gene therapy,[18] antibacterial response, as well as extraction of biomolecules,[19] and so on. However, the intracellular level response includes apoptosis and variation in the intracellular calcium ion concentration across the membranes of various intracellular organelles, which play a crucial part in the alteration of cytoskeleton as well as cellular differentiation.[20–22]

The electroporation phenomenon in the cell membrane can be reversible or irreversible which depends upon the applied electrical stimulation parameters.[23–27] For example, the E-field strength of ~5 kV/cm with a pulse duration ranging from 0.4 to 1 μs or several (~100) milliseconds can be applied to obtain reversible electroporation. For irreversible electroporation, pulses in the range of milliseconds to seconds are required for a similar strength of E-field. However, E-field pulses in nanoseconds duration range (600–10 ns) with higher strength (10–300 kV/cm) are required for intracellular effects. The applied pulsed E-field <1 microsecond also form nanopores in the cell membrane.[28] When the E-field is applied across the living cells, it disturbs the equilibrium which sometimes results in a substantial increase in the movement of ions across the cellular membrane and consequently influences the transmembrane potential. The applied E-field can also lead to conformational or structural changes in proteins, lipids and enzymes.[9]

The mechanism of interaction of living cells with external electrical stimulation is not fully explored. However, a number of attempts have been made to develop a theoretical understanding for the interaction of cells with E-field with the help of electrical equivalent circuits of cells/membranes.[29] Deng et al.[30] proposed an electrical equivalent circuit of a single living cell. The equivalent circuit of the living cell has been reported in other studies as well to understand its behavior under an external E-field.[31–33] In general, while considering the electrical equivalent circuit, the cell and nuclear membranes were considered to be perfect dielectrics which is far from the realistic case. This is due to the fact that the cell membrane consists of ionic channels, pumps, transporters, and other channels, which mediate the flow of ions. Similarly, the nuclear membrane consists of pores, which passes selective molecules across the cytoplasm and nucleoplasm. Following this, Ellappan and Sundararajan[34] proposed the electrical equivalent circuit of a living cell, by taking into account the cell membrane as a leaky dielectric and nuclear membrane to be a perfect dielectric. In this study, the shape of the cell was assumed to be a cuboid.

However, these studies highlighted the dielectric properties of the membranes and conductive properties were feebly analyzed. The membrane conductive proteins allow different ionic currents across the cell membrane, while pores in the nuclear membrane permit the selective molecules to pass. The change in the behavior of the currents can considerably influence the cellular functionalities.

The equivalent circuit of the cell membrane is supposed to consist of a resistor and capacitor where the resistor-like properties are perceived to mimic the behavior of voltage-gated channels and the capacitor represents the behavior of the lipid bilayer. The nuclear membrane is also represented in a similar form with the resistance depicting the behavior of pores and the capacitance representing the behavior of the lipid bilayer.

Combining these, the electrical equivalent circuit of a single cell is represented under the assumption that on application of an E-field, ions traverse the least resistance path. The time constant of the circuits has been evaluated and its variation with various cellular adaptation processes is elaborately discussed. The strength of the E-field needed to electroporate the cells has also been provided, assuming various critical transmembrane potentials (TMPs). Using MATLAB simulations, models were also analyzed with different pulsed voltage input signals of various durations.[35]

9.5.2 Model Evolution

The polarized nature of the cell membrane gives rise to the endogenous E-field. In addition, the nuclear membrane also consists of pores, which allow the passage of selected ions/molecules across the cytoplasm and nucleoplasm. Though, with the presence of channels and pores in both the membranes, the nuclear membrane impedance is comparatively higher (~5–10 times) than that of the cell membrane.[35] As stated previously, the cell membrane can be represented as a combination of resistors and capacitors. Moreover, in the complete electrical equivalent circuit of a cell, the presence of an extracellular matrix (ECM) is also depicted with the presence of a resistor and capacitor.[20] In addition, the cytoplasm and nucleoplasm are depicted as resistors.

Application of E-field across the cell stimulates the numerous voltage-gated channels, resulting in depolarization of the cell. Thereby, transmembrane potential (TMP) increases to the critical value (0.7–1 V). On reaching the critical value, structural changes in the cellular membrane lead to the formation of pores in the membrane, which is generally known as electroporation.[36]

Initially, under the influence of E-field, the conductive currents flow across the various channels and pores of the cell as well as nuclear membrane and fluids of cytoplasm and nucleoplasm, while the displacement current flows across the dielectric portion (lipids in the membrane) of the membranes. The stability of the pores formed in the cell membrane depends upon the E-field parameters as well as electrophysical state of the cell.[24,29] The pores can be resealed and overall the structure of the cell remains unhindered or it can lead to permanent damage to the structure, which results in cell death. The former process is known as reversible electroporation, while latter is irreversible electroporation.[29] The E-field induced transmembrane potential also influences other cellular processes, such as, proliferation, differentiation, migration, apoptosis, and so on.[37,38]

Figure 9.11a represents the electrical equivalent circuit of the cell by assuming the cell membrane to act as a leaky dielectric in the earlier proposed model.[34] C_m and C_n represent the capacitances of cell and nuclear membranes, respectively. R_{C1}, R_{C2}, R_{C3} are cytoplasmic resistances and R_n represents the resistance of the nucleoplasm. In this study, the voltage-gated channels are taken into account which equivalently can be considered as a combination of resistance R_m and the voltage source E_m. R_S and C_S represent the resistance and capacitance for the ECM, respectively.

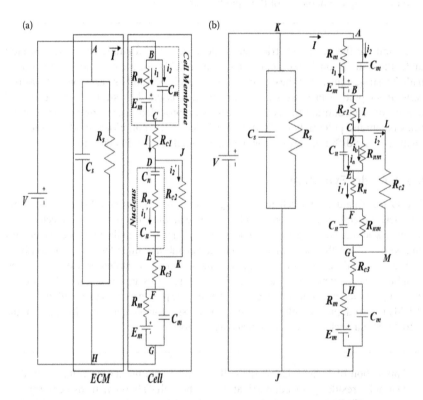

FIGURE 9.11 Electrical equivalent circuits for a living cell by assuming (a) cell membrane to be a leaky dielectric and the nuclear membrane to be a perfect dielectric and (b) considering the cell as well as the nuclear membrane to be leaky dielectric. The arrow indicates the assumed ionic path under the applied electrical stimulation. The parameters such as R_s and C_s represent the resistance and capacitance for the ECM, respectively. The membrane potential is represented by E_m. The cell and nuclear membrane capacitance is represented by C_m and C_n, while, R_{c1}, R_{c2}, and R_{c3} represent the cytoplasmic resistances and R_n is the nucleoplasm resistance. R_{nm} represents the resistance of the nuclear membrane. (From Saxena A., A. K. Dubey, *J. Phys. D: Appl. Phys.*, 2019, 52, 015401. Reproduced with permission from IOP Publishing.)

Generally, the response of the living cell under an applied electrical stimulation is evaluated using the Laplace equation ($\nabla^2 V = 0$) with certain assumptions (cell size and shape) and boundary conditions.[39] However, if any of the intracellular structures are considered, the analysis becomes more complex. The electrical equivalent circuit is however a simpler model and study of the interaction of a living cell with an E-field can be analyzed more deeply.[35]

Figure 9.11a shows the second-order circuit where the cell and nuclear membranes are considered to be leaky and perfect dielectrics, respectively. On evaluating, it renders the second-order differential equation as,

$$
\left[\frac{C_m}{2}(R_{C1} + R_{C3}) + \frac{C_m R_n}{2 R_{C2}}(R_{C1} + R_{C3}) + \frac{C_m R_n}{2} \right] \frac{d^2 i_1'}{dt^2}
$$

$$
+ \left[\begin{array}{l} \dfrac{(R_{C1} + R_{C3})}{2 R_m} + \dfrac{R_n(R_{C1} + R_{C3})}{2 R_m R_{C2}} + \dfrac{R_n}{2 R_m} \\[2mm] + \dfrac{C_m(R_{C1} + R_{C3})}{C_n R_{C2}} + \dfrac{C_m}{C_n} + \dfrac{R_n}{R_{C2}} + 1 \end{array} \right] \frac{di_1'}{dt}
$$

$$
+ \left[\frac{(R_{C1} + R_{C3})}{R_m C_n R_{C2}} + \frac{1}{R_m C_n} + \frac{2}{C_n R_{C2}} \right] i_1' = 0
$$

(9.5)

Using the above equation, the time constant can be obtained using the general solution of the linear second-order homogeneous differential equation.

The specific capacitance (per unit area) for both the membranes has been considered to be 10^{-2} F/m².[30,31] The resistivity of cytoplasmic and nucleoplasmic fluids has been taken as 100 Ωcm.[30] It has been reported that the cell membrane resistivity lies between 0.01 and 1 Ωm², depending on the cell type and physiological conditions.[40] Here, the membrane resistivity has been taken to be 1 Ωm² because it is mainly associated with spherical/spheroidal shaped cells. The cytoplasm and nucleoplasm resistances have been considered from the reported literature work.[31]

The substitution of these numerical values in Equation 9.5 provides the value of time constant as,

$$
\tau_1 = 3.26 \times 10^{-6}\,s, \tau_2 = 1.535 \times 10^{-6}\,s.
$$

Further, the additional circuit (Figure 9.11b) considers both cell and nuclear membranes as leaky dielectrics. This is a third-order circuit which, on evaluation, gives a third-order differential equation as,

$$
\left[\frac{C_m R_{nm} C_n (R_{C1} + R_{C3})}{2} + \frac{C_m C_n R_{nm} R_n (R_{C1} + R_{C3})}{2 R_{C2}} + \frac{R_n C_m R_{nm} C_n}{2} \right] \frac{d^3 i_b}{dt^3}
$$

$$
+ \left[\begin{array}{l} \dfrac{R_{nm} C_n (R_{C1} + R_{C3})}{2 R_m} + \dfrac{R_n R_{nm} C_n (R_{C1} + R_{C3})}{2 R_m R_{C2}} + \dfrac{R_n R_{nm} C_n}{2 R_m} + \dfrac{C_m (R_{C1} + R_{C3})}{2} \\[2mm] + \dfrac{C_m R_{nm}(R_{C1} + R_{C3})}{R_{C2}} + \dfrac{C_m R_n (R_{C1} + R_{C3})}{2 R_{C2}} + C_m R_{nm} + \dfrac{R_n C_m}{2} + R_{nm} C_n + \dfrac{R_{nm} R_n C_n}{R_{C2}} \end{array} \right] \frac{d^2 i_b}{dt^2}
$$

$$
+ \left[\frac{(R_{C1} + R_{C3})}{2 R_n} + \frac{R_{nm}(R_{C1} + R_{C3})}{R_m R_{C2}} + \frac{R_n(R_{C1} + R_{C3})}{2 R_m R_{C2}} + \frac{R_{nm}}{R_m} + \frac{R_n}{2 R_m} + 1 + \frac{2 R_{nm}}{R_{C2}} + \frac{R_n}{R_{C2}} \right] \frac{di_b}{dt} = 0
$$

(9.6)

In this equation, the resistivity of the nuclear membrane is considered to be similar to that for the cell membrane and therefore, the resistance of the nuclear membrane is evaluated to be

$$R_{nm} = 591390.25\,k\Omega$$

The numerical values of resistances and capacitances of the above cellular components provide the value of time constant for the electrical equivalent circuit of a living cell, as proposed in Figure 9.11b, can be obtained as,

$$\tau_1 = 3.26 \times 10^{-6}\,s,\ \tau_2 = 1.535 \times 10^{-6}\,s$$

Further, the variation of time constants for the electrical equivalent circuit has been analyzed in terms of the various cellular adaptation processes.

Equation 9.6 suggests that the time constant is the function of a number of cellular/intracellular parameters such as the radius of the cell, the resistance of cytoplasm and nucleoplasm, and the resistance and capacitances of cell and nuclear membranes. It can be observed that the time constant evaluated for both the equivalent circuits of Figure 9.11a and b are the same. It is due to the fact that the nuclear membrane resistance is quite large. It can, therefore, be considered to be close to a perfect dielectric. In addition, the variation of time constants with the cellular adaptation processes exhibit similar responses for both the equivalent models.

On observing the time constant values, it is indicated that the response of the cell toward externally applied electrical stimulation is instantaneous.[35]

9.5.3 Model Implications

Figure 9.12 demonstrates the variation of time constant with size of the cell for different size ratios of the nucleus to the cell. It can be observed in all the cases that the time constant is linearly dependent on the cell size. In addition, as the size of the nucleus increases with respect to the cell, the steepness of the curve increases. Overall, Figure 9.12 suggests that cells having different nuclear sizes respond differently under the applied electrical stimulus. It is well known that in the presence of internal or external stimulus, cells change their size (atrophy and hypertrophy) as well as electrochemical (concentration gradients of ions across the cellular membrane) and electrophysical (transmembrane potential and size of the nucleus) parameters. These functionalities facilitate the cells to adapt to different physiological environments and are called cellular adaptation processes. In addition, during interaction of cells with biomaterial surfaces, they change their size accordingly.[41] Therefore, the dependence of time constant on cell size indicates that there would be different responses of cell as their sizes vary under the applied stimulus.

Figure 9.13 illustrates the variation of time constant with the capacitance of cell and nuclear membranes. These capacitances are shown to vary linearly with time constant. In addition, the variation of the nuclear membrane capacitance is steeper than the cell membrane capacitance. The figure is indicative of the comparative charging characteristics of both the membranes, in which nuclear membrane charges more promptly than the cell membrane. In addition, large sized cells would have larger capacitance. Therefore, their response is slow under an externally applied E-field.

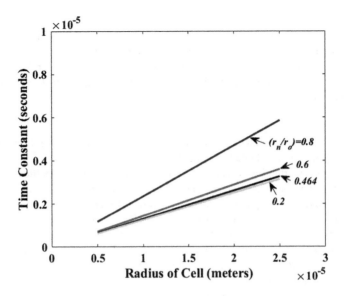

FIGURE 9.12 Time constant variation with cell size for varying nucleus to cell size (r_n/r_o) ratio. (From Saxena A., A. K. Dubey, *J. Phys. D: Appl. Phys.*, 2019, 52, 015401. Reproduced with permission from IOP Publishing.)

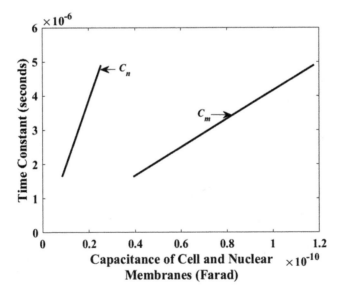

FIGURE 9.13 Time constant variation with capacitance of cell and nuclear membranes. (From Saxena A., A. K. Dubey, *J. Phys. D: Appl. Phys.*, 2019, 52, 015401. Reproduced with permission from IOP Publishing.)

> Physically, the time constant is a measure of duration for which the entire cell is exposed to the external electrical stimulation. In addition, time constant is also the threshold time until the cell induces physically measurable effects such as the formation of pores in the membrane, modification in membrane conductivity, etc.

In other words, time constant is a measure of duration, required for charging the capacitors across cell and nuclear membranes.

Figure 9.13 is plotted for the constant cell size by changing the specific capacitance per unit area. The capacitance of the membrane per unit area is directly proportional to the induced transmembrane potential. Therefore, different cell types having varying capacitance values per unit area would induce different transmembrane potentials in response to the externally applied E-field. The transmembrane potential of cancerous cells differs from that of normal cells. Therefore, the E-field response is expected to be different for both cell types and in the process it can normalize cell proliferation as well as kill the cancerous cells selectively.[32,42,43]

Figure 9.14 demonstrates the time constant variation as a function of cytoplasm and nucleoplasm resistance, which is illustrated to be hyperbolic in nature. This behavior indicates that as the resistances of various volumetric regions of the cytoplasm and nucleoplasm increases, the time constant decreases. The cytoplasmic resistance shows quite steep response than nucleoplasmic resistance. The variation of time constant as a function of cell as well as nuclear membrane resistance is illustrated in Figure 9.15. The non-linearity of the curve indicates that a cell with a high value of membrane

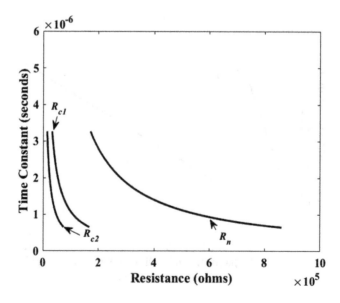

FIGURE 9.14 Time constant variation with the cytoplasmic and nucleoplasmic resistances. (From Saxena A., A. K. Dubey, *J. Phys. D: Appl. Phys.*, 2019, 52, 015401. Reproduced with permission from IOP Publishing.)

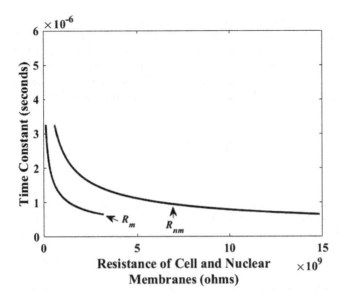

FIGURE 9.15 Time constant variation with cell and nuclear membrane resistances. (From Saxena A., A. K. Dubey, *J. Phys. D: Appl. Phys.*, 2019, 52, 015401. Reproduced with permission from IOP Publishing.)

resistance responds promptly under the applied electrical stimulus. In addition, the nuclear membrane resistance also plays a crucial role for cells to respond under an applied electrical stimulus. A higher membrane resistance is analogous to the high dielectric characteristics. In the simple electrical equivalent circuit of the living cell, the combination of resistance and capacitance represents the heterogeneous electrical properties of the cell and nuclear membranes, while the resistance represents the conductive nature of the fluids present inside the cell as well as nucleus. It is to emphasize that various cellular adaptation processes represented here are similar for both equivalent circuits (Figure 9.11a and b). The validation of the electrical equivalent circuit (Figure 9.11a) of the living cell has been carried out with the help of the well-recognized electroporation method. In addition, the required E-field to electroporate the cells has been analyzed and the results are compared with those of the clinically used values.

9.5.4 Electroporation

The applied E-field leads to the structural changes in the cell membrane, which results in the increased permeability of the membrane, which leads to electroporation. Apart from it, various non-linear phenomena are responsible for the pore formation in the cell membrane. Overall, the gradual depolarization of the membrane is the primary reason for electroporation to occur. The pores in the membranes are created at the instant when the E-field is applied. However, they are observable when the membrane potential reaches the critical value. Therefore, the exact threshold potential of the cell membrane is hard to determine. In this perspective, the approximate range of the critical potential is specified in literature studies. In addition, electroporation depends on the cell size and parameters of the applied electrical stimulation such as field strength, duration, and number of pulses applied.[33,44] In addition, the cellular parameters such as capacitances/resistances of the cell and nuclear membranes and the impedance of matrix/conductive fluids, as discussed, play an important role as far as the electric field stimulated electroporation phenomenon is concerned. The E-field parameters to electroporate a cell can be evaluated using Figure 9.11a as,

$$E = \frac{V_m C_o}{2R_{cell}(\alpha\tau_1 + \beta\tau_2 - \alpha\tau_1 e^{-t_0/\tau_1} - \beta\tau_2 e^{-t_0/\tau_2})} \tag{9.7}$$

Where

$$V_m = \frac{V}{C_o}\left\{ \begin{array}{c} \overbrace{\left[\left[\frac{k_1(R_{C1}+R_{C3})}{2} + \frac{R_n k_1(R_{C1}+R_{C3})}{2R_{C2}} - \frac{(R_{C1}+R_{C3})}{C_n R_{C2}} - \frac{1}{C_n} + \frac{R_n k_1}{2}\right]\left[\frac{e^{-k_1 t}}{-k_1}\right]_0^{t_0}\right]}^{\alpha} + \\ \underbrace{\left[\left[\frac{(R_{C1}+R_{C3})}{C_n R_{C2}} - \frac{R_n(R_{C1}+R_{C3})}{2R_{C2}} - \frac{k_2(R_{C1}+R_{C3})}{2} + \frac{1}{C_n} - \frac{R_n k_2}{2}\right]\left[\frac{e^{-k_2 t}}{-k_2}\right]_0^{t_0}\right]}_{\beta} \end{array} \right\} \tag{9.8}$$

$$C_o = \left[k_1(R_{C1}+R_{C3}) + \frac{R_n k_1(R_{C1}+R_{C3})}{R_{C2}} - \frac{2(R_{C1}+R_{C3})}{C_n R_{C2}} - \frac{2}{C_n} + R_n k_1\right]\left[\frac{e^{-k_1 t_0}}{-k_1} - \frac{1}{k_1}\right] +$$
$$\left[\frac{2(R_{C1}+R_{C3})}{C_n R_{C2}} - \frac{R_n(R_{C1}+R_{C3})}{R_{C2}} - k_2(R_{C1}+R_{C3}) + \frac{2}{C_n} - R_n k_2\right]\left[\frac{e^{-k_2 t_0}}{-k_2} + \frac{1}{k_2}\right]$$
$$+ \frac{2}{C_n}\left[\frac{e^{-k_1 t_0}}{-k_1} + \frac{e^{-k_2 t_0}}{k_2} + \frac{1}{k_1} - \frac{1}{k_1}\right] + (R_{C1}+R_{C3})\left[\begin{array}{c}\left[1 + \frac{R_n}{R_{C2}} - \frac{2}{C_n R_{C2} k_1}\right]e^{-k_1 t_0} + \\ \left[\frac{2}{C_n R_{C2} k_2} - \frac{R_n}{R_{C2}} - 1\right]e^{-k_2 t_0}\end{array}\right]e^{-k_2 t_0} \tag{9.9}$$

where $\dfrac{1}{k_1} = \tau_1, \dfrac{1}{k_2} = \tau_2$

The above equation is similar to the experimentally verified hypothesis that for electroporation to occur the strength of E-field is inversely related to cell size ($V_m = 1.5\, ER \cos\varphi$), where R represents the cell radius and φ is the angle between E-field and surface area vectors.[45] In addition, Schwan's equation also suggests a similar relationship with the applied electrical stimulation and cell size. The application of external stimulation induces a potential (V_m) across the cell as,

$$V_m = fER\cos\varphi(1 - e^{t/\tau}) \tag{9.10}$$

where f, t, and τ are the shape factor, time elapsed for the onset of the E-field and time constant for membrane charging, respectively. The cellular adaptation processes such as atrophy and hypertrophy can be suggested to provide the major relevance to such kind of analytical studies.

It is observed that when the biomaterial is placed in the cell culture medium, proteins get adhered on the surface and therefore, cells first interact to the adhered proteins on the surface and in the process, cells change their shape and size. Thus, cells adhered on the biomaterial surface would have different electrical responses with that of non-adhered cells under the similar electrical stimulus.

Weaver et al.[44] proposed the E-field strength duration map, which suggests that for conventional (reversible) electroporation to occur, the field strength of 0.1–100 kV/cm with a pulse duration of 1 µs to 1 s is required. Earlier, Neumann et al.[16] reported the insertion of DNA molecules into the mouse lyoma cells under the application of E-field with a strength of 8 kV/cm and a duration of 5 µs, which was the first reported *in vivo* application of electroporation. In electrochemotherapy (reversible electroporation) as well as necrosis (irreversible electroporation), an E-field strength of 1.5 kV/cm with a pulse duration of 100 µs are generally utilized. While for tumor treatment, a large strength of E-field (40 kV/cm) and short duration pulses (~10 ns) are usually applied.[46] Therefore, taking the E-field strength—duration map into consideration, pulse duration in the range of 100 µs to 1 s is required for reversible electroporation to occur. The number of medical treatment such as, gene therapy (0.1–1 kV/cm), electro-chemotherapy (0.1–1 kV/cm), etc. used electrical stimulation in such ranges. On the other hand, for irreversible electroporation, an E-field duration of 1 ms to 1 s is desirable with the strength, ranging from 1 to 10 kV/cm. Tissue ablation (1–10 kV/cm), necrosis (1–10 kV/cm), bacterial decontamination, and so on are few of the examples of irreversible electroporation processes.[44] Accordingly, it can include various applications of irreversible and reversible electroporation. Consequently, it was reported that the application of E-field strength of 4.5–8.1 kV/cm with 40 µs pulses results in the apoptosis of Jurkat T-lymphoblasts and HL-60 cells.[47]

Due to the enormous amount of reported literature on E-field parameters applied to different cell culture medium in order to achieve the desired effect (pore formation in the cell membrane and membranes of intracellular organelles), the E-field strength–duration map is an approximate assumption which is based on the experimental studies. Table 9.2 summarizes the electrical stimulation parameters, such as field strength and pulse duration used to electroporate the cells (*in vitro* and *in vivo*) for various applications.[35] The E-field strength for the onset of electroporation depends upon the electrophysical conditions of cells in addition to the cell size. A comparison of experimental values of electrical stimulation parameters (which are currently being utilized for various applications) with those of the values, calculated using electrical equivalent circuits suggests that these values are of the same order. This proves the validity of the electrical equivalent circuit of the cell.

9.5.5 Induced Current/Voltage across Cell/Nuclear Membranes

Figure 9.16 represents the variation of current and voltage, induced across the cell and nuclear membranes after the application of electrical stimulation of strength and duration of 1 V and 1 ns, respectively. This analysis was performed for the electrical equivalent circuit, as shown in Figure 9.11(b). The current response due to the pulsed E-field stimulus is observed to be in the microampere range. The response characteristics behavior of current is similar to the pulsed input. This behavior is similar to the step response of the RC network which is assumed to be of membranes.

TABLE 9.2

Electrical Stimulation Parameters Used to Electroporate the Cells for Various Applications

S.No.	E-Field	Pulse Duration	Type of Cells	Application	References
1	5–10 kV/cm	5–10 μs	Mouse muscle cells	Gene transfer (*in vivo*)	Wong and Neumann[27] (1982)
2	5 kV/cm	2 ms	Mouse tumor cells	Administration of blomycein drug to cancerous cells (*in vivo*)	Okino and Mohri[48] (1987)
3	0.3 kV/cm	1 ms	Rat skeletal muscle cells	Electrotransfer of DNA (*in vivo*) gene therapy)	Mir et al.[49] (1999)
4	5.3 MV/m	60 ns	Human eosinophil cells	Intracellular effects (apoptosis induction, gene delivery, etc.) (*in vitro*)	Schoenbach et al.[22] (2001)
5	≤300 kV/cm	10–300 ns	Human Jurkat cells and fibrosarcoma in C57B1/6 mice	Controlled cell death and apoptosis induction (*in vitro*)	Beebe et al.[25] (2003)
6	0.8–1.3 kV/cm	100 μs	Melanomas, breast, head, and neck cancerous cells	Electrochemotherapy (*ex vivo*)	Gothelf et al.[50] (2003)
7	80–240 kV/cm	10 ns	1,2-Di-oleoyl-*sn*-glycero-3-phosphocholine phospholipid vesicles, COS-7 cells	Permeabilization of intracellular vacuoles (*in vitro*)	Tekle et al.[51] (2005)
8	1 kV/cm	20 ms	Liver cells of male Sprague–Dawley rats	Tissue ablation (*in vivo*)	Edd et al.[52] (2006)
9	≥20 kV/cm	360 ns	Murine melanoma B16-F10 cells	Tumor growth inhibition (*in vitro*)	Nuccitelli et al.[53]
10	320–550 V with an electrode gap of 5.7 mm and 4.4 mm	100 μs	Rat skeletal muscle cells and liver cells	Electrotransfer of DNA (*in vivo*) gene therapy)	Cukjati et al.[54] (2007)
11	2.5 kV/cm	100 μs	Tumor cells in mice	Tissue ablation (*in vivo*)	Al-Sakere et al.[55] (2007)
12	60 kV/cm	60 ns/300 ns	HCT116p53[+/+], HCT116p53[−/−] colon carcinoma cells	Apoptosis (*in vitro*)	Hall et al.[56] (2007)
13	80 kV/cm	1 μs	Bacteria found in wastewater (*Pseudomonas aeruginosa* and *Enterococcus faecium*)	Bacterial inactivation (*in vivo*)	Rieder et al.[57] (2008)

(Continued)

TABLE 9.2 (*Continued*)

Electrical Stimulation Parameters Used to Electroporate the Cells for Various Applications

S.No.	E-Field	Pulse Duration	Type of Cells	Application	References
14	25–100 kV/cm	60 ns	Jurkat cells	Regulation of intracellular calcium ions (*in vitro*)	Scarlett et al.[58] (2009)
15	0–60 kV/cm	300 ns	E4 squamous carcinoma cells	Apoptosis induction (*in vitro*)	Ren and Beebe[59] (2011)
16	Combination of high and low-voltage pulses (500 V and 45 V with 0.7 cm electrode gap)	500 μs and 250 ms	Dermatomed pig cells	Transdermal drug delivery (*in vivo*)	Zorec et al.[60] (2013)
17	1.3 kV/cm	50–50 μs (biphasic pulse)	Human alveolar basal epithelial carcinoma cell line A549	Electrochemotherapy (*in vitro*)	Spugnini et al.[61] (2014)
18	3 kV with 2.5 cm electrode gap	70–80 μs	Pancreatic adenocarcinoma cancerous cells	Tumor ablation (*in vivo*)	Tasu et al.[62] (2016)
19	0.5 kV/cm, 1.2 kV/cm and 35 kV/cm	100 μs, 240 μs and 60 ns	MDA-MB-231 human breast cancer cells	Cancer cells viability tests, electroporation and electrochemotherapy (*in vitro*)	Mittal et al.[63] (2017)
20	0.1–0.9 kV/cm	100 μs	Human mesenchymal stem cells	Control of differentiation and proliferation due to Ca^{2+} oscillations (*in vitro*)	Hanna et al.[64] (2017)
21	32–64 V/mm	50 ms	Human T lymphoma cell line derived from T cell leukemia (Jurkat cells)	Transfection (*in vitro*)	Im and Jeong[65] (2017)
22	400 V with linear array of needle electrodes	0.1 ms (8 pulses with 5 kHz frequency)	Cutaneous metastases from breast cancer and malignant melanoma	Calcium electroporation with electrochemotherapy (*in vivo*)	Falk et al.[66] (2018)
23	10 kV Electrode gap of 3–6 mm	60 ns	Osteoblast-like cells (MG63) and cardiomyocytes	Calcium mobilization leading to new bone formation (*in vitro*)	Zhou et al.[67] (2018)
24	1.5 kV/cm	90 μs	Hepato-pancreatico biliary tumor cells	Tumor ablation (*in vivo*)	Ruarus et al.[68] (2018)
25	0.4–1.2 kV/cm	100 μs	Human malignant melanoma cell lines (Me45 and MeWo)	Inhibition of tumor growth (*in vivo*)	Choromanska et al.[69] (2018)

Source: Reproduced with permission from Saxena A., A. K. Dubey, *J. Phys. D: Appl. Phys.*, 2019, 52, 015401. IOP Publishing.

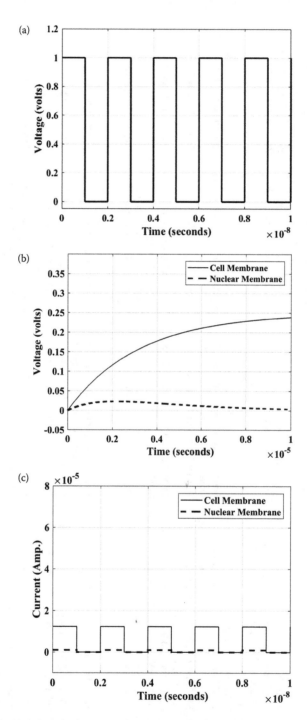

FIGURE 9.16 Variation of voltage (b) and current (c) across both, the cell and nuclear membranes with the application of (a) nanosecond duration pulse of intensity 1 V. (From Saxena A., A. K. Dubey, *J. Phys. D: Appl. Phys.*, 2019, 52, 015401. Reproduced with permission from IOP Publishing.)

The pulsed voltage being considered here is of very short duration (1 ns). The decrease in the pulse duration (1 ms to 1 ns) decreases the rise/fall time of the applied pulses. Therefore, the current during the rise and fall phases of the applied pulses would be higher as compared to the constant phase of the applied pulse. This is due to the fact that the current across the capacitive component of the membrane is the time derivative of applied voltage. As duration of the pulse decreases, the time derivative of the voltage increases and therefore, the current would increase. It is important to note here that such analysis will be valid prior to reaching the critical value of the transmembrane potential (1 V) which otherwise results in a non-linear increase in membrane conductivity. In general, a gradual increase in transmembrane potential is observed with the application of pulsed electrical stimulation. Overall, the analytical computation indicates that the duration of the applied pulses plays a critical role, as far as the pulsed electrical stimulation induced cellular response is concerned.

9.6 Closure

The inherent nature of the living cell as an electrically active unit has motivated the researchers to understand as to how cellular fate processes can be altered/controlled under the external E-field stimulation. Such understanding requires integrated approach of physics and biology and therefore, the importance of interdisciplinary approach can be realized. A number of studies have been carried out to probe into the interaction of E-field with a single living cell. To discuss this aspect, two different electrical equivalent circuits or models for living cells have been considered, based on the dielectric characteristics of the membranes (cell and nuclear) and considering specific ionic pathways under an applied electrical potential. The considered models are based on two-fold assumptions, that is, by assuming (a) leaky dielectric behavior of the cell membrane only and (b) leaky dielectric behavior of both the cell as well as nuclear membrane. The evaluation of time constant is further followed by understanding its variation with the cell size, cytoplasmic and nucleoplasmic resistances, and cell/nuclear membrane resistances and capacitances. The proposed electrical equivalent models have been validated by comparing with the experimentally established electroporation phenomenon. Furthermore, it is shown that the results are consistent with the experimentally observed values. In addition, the influence of the pulse duration as well as the strength of electrical stimulation on the flow of current and voltage, induced across the cell/nuclear membranes, have also been presented in this chapter. Overall, this chapter clearly indicates that how an interdisciplinary approach can effectively provide the solution of such a technologically relevant problem.

REFERENCES

1. Walter B. F., B. L. Emile, *A Medical Physiology: A Cellular and Molecular Approach*, 2012, ISBN: 978-1-4377-1753-2.
2. Kandel E. R., J. H. Schwartz, T. M. Jessell, S. A. Siegelbaum, A. J. Hudspeth, *Principles of Neural Science*, 5th edn, McGraw-Hill Medical, 2013, ISBN: 978-0-07-181001-2.
3. Fick A., Uber diffusion, *Ann. Phys. Chem.*, 1855, 94, 59–86.

 4. Sterratt D., B. Graham, A. Gillies, D. Willshaw, *Principles of Computational Modelling in Neuroscience*, Cambridge University Press, ISBN-978-0-521-87795-4.
 5. Gandhi O. P., X. B. Chen, Specific absorption rates and induced current densities for an anatomy-based model of the human for exposure to time-varying magnetic fields of MRI, *Magn. Reson. Med.*, 1999, 41, 816–823.
 6. Gabriel C., S. Gabriel, http://www.brooks.af.mil/AFRL/HED/hedr/reports/dielectric/Report/Report.html, 2002.
 7. Bencsik M., R. Bowtell, R. Bowley, Electric fields induced in the human body by time-varying magnetic field gradients in MRI: numerical calculations and correlation analysis, *Phys. Med. Biol.*, 2007, 52, 2337–2353.
 8. Zhou S.-A., M. Uesaka, Bioelectrodynamics in living organisms, *Int. J. Eng. Sci.*, 2006, 44(1–2), 67–92.
 9. Gowrishankar T. R., J. C. Weaver, *Proc. Natl. Acad. Sci.*, 2003, 100(6), 3203–3208.
10. Boron W. F., E. L. Boulpaep, *A Medical Physiology: A Cellular and Molecular Approach*, Saunders Elsevier, Philadelphia, 2012, pp. 147–165.
11. V-Acosta J. C., PhD thesis, TechnischeUniversiteit Eindhoven, 2015.
12. Nuccitelli R., in *Current Topics in Developmental Biology*, ed. G. P. Schatten, Elsevier, Amsterdam, 1st edn, 2003, vol. 58, ch. 1, pp. 1–26.
13. Nuccitelli R., *Radiat. Prot. Dosim.*, 2003, 106(4), 375–383.
14. McCaig D., Colin, M. Zhao, *Bio. Essays*, 2005, 19(9), 819–826.
15. Fröhlich F., D. A. McCormick, *Neuron*, 2010, 67(1), 129–143.
16. Neumann E., M. Schaefer-Ridder, Y. Wang, P. H. Hofschneider, *EMBO J.*, 1982, 1(7), 841–845.
17. Dubey A. K., K. Balani, B. Basu, in *Nanomedicine: Technologies and Applications*, ed. T. J. Webster, Woodhead Publishing, Cambridge, 1st edn, 2012, ch. 18, pp. 537–570.
18. Yarmush M. L., A. Golberg, G. Serša, T. Kotnik, D. Miklavčič, *Annu. Rev. Biomed. Eng.*, 2014, 16(1), 295–320.
19. Batista T., Napotnik, D. Miklavčič, *Bioelectrochemistry*, 2018, 120, 166–182.
20. Schoenbach K. H., S. Katsuki, R. H. Stark, E. S. Buescher, S. J. Beebe, *IEEE Trans. Plasma Sci.*, 2002, 30(1), 293–300.
21. Batista T., Napotnik, M. Reberšek, P. T. Vernier, B. Mali, D. Miklavčič, *Bioelectrochemistry*, 2016, 110, 1–12.
22. Schoenbach K. H., S. J. Beebe, E. S. Buescher, *Bioelectromagnetics*, 2001, 22(6), 440–448.
23. Robinson K. R., *J. Cell Biol.*, 1985, 101(6), 2023–2027.
24. Tsong T. Y., *Biophys. J.*, 1991, 60(2), 297–306.
25. Beebe S. J., P. M. Fox, P. M, L. J. Rec, K. Somers, R. H. Stark, K. H. Schoenbach, *IEEE Trans. Plasma Sci.*, 2002, 30(1), 286–292.
26. Zimmermann U., J. Vienken, *J. Membr. Biol.*, 1982, 67(1), 165–182.
27. Wong T.-K., E. Neumann, *Biochem. Biophys. Res. Commun.*, 1982, 107(2), 584–587.
28. Beebe S. J., *J. Nanomed. Nanotechnol.*, 2013, 4(2), 1–8.
29. Schoenbach K. H., F. E. Peterkin, R. W. Alden, S. J. Beebe, *IEEE Trans. Plasma Sci.*, 1997, 25(2), 284–292.
30. Deng J., K. H. Schoenbach, E. S. Buescher, P. S. Hair, P. M. Fox, S. J. Beebe, *Biophys. J.*, 2003, 84(4), 2709–2714.
31. Dubey A. K., S. Dutta-Gupta, R. Kumar, A. Tewari, B. Basu, *J. Appl. Phys.*, 2009, 105(8), 084705.
32. Dubey A. K., M. Banerjee, B. Basu, *J. Biol. Phys.*, 2011, 37(1), 39–50.

33. Dubey A. K., R. Kumar, M. Banerjee, B. Basu, *J. Comput. Theor. Nanosci.*, 2012, 9(1), 137–143.
34. Ellappan P., R. Sundararajan, *J. Electrostat.*, 2005, 63(3), 297–307.
35. Saxena A., A. K. Dubey, *J. Phys. D: Appl. Phys.*, 2019, 52, 015401.
36. Weaver J. C., *IEEE Trans. Plasma Sci.*, 2000, 28(1), 24–33.
37. Sundelacruz S., M. Levin, D. L. Kaplan, *Stem Cell Rev. Rep.*, 2009, 5(3), 231–246.
38. Ly J. D., D. R. Grubb, A. Lawen, *Apoptosis*, 2003, 8(2), 115–128.
39. Kotnik T., D. Miklavčič, *Biophys. J.*, 2000, 79(2), 670–679.
40. Foster K. R., H. P. Schwan, in *Handbook of Biological Effects of Electromagnetic Fields*, eds. C. Polk, E. Postow, CRC Press, New York, 2nd edn, 1995, vol. 45, ch. 1, pp. 3–16.
41. Ratner B., A. S. Hoffman, F. J. Schoen, J. E. Lemons, *Biomaterials Science: An Introduction to Materials in Medicine*, Academic Press, New York, 1996.
42. Yang M., W. Brackenbury, *Front. Physiol.*, 2013, 4(185). https://doi.org/10.3389/fphys.2013.00185.
43. Vodovnik L., D. Miklavčič, G. Serša, *Med. Biol. Eng. Comput.*, 1992, 30(4), CE21–CE28.
44. Weaver J. C., K. C. Smith, A. T. Esser, R. S. Son, T. R. Gowrishankar, *Bioelectrochemistry*, 2012, 87, 236–243.
45. Dev S. B., D. P. Rabussay, G. Widera, G. A. Hofmann, *IEEE Trans. Plasma Sci.*, 2000, 28(1), 206–223.
46. Son R. S., K. C. Smith, T. R. Gowrishankar, P. T. Vernier, J. C. Weaver, *J. Membr. Biol.*, 2014, 247(12), 1209–1228.
47. Hofmann F., H. Ohnimus, C. Scheller, W. Strupp, U. Zimmermann, C. Jassoy, *J. Membr. Biol.*, 1999, 169(2), 103–109.
48. Okino M., H. Mohri, *Jpn. J. Cancer Res.: Gann*, 1987, 78(12), 1319–1321.
49. Mir L. M., M. F. Bureau, J. Gehl, R. Rangara, D. Rouy, J.-M. Caillaud, P. Delaere, D. Branellec, B. Schwartz, D. Scherman, *Proc. Nat. Acad. Sci. USA*, 1999, 96(8), 4262–4267.
50. Gothelf A., L. M. Mir, J. Gehl, *Cancer Treat. Rev.*, 2003, 29(5), 371–387.
51. Tekle E., H. Oubrahim, S. M. Dzekunov, J. F. Kolb, K. H. Schoenbach, P. B. Chock, *Biophys. J.*, 2005, 89(1), 274–284.
52. Edd J. F., L. Horowitz, R. V. Davalos, L. M. Mir, B. Rubinsky, *IEEE Trans. Biomed. Eng.*, 2006, 53(7), 1409–1415.
53. Nuccitelli R., U. Pliquett, X. Chen, W. Ford, R. J. Swanson, S. J. Beebe, J. F. Kolb, K. H. Schoenbach, *Biochem. Biophys. Res. Commun.*, 2006, 343(2), 351–360.
54. Cukjati D., D. Batiuskaite, F. André, D. Miklavčič, L. M. Mir, *Bioelectrochemistry*, 2007, 70(2), 501–507.
55. Al-Sakere B., F. André, C. Bernat, E. Connault, P. Opolon, RV Davalos RV, *PLoS ONE*, 2007, 2(11), e1135.
56. Hall E. H., K. H. Schoenbach, S. J. Beebe, *Apoptosis*, 2007, 12(9), 1721–1731.
57. Rieder A., T. Schwartz, K. Schön-Hölz, S. M. Marten, J. Süß, C. Gusbeth, W. Kohnen, W. Swoboda, U Obst, W. Frey, *J. Appl. Microbiol.*, 2008, 105(6), 2035–2045.
58. Scarlett S. S., J. A. White, P. F. Blackmore, K. H. Schoenbach, J. F. Kolb, *Biochim. Biophys. Acta (BBA) – Biomembr.*, 2009, 1788(5), 1168–1175.
59. Ren W., S. J. Beebe, *Apoptosis*, 2011, 16(4), 382–393.
60. Zorec B., S. Becker, M. Reberšek, D. Miklavčič, N. Pavšelj, *Int. J. Pharm.*, 2013, 457(1), 214–223.
61. Spugnini E. P., A. Melillo, L. Quagliuolo, M. Boccellino, B. Vincenzi, P. Pasquali, A. Baldi, *J. Cell. Phys.*, 2014, 229(9), 1177–1181.

62. Tasu J. P., G. Vesselle, G. Herpe, J. P. Richer, S. Boucecbi, S. Vélasco, M. Carretier, B. Debeane, D. Tougeron, *Diagn. Interv. Imag.*, 2016, 97(12), 1297–1304.
63. Mittal L., V. Raman, I. G. Camarillo, A. L. Garner, A. J. Fairbanks, G. A. Dunn, R. Sundararajan, *IEEE Conference on Electrical Insulation and Dielectric Phenomenon (CEIDP)*, 2017, pp. 596–599.
64. Hanna H., F. M. Andre, L. M. Mir, *Stem Cell Res. Ther.*, 2017, 8(1), 91.
65. Im D. J., S.-N. Jeong, *Biochem. Eng. J.*, 2017, 122, 133–140.
66. Falk H., L. W. Matthiessen, G. Wooler, J. Gehl, *ActaOncologica*, 2018, 57(3), 311–319.
67. Zhou P., F. He, Y. Han, B. Liu, S. Wei, *Bioelectrochemistry*, 2018, 124, 7–12.
68. Ruarus A. H., L. G. P. H. Vroomen, R. S. Puijk, H. J. Scheffer, B. M. Zonderhuis, G. Kazemier, M. P. van den Tol, F. H. Berger, M. R. Meijerink, *Can. Assoc. Radiol. J.*, 2018, 69(1), 38–50.
69. Choromanska A., S. Lubinska, A. Szewczyk, J. Saczko, J. Kulbacka, *Bioelectrochemistry*, 2018, 123, 255–259.

10

Physical Laws of Solar–Thermal Energy Harvesting

In this chapter, we review the necessary fundamentals related to the underlying principles involved in harvesting solar–thermal energy using optical reflector and spectrally selective absorber materials. This will start with the recapitulation of the characteristics of the electromagnetic spectrum of solar irradiation. This will be followed by revisiting the physical principles of reflection phenomenon. In particular, reference to specular reflectance, the role of surface plasmon resonance, and plasma edge properties will be discussed. The discussion in this chapter is restricted to concentrated solar power (CSP) technology, an inter-disciplinary research area, which is currently gaining more attention in the energy sector. More details of the concepts presented in this chapter are available in many papers published by the research group of one of the authors of this book.[1–10]

10.1 Electromagnetic Spectrum and Solar Range

Capturing the solar energy using technologically feasible approaches and harvesting it have been the central paradigm of renewable energy generation. Figure 10.1(a) presents the electromagnetic spectrum of light, revealing the wide range of wavelengths and frequencies of electromagnetic radiation. Two specific wavelength intervals[11] are identified for the appropriate utilization of solar spectrum. The wavelength range from 280 to 2500 nm comprises the entire solar radiation coming to the earth's surface. This wavelength range can be exploited for harnessing solar energy. Simultaneously, special attention is required to avoid thermal heat loss through infrared radiation. The spectral radiation of more than >2500 nm causes such heat radiation.

The spectral distribution in solar spectrum depends on the atmospheric conditions. A small portion of solar radiation (approximately 2%–3%) consists of ultraviolet radiation. On the same spectrum, the visible light spans in the range of 400–700 nm and comprises around 45%–46% of the spectrum. Finally, infrared radiation (700–2500 nm) contributes to remaining 50%–52% of the solar spectrum.

Quantification of solar radiation is performed by determining the solar constant, which represents mean incident irradiation passing an unit cross-sectional area. The entire solar spectrum is used to measure the solar constant, which comes out to be 1.361 kilowatts per square meter (kW/m²). Solar irradiance is an important parameter to be taken into consideration to infer the magnitude at one astronomical unit (AU), thereby to find the solar constant. AM stands for air mass, which is the relative optical mass calculated using the density of air as a function of altitude. The standard spectrum used for space-related applications is AM 0, which obtains a total power of

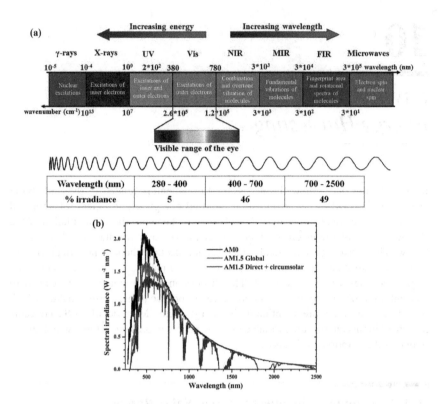

FIGURE 10.1 (a) Electromagnetic spectrum of light 13 and (b) standard solar spectra for space and terrestrial use. (Adapted from Ref. 23).

1366.1 W/m². There are two standards, which is known to be applied for terrestrial use. The primary one is the AM 1.5 global spectrum that is used for flat plate collectors having an integrated power of 1000 W/m² (100 mW/cm²). The other one is the AM 1.5 direct (+ circumsolar) spectrum, which is designed for the system operation using concentrated solar radiation. The latter one has an integrated power density of 900 W/m² and it involves the direct beam from the sun along with the circumsolar component in a disk 2.5° around the sun. The standard solar spectra can be generated employing SMARTS (Simple Model of the Atmospheric Radiative Transfer of Sunshine) program. The same program is also useful to generate other spectra, as required.

10.2 Physics of Reflection Phenomenon

In this section, we will revisit the necessary fundamentals to explain the reflectance phenomenon, followed by physical principles related to specular reflectance.

10.2.1 Interaction of Light with Matter

The three widely known major optical phenomena, namely absorption (A), transmission (T), and reflection (R), take place when there is a light–matter interaction (see Figure 10.2).[12]

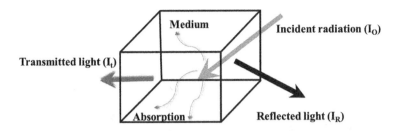

FIGURE 10.2 Interaction of electromagnetic radiation with a solid.[24]

$$A + T + R = 1 \qquad (10.1)$$

Electromagnetic waves with different wavelengths have respective frequencies. Interestingly, a specific color can be represented by a distinct frequency in the visible spectra. A specific frequency at which an electron vibrates while interacting with an EM wave is known as natural frequency. The absorption, transmission, or reflection of the EM wave by a material is determined by the combination of electrons' natural frequency and the incident radiation.

10.2.1.1 Absorption

When frequency of an EM wave interacting with a matter matches with an electron's natural frequency, the electrons start to vibrate. It results in the absorption of the incident radiation leading to a vibrational motion of electrons. The electrons, in succession, bump up against neighboring atoms. This phenomenon contributes to thermal energy. Absorbance (A) is the amount of light absorbed by a sample.

10.2.1.2 Reflection

Reflectance (R) is the ratio of the radiant flux (φ_r) reflected from a material to that of the incident flux (φ_i). This phenomenon takes place, when there is a difference between the frequency of the incoming light and electrons' natural frequency. The reflection is a material surface property, which is determined by surface characteristics (roughness) and composition. The reflectance property therefore can be tailored in a tandem coating structure with layers having different compositions and thicknesses.

In the following, the broader classification of reflectance together with physical mechanisms is explained. There are mainly two types of reflectance, namely specular reflectance and diffuse reflectance.

10.2.1.2.1 Specular Reflection

The law of reflection states that in case of specular reflection, the incident angle is equal to the angle of reflectance. Typically, the example of specular reflection is the interaction of an EM wave with a mirror, where the major amount of incident wavelength gets reflected back into the acceptance angle (φ).

In Figure 10.3(a), the light incidents on the mirror, which is known as the incident ray (labeled I in Figure 10.3(a)). The ray of light, which comes out of the mirror, is known as the reflected ray (labeled R in Figure 10.3(a)). At the point of incidence where

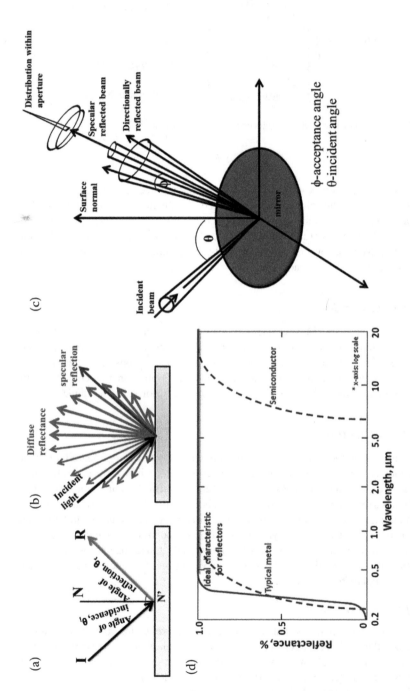

FIGURE 10.3 Schematic representing the (a) law of reflection and (b) diffuse reflectance. (c) Plasma edge: Rationale for selection of metals as reflectors and (d) schematic explaining specular reflectance measurements. (Adapted from Ref. 23).

the incident radiation interacts with the mirror, a line can be drawn perpendicular to the surface of the mirror (labeled N in Figure 10.3(a)). The normal line separates the angle between the incident ray (θ_i) and the reflected ray (θ_r) into two equal angles. The angle between the incident ray and the normal line is known as the angle of incidence.

10.2.1.2.2 Diffuse Reflectance

Diffuse reflectance is represented by the amount of the incident radiation reflected in all directions, as shown in Figure 10.3(b). Free powders are examples of the source of diffuse reflectance. Diffuse reflectance consists of diffusely scattered light in all directions and is dependent on the wavelength (λ) and the angle of incidence θ. An integrating sphere is required to measure the total or diffuse reflectance.

Specular and diffuse reflectance take place simultaneously, while the EM wave interacts with materials. As far as the underlying mechanisms for reflection are concerned, both surface plasmon resonance and plasma edge properties play a key role in understanding the reflectance property of a surface.

10.2.1.2.3 Surface Plasmon Resonance

It is well known that there are four fundamental states of matter. One of these states is plasma in which matter exists in the form of ions. Plasma is an electrically neutral medium of unbound positive and negative particles, that occur at very high temperature and low particle density. Plasmon, a quantum of plasma oscillations, is considered to be the collective oscillation of the free electron gas density. The coherent excitation of conduction electrons near the interface of metal and dielectric surfaces is known as surface plasmon. Specific electromagnetic waves can be generated within a system with proper interactions between electrons and incident photons. Surface plasmon polaritons (SPPs) are basically propagating surface plasmons in/on extended structures including metal films or metal nanostructures. In contrast, localized surface plasmons (LSPs) are another type of surface plasmons, which are often in the form of standing waves, in metal nanoparticles.

When electromagnetic radiation (e.g., solar radiation) incidents on the surface of any material, internal electric and magnetic fields are generated due to the presence of electric charges inside the material. This is because, the center of mass of the nuclei and that of the electrons coincide. However, when they shift with respect to each other, an oscillation occurs as the Coulomb force restricts the movement by restoring their positions. The frequency at which these oscillations resonate with the incident wavelength is known as "plasma frequency" (ω_p), which has an important role in the propagation of electromagnetic waves. The plasma frequency is calculated using the numerical values of density of electrons, material's density and molecular mass. The plasma frequency of non-plasma materials can be determined from the natural collective oscillation frequency of the "electron cloud." The plasma frequency is given by,

$$\omega_p = \sqrt{\frac{ne^{10}}{m\epsilon_o}} \tag{10.2}$$

where n is the free electron density, e is the elementary charge (1.602×10^{-19} C), m is the electron mass 9.1×10^{-31} kg, and ϵ_o is the permittivity of free space 8.854×10^{-12}.

From Equation 10.2, it can be seen that more is the free electron density, the larger will be the plasma frequency. The free electron density (n) can be calculated by multiplying the number of atoms per unit volume (N_v) and electrons per atom (e/a).

$$n = N_v \times (e/a) \tag{10.3}$$

The electron concentration (e/a) for an intermetallic compound A_aB_b is calculated using Hume–Rothery rule,

$$(e/a) = \frac{ax + by}{a + b} \tag{10.4}$$

where x and y are the respective group numbers of elements A and B in the periodic table, and a and b are the number of atoms in the molecular formula, respectively.

Number of atoms per unit volume,

$$N_v = (N_A \times \rho)/M \tag{10.5}$$

where N_A is the Avogadro number, ρ is the density, and M is the atomic mass.

Substituting Equations 10.3, 10.4, and 10.5, into 10.2, the expression for plasma frequency comes out as,

$$\omega_p = \sqrt{\frac{(N_A \times \rho) \times (ax + by)e^2}{Mm\epsilon_o(a + b)}} \tag{10.6}$$

From Equation 10.6, it is evident that the plasma frequency is directly proportional to the square root of the ratio of the density to its molecular weight of the material and also to the square root of electron concentration. Thus, a material with a combination of high electron concentration, high density, and a low molecular weight, can be chosen as the reflector material.

10.2.1.2.4 Plasma Edge Properties

From the perspective of fundamental physics, the interaction of EM waves with a surface of material introduces small oscillations, causing a radiation of secondary waves from each atom in all directions (dipole antenna). All such radiations are the reason of reflection and refractions. As mentioned earlier, an interesting phenomenon in solids is the vibration of free electrons collectively, which tends to give a sharp edge in EM spectra known as "plasma edge."[13] The reflectance of a material drops at a particular wavelength, which is associated with the plasma edge. The location of plasma edge depends on the number of free carriers. The location of plasma edge is different for different metals and alloys and as a result of this the reflectance properties of each material change (Figure 10.2(c)).

10.2.1.2.5 Solar Specular Reflectance

It can be reiterated once more that the reflectance (R) is the ratio of the intensity of reflected (I_R) radiation from a surface to that of the incident radiation (I_o).

$$R = \frac{I_R}{I_o} \tag{10.7}$$

The average specular reflectance (see Figure 10.2(d)) is obtained by calculating the average value obtained in the solar range,

$$\rho_s\left(\lambda,\theta,\phi\right)=\sum_{i=280}^{2500}\rho_s\left(\lambda_i,\theta,\phi\right)$$ (10.8)

where λ is the incident wavelength, θ is the angle of incidence, and ϕ is the acceptance angle. The solar-weighted specular reflectance $\rho_s(sw,\ \theta,\ \phi)$ of a solar mirror[14] is determined using the values of the measured specular reflectance spectrum $\rho_s(\lambda,\ \theta,\ \phi)$ and the direct solar irradiance spectrum (E_λ) at wavelength intervals of $\Delta\lambda_i$ with the following formula,

$$\rho_s\left(sw,\theta,\phi\right)=\frac{\sum_{i=280}^{2500}\rho_s\left(\lambda_i,\theta,h\right)\cdot E_\lambda\left(\lambda_i\right)\cdot\Delta\lambda_i}{\sum_{i=280}^{2500}E_\lambda\left(\lambda_i\right)\cdot\Delta\lambda_i}$$ (10.9)

The solar weighted hemispherical reflectance is determined by the following expression,

$$\rho_h\left(sw,\theta,h\right)=\sum_{i=280}^{2500}\rho_h\left(\lambda_i,\theta,h\right)\cdot F(\lambda_i)$$ (10.10)

where h stands for the hemispherical region of the integrating sphere, $F(\lambda_i)$ stands for the weighting factor, and sw stands for solar weighted values. In case, the measurements of specular reflectance is carried out with a reflectometer at a particular wavelength λ_{meas}, the value of solar weighted specular reflectance $\rho_s(sw,\ \theta,\ \phi)$ can be approximated with the formula,

$$\rho_s\left(sw,\theta,\phi\right)=\frac{\rho_s\left(\lambda_{meas},\theta,\phi\right)}{\rho_h\left(\lambda_{meas},\theta,h\right)}\cdot\rho_h\left(sw,\theta,h\right)$$ (10.11)

From the above equations, it is evident that the solar weighted specular reflectance is a function of wavelength of incident wavelength, angle of incidence, and angle of acceptance. For achieving higher accuracy, the measurements should be taken at several narrow wavelength intervals of the solar spectrum for obtaining a maximum amount of reflectance data. Furthermore, a minimum angle of acceptance reduces the loss of incident radiation by restricting the diffuse reflectance and a constant angle of incidence throughout the surface allows maximum focussing of reflected radiation onto the receiver.

10.2.1.3 Transmission

Transmittance (T) is represented by the fraction of incident radiation, which travels through a sample and reaches the other end. In case of transmission, there is also a significant difference between two frequencies, which includes incident radiation and natural frequency of electrons. However, transparent or semi-transparent objects are required for transmission to occur. The incident wave causes the atoms to vibrate at small amplitudes, thereby transfer of such vibration helps the EM wave to pass through the material and exit through the opposite side.

Metals contain a large number of free electrons and the periodic potential energy of the lattice makes the electrons to move around the lattice nuclei without being bound to a single lattice. This delocalized collection of electrons is known as "electron cloud." The sea of electrons flows freely around the lattice to conduct electricity in metals. In the presence of electric potential applied across the metal, electrons flow rapidly to the positive charge and cancel it out across the metal surface.

In insulators, the electrons are firmly bound to the lattice nuclei and can only oscillate around their original position. This oscillation influences the propagation of light, resulting in a reduced wave velocity and a small loss of energy. However, in case of metals, the free electrons travel large distances before the energy is dissipated due to damping.

10.3 Spectrally Selective Optical Properties

A remarkable revolution in scientific community happened in 1865, while Maxwell first introduced the classical theory of electromagnetism. Using his theory, the light was predicted as electromagnetic radiation.[15] There are different classifications of electromagnetic spectrum depending on the wavelength range. The spectrum starts from cosmic and gamma rays (wavelength $\sim 10^{-8}$ μm) and continues till long radio waves (wavelength $\sim 10^{10}$ μm). The sun is considered as an electromagnetic source and the radiation emitted from the sun reaching the atmosphere of earth is known as solar spectrum, which lies in the wavelength range of 0.3–2.5 μm. In a similar way, a hot object kept at a temperature of 200°C and above is also an example of an EM wave radiation source, which generates the wavelength in the infrared regime, particularly in 2.5–50 μm. These two parts of electromagnetic spectrum, namely solar radiation and infrared range, are extremely important for a solar selective absorber coating.

10.3.1 Solar Absorptance

When light, an electromagnetic (EM) wave, incidents on a material, an interaction between the incident wave and the atoms, ions, and/or electrons takes place in the material. The light–matter interaction depends mainly on two factors including the frequency of the light and the atomic structure of the material. As mentioned earlier, this interaction process causes three common physical phenomena such as absorption, reflection, and transmission (see Figure 10.2). In typical interactions, the incident EM wave while passing through the material completely dissipates. Such interactions cause absorption of incident radiation in matter. The exact match of the energy of an incident photon with the band gap of a material leads to a high absorption process.

The absorptance can be represented as the ratio of the absorbed radiation by any material to the incident flux. The following equation expresses the absorptance as,

$$\alpha_\lambda(\lambda) = \frac{G_{\lambda,abs}(\lambda)}{G_\lambda(\lambda)} \tag{10.12}$$

where $G_{\lambda,abs}(\lambda)$ and $G_\lambda(\lambda)$ are the absorbed and incident radiations, respectively.

According to the conservation of energy, the sum of the transmission, reflection, and absorption of the incident flux is equal to 1.

$$\alpha_\lambda + \rho_\lambda + \tau_\lambda = 1 \tag{10.13}$$

where α_λ, ρ_λ, and τ_λ are the absorptance, reflectance, and transmittance at wavelength, λ. There are some specific materials which do not allow the penetration of the incident radiation through it. In that case, the transmittance is considered as zero ($\tau_\lambda = 0$) and the conservation of energy equation is modified as follows:

$$\alpha_\lambda + \rho_\lambda = 1 \tag{10.14}$$

$$\alpha_\lambda = 1 - \rho_\lambda \tag{10.15}$$

Using Equations 10.12 and 10.15, the absorptance can be evaluated as a function of reflectance using the equation below

$$\alpha(\theta) = \frac{\int_{\lambda_1}^{\lambda_2} [1 - \rho_\lambda] G(\lambda) d\lambda}{\int_{\lambda_1}^{\lambda_2} G(\lambda) d\lambda} \tag{10.16}$$

where λ is the wavelength, θ is the incident angle of light, and λ_1 and λ_2 are minimum and maximum solar wavelengths, respectively. The standard UV–Vis–NIR spectrophotometer measures diffuse reflectance spectra of solar selective absorbers and the data are used to calculate solar absorptance. The reflectance measurements are carried out throughout the solar spectrum (0.3–2.5 μm) and at near normal angle of incidence ($\theta = 0°$).

10.3.2 Thermal Emittance

The thermal emittance of any object is related to radiation emitted from a body in the infrared regime. However, before going into definition, it would be better to clarify the idea on blackbody radiation.[16] Under hypothetical conditions, an object which absorbs all the radiation incident on it is known as an ideal blackbody. Max Planck first established the radiation properties of a perfect blackbody using quantum physics. The expression of blackbody radiation is as follows,

$$E_{b\lambda}(\lambda, T) = \frac{8\pi hc}{\lambda^5 \exp\left((hc/\lambda k_B T) - 1\right)} \tag{10.17}$$

where c is the hand, k_B are the speed of light (m/s), Planck's constant (J s) and Boltzmann constant (J/K), respectively.

An integration of blackbody spectra over the entire wavelength range will provide the amount of radiation emitted from a blackbody, which can be represented using Stefan–Boltzmann law.[17]

$$E_b(T) = \int_0^\infty E_{b\lambda}(\lambda,T)d\lambda = \sigma T^4 \qquad (10.18)$$

where σ is the Stefan–Boltzmann constant.

It is important to observe that the above equation indicates a fact that the total blackbody radiation is proportional to the fourth power of its temperature. The equation also describes that a small change in the temperature of a blackbody will increase the thermal radiation from a black body significantly.

The emittance (ε) of a surface is a dimensionless number which ranges from 0 to 1, that is $0 \le \varepsilon \le 1$. The emittance is the representation of the ratio of radiation emitted by the object at a given temperature (E_λ) to the radiation emitted by a perfect blackbody ($E_{b\lambda}$) at the same temperature. The emittance is a measure, which gives an idea about the comparison of the radiative properties of any material and a perfect blackbody. There are various factors of an object, which determines the amount of emission. These factors include temperature, chemical composition, surface roughness, intrinsic geometrical structure, observing angle, etc. The following equation represents the total hemispherical emittance

$$\varepsilon_\lambda(\lambda,T) = \frac{E_\lambda(\lambda,T)}{E_{b\lambda}(\lambda,T)} \qquad (10.19)$$

Apart from Planck's law on blackbody radiation, there are some other representation, which relates the light–matter interaction, For example, according to Kirchhoff's law, the total hemispherical emittance (λ) of a surface at temperature T is equal to its total hemispherical absorptance $\alpha(\lambda)$ from a blackbody at the same temperature. Hence, the emissivity can also be expressed in terms of absorptance, and thereby the reflectance of a material using the equation below,

$$\varepsilon_\lambda(\lambda) = \alpha_\lambda(\lambda) = 1 - \rho_\lambda(\lambda) \qquad (10.20)$$

The combination of Equations 10.19 and 10.20 will give an overall idea on the emittance of a material at a temperature T as a function of reflectance

$$\varepsilon(T) = \frac{\int_{\lambda_1}^{\lambda_2} [1 - \rho_\lambda(\lambda)] E_{b\lambda}(\lambda)d\lambda}{\int_{\lambda_1}^{\lambda_2} E_{b\lambda}(\lambda)d\lambda}, \qquad (10.21)$$

where $E_b(\lambda)$ is the blackbody radiation spectrum. An object with a temperature of about 300°C emits radiation in the wavelength range from 2.5 to 25 μm, which can be used as λ_1 and λ_2, respectively in equation above. Using a FTIR spectrophotometer with an integrated sphere set-up is the routine procedure to measure the emittance of spectrally selective thin films between 400 and 4000 cm^{-1} (2.5–25 μm).

10.3.3 Solar Selectivity

One of the most important criteria for solar absorbers used in the CSP system is to optimize their spectrally selective properties by enhancing the absorptance (α) >0.95 in solar spectrum (0.25–2.5 μm) and minimizing thermal emittance (ε) <0.05 in thermal radiation range (2.5–25 μm). The metric used in assessing the optical properties of solar absorbers is spectral selectivity which is the ratio of solar absorptance to thermal emittance. Higher the value of spectral selectivity better would be the performance of the receiver tube coated with absorbers. To achieve high selectivity, appearance of a step function in the reflectance spectrum of absorber is desired in between the solar spectrum and the infrared regime. The particular wavelength at which the step arises is known as the cutoff wavelength or plasma wavelength, λ_{plasma}.

10.4 Performance Evaluation

In recent years, researchers paid their attention to evaluate some other parameters apart from the routinely measured solar absorptance and thermal emittance to understand the properties and performance of spectrally selective absorbers. The following section describes few such functions or parameters.

10.4.1 Merit Function and Absorber Efficiency

The merit function indicates how efficiently an absorber can utilize the absorbed radiation by minimizing the heat loss. The concept of merit function was first proposed by Sergeant et al.[18] They have demonstrated a simple equation which establishes the correlation between the merit function $F(T)$ and absorptance (α), emittance (ε), and the operating temperature (T).

$$F(T) = \alpha[1 - \varepsilon T] \tag{10.22}$$

From the equation, it would be interesting to note that maximization of α and minimization of ε would be the key to achieve the maximum $F(T)$. A higher merit function indicates a better spectral selectivity of the material. Another important function, cut off wavelength, can also be estimated using the value of the merit function.

The role of the receiver tube is to convert most of the absorbed solar radiation into a useful form of energy, that is, thermal energy. Therefore, the evaluation of the conversion of solar to thermal energy appears to be an essential task. Solar to thermal conversion efficiency, $\eta_{sol-the}$, at a particular temperature (T) which was first discussed by Cindrella[19] and Ho et al.[20] is,

$$\eta_{sol-the} = \alpha - \frac{\varepsilon(T)\sigma T^4}{CI_s} \tag{10.23}$$

where C is the concentration ratio which is usually in the order of 10–1000, I is the solar flux density (W/m²), and σ is Stefan–Boltzmann constant.[21] It can be estimated that a combination of higher concentrations and lower operating temperatures would result in a low conversion efficiency. Another way of achieving high efficiency is to increase the absorptance along with the concentration factor. For a specific solar selective absorber, it is also necessary to maximize the overall solar–thermal efficiency by considering a specific structure, material, working temperature, and the concentration factor.[22]

10.5 Closure

This chapter introduces the concepts of the fundamental material properties of relevance for a solar energy system, particularly for CSP. It is believed that these fundamental concepts would allow one to select a combination of materials for the solar reflector–absorber system. This chapter also highlights the importance of interdisciplinary concepts, particularly drawn from physical science domain, toward the development of new materials for the CSP system.

REFERENCES

1. Dan A., A. Soum-Glaude, A. Carling-Plaza, C. Ho, K. Chattopadhyay, H. Barshilia, B. Basu, Photothermal conversion efficiency, temperature and angle dependent emissivity and thermal shock resistance of W/WAlN/WAlON/Al2O3-based spectrally selective absorber, *ACS App. Energy Mater.*, 2019, 2(8), 5557–5567.
2. Alex S., R. Kumar, K. Chattopadhyay, H. C. Barshilia, B. Basu, Thermally evaporated Cu–Al thin film coated flexible glass mirror for concentrated solar power applications, *Mater. Chem. Phys.*, 2019, 232, 221–228.
3. Dan A., B. Basu, T. Echániz, I. González de Arrieta, G. A. López, H. C. Barshilia, Effects of environmental and operational variability on the spectrally selective properties of W/WAlN/WAlON/Al2O3–based solar absorber coating, *Solar Energy Mater. Solar Cells*, 2018, 176, 157–166.
4. Dan A., A. Biswas, P. Sarkar, S. Kashyap, K. Chattopadhyay, H. C. Barshilia, B. Basu, Enhancing spectrally selective response of WWAlN/WAlON/Al$_2$O$_3$-based nanostructured multilayer absorber coating through graded optical constants, *Solar Energy Mater. Solar Cells*, 2018, 176, 157–166.
5. Alex S., K. Chattopadhyay, B. Basu, Tailored specular reflectance of Cu-based novel intermetallic alloys, *Solar Energy Mater. Solar Cells*, 2016, 149, 66–74.
6. Dan A., J. Jyothi, H. C. Barshilia, K. Chattopadhyay, B. Basu, Spectrally selective absorber coating of WAlN/WAlON/Al2O3 for solar thermal applications, *Solar Energy Mater. Solar Cells*, 2016, 157, 716–726.
7. Dan A., H. C. Barshilia, K. Chattopadhyay, B. Basu, Angular solar absorptance and thermal stability of WAlN/WAlON/Al$_2$O$_3$-based solar selective absorber coating, *Appl. Therm. Eng.*, 2016, 109B, 997–1002.
8. Alex S., S. Sengupta, U. K. Pandey, B. Basu, K. Chattopadhyay, Electrodeposition of δ-phase based Cu–Sn mirror alloy from sulfate-aqueous electrolyte for solar reflector application, *Appl. Therm. Eng.*, 2016, 109B, 1003–1010.
9. Dan A., K. Chattopadhyay, H. C. Barshilia, B. Basu, Colored selective absorber coating with excellent durability, *Thin Solid Films*, 2016, 620, 17–22.
10. Dan A., H. C. Barshilia, K. Chattopadhyay, B. Basu, Solar energy absorption mediated by surface plasma polaritons in spectrally selective dielectric-metal-dielectric coatings: A critical review, *Renew. Sustain. Energy Rev.*, 2017, 79, 1050–1077.

11. Roos A., Use of an integrating sphere in solar energy research, *Solar Energy Mater. Solar Cells*, 1993, 30, 77–94.

12. https://segoianphysics.wordpress.com/2010/01/15/how-does-light-interact-with-matter/.

13. Dixon A. E., J. D. Leslie, *Solar energy conversion—An introductory course*, 1979, 10, 87–329, eBook ISBN: 9781483189284.

14. Guidelines, Measurement of solar weighted reflectance of mirror materials for concentrating solar power technology with commercially available instruments, Solar PACES interim Version 1.1, May 2011.

15. Jackson J. D., *Classical Electrodynamics*, Wiley, 1999.

16. Kuhn T. S., *Black-body Theory and the Quantum Discontinuity, 1894–1912*, University of Chicago Press, 1978.

17. Cardy J., The ubiquitous 'c': from the Stefan–Boltzmann law to quantum information Boltzmann Medal Lecture, Statphys24, Cairns, July 2010, *J. Stat. Mech.: Theory Exp.*, 2010, 2010, P10004.

18. Sergeant N. P., O. Pincon, M. Agrawal, P. Peumans, Design of wide-angle solar-selective absorbers using aperiodic metal-dielectric stacks, *Opt. Express*, 2009, 17, 22800–22812.

19. Cindrella L., The real utility ranges of the solar selective coatings, *Solar Energy Mater. Solar Cells*, 2007, 91, 1898–1901.

20. Ho C. K., A. R. Mahoney, A. Ambrosini, M. Bencomo, A. Hall, T. N. Lambert, Characterization of Pyromark 2500 for high-temperature solar receivers, *Am. Soc. Mech. Eng.*, 2014, 136(1), 509–518.

21. Bermel P., J. Lee, J. D. Joannopoulos, I. Celanovic, M. Soljacie, Selective solar absorbers, *Ann. Rev. Heat Transfer*, 2012, 15.

22. Li P. et al., Large-scale nanophotonic solar selective absorbers for high-efficiency solar thermal energy conversion. *Adv. Mater.*, 2015, 27, 4585–4591.

23. Alex S., Development of Cu-based intermetallic reflector materials for Concentrated Solar Power Application, PhD thesis, Indian Institute of Science, Bangalore, 2017.

24. Dan A., Spectrally selective tandem absorbers for photothermal conversion in high temperature solar thermal systems, PhD thesis, Indian Institute of Science, Bangalore, 2019.

Section II

Applications

11

Environmental/Societal Needs of Alternate Energy and Energy Storage

For the society to progress further, demand for energy in the useful form (i.e., electricity) is bound to go up very rapidly. A rough estimation indicates that the world needs to double its energy supply by 2050. It is believed that the ability to fulfill this steeply increasing demand, primarily in the form of electricity, will form the benchmark for progress of human civilization. However, the conventional sources of energy, namely, fossil fuel, coal, nuclear, natural gas, and so on are usually termed as "non-renewable" sources. Already, as of date, there is considerable scarcity for these sources. Additionally, the environmental pollution, primarily caused by excessive usage of fossil fuels as energy sources for automobiles and other applications, has also been on a very steep increase. These are some of the reasons for trying to move away from the "fossil fuels" and even "hydroelectric systems," and advance toward the "renewable" (or sustainable) and "green" energy sources. Accordingly, there has been significant impetus toward efficient "harvesting" and utilization of the "renewable" energy sources such as solar, wind, biomass, geothermal and so forth for generating useful forms of energy. For instance, in the Indian scenario, the government is presently targeting 175 GW of renewable energy by 2022; with 100 GW from solar, 60 GW from wind, 10 GW from biomass, and 5 GW from hydro. Additionally, efficient utilization of energy "harvested" from such sources, most of which are intermittent in nature, necessitates the development of advanced energy storage technologies. With regard to automobiles, having them operate solely based on such advanced energy storage technologies, like electrochemical energy storage (batteries and supercapacitors), will also contribute immensely toward the control of environmental pollution. In these contexts, this chapter will demonstrate the usage of some basic scientific concepts toward understanding and developing technologies to harvest and efficiently utilize energy from these "alternate" sources.

11.1 Science and Technology toward Efficient Energy "Harvesting" from the Sun and Wind

11.1.1 Harvesting the Solar Energy via Photovoltaics

Solar energy can be utilized either directly in the form of heat energy or as electrical energy; with the latter being the more preferred and predominant form for technological usage. The very basic principle governing the conversion of solar energy to electrical energy (i.e., photovoltaic science) involves the excitement of electrons to the conduction band, along with a simultaneous creation of holes, upon absorption of solar energy (i.e., photons of different wavelengths/energy) by the material used as the "photovoltaic." This leads to the direct conversion of solar energy to electrical energy. Some of the scientific aspects governing the performance and efficiency (i.e., energy conversion efficiency) of a photovoltaic system (neglecting the various electrical losses) are absorption of a greater range of solar spectrum which contain electromagnetic radiations having energies between ~0.5 and ~2.9 eV, prevention/minimization of the recombination of electrons and holes generated by the absorbed photons, and optimization of the current/voltage generated upon absorption of photons leading to maximization of the output power (which is directly proportional to the product of current and voltage). For more detailed understanding on these aspects, the reader may refer to References [1–3].

The very material used for the "photovoltaic effect" has influence on all the above three. A material with a very low bandgap can improve the efficiency of absorbance by utilizing most of the photons received from the sun (that possesses energy in the range mentioned above) for the generation of the electron–hole pair. This would, in turn, lead to enhancement in the current produced by the solar cell, however, at the cost of voltage generated (which, in turn, increases with the band-gap of the "photovoltaic material"). For example, crystalline Si (the conventional and still the most commonly used "photovoltaic material," ever since 1954) has a bandgap of ~1.1 eV and can use photons of solar radiations having a wavelength $\geq$ ~1.1 eV for "photovoltaic effect." By contrast, the same material, that is, Si in the amorphous form has a greater bandgap of ~1.7 eV, which reduces the current efficiency, but increases the voltage output of the cell. Accordingly, amorphous Si is now gaining increasing popularity, especially upon doping with H for reducing dangling bonds as defect sites.

Overall, materials having a bandgap between 1 and 1.8 eV qualify as "photovoltaic materials," with those possessing a bandgap of ~1.3–1.5 eV usually result in the best efficiency in terms of output power (i.e., "Shockley–Queisser Efficiency Limit" or "SQ Limit" of up to ~34%).

In addition to Si-based materials (presently, i.e., in early 2017, showing efficiencies of ~25%–30%), the materials that are becoming increasingly popular for research and development are crystalline GaAs, InP, copper indium gallium selenide (or CIGS), CdTe, perovskite-structured materials, a few organic semiconductors, etc. due to the range of bandgaps (especially for the first few), processing simplicity compared to Si (especially for the organic–inorganic materials, including the perovskites), and also structural/design flexibility (especially for the organic materials, despite efficiency usually being poor) offered by them. A summary of the materials that are being used/investigated in terms of their bandgaps and SQ energy conversion efficiencies are presented in Figure 11.1.[4] On a different note, with respect to stability upon exposure to sunlight over a long duration, the more "novel" inorganic materials fare better as compared to Si-based photovoltaics, which is known to have reduced efficiency by 10%–20% over time primarily due to the creation of defect sites in the lattice itself upon continued interactions with the solar photons.

Of course, barring the stability issues, the efficiencies can only be approached to the best extent possible if optical losses due to reflections, shading by grid patterns, and electrical losses at the contacts are minimized and also if the thickness of the concerned photovoltaic material is sufficient to absorb the useful incident photons.

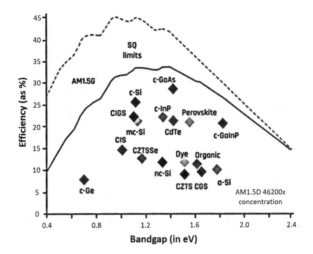

FIGURE 11.1 Shockley–Queisser (SQ) solar energy conversion efficiency limits under global sunlight (AM1.5G), versus band gap, in comparisons with the energy absorption threshold (the solid line), highest experimentally recorded efficiencies for the different photovoltaic materials (the specific data points), and also the limits for direct AM1.5D sunlight conversion under maximum possible concentration of 46,200 suns (dashed line), all at cell temperature of 298 K (Here, CIS, copper indium selenide; CGS, copper gallium selenide; CIGS, copper indium gallium selenide; CZTS, Cu_2ZnSnS_y.). (Adapted from Green M. A., S. P. Bremner, *Nat. Mater.*, 2017, 16, 23–24.)

All the above stress upon not only on the material selection but also on profound importance of the engineering design aspects. Table 11.1 provides a summary of the other features, including the bottlenecks associated with some of the more investigated "photovoltaic materials." Another more recent scientific/technological development is the use of multi-junction solar cells which, in very simple terms, is a stack of more than one cells (p–n junctions) based on materials having different band gaps (i.e., from higher to lower downwards) to allow net absorption of a wider range of solar spectrum and hence increment in the overall efficiency. However, tuning such structures and rendering them suitable for widespread practical applications need further developmental work at this stage.

Not just the bandgap, but also the very structure and defect nature of the materials used influence the performance. Here, one of the important aspects is minimizing the recombination of photo-generated electrons and holes, which, if "left alone" would recombine in almost no time (say a millionth of a second) and, thus, would not contribute toward electrical current (known as "recombination losses"). The separation or "sorting out" of the electrons and holes (i.e., the majority and minority carriers, as per the material concerned) is enabled by the electric field set-up at the p–n junction. However, given their propensity to recombine, the material used must possess sufficient mobility for the photogenerated electrons and holes, and must provide only minimal obstructions to the same for traversing a "considerable distance" within almost "no time," as effected by the electric field. Accordingly, a less defected lattice structure is preferred, which is one of the reasons why amorphous Si, having a disordered arrangement of atoms compared to crystalline Si and also dangling bonds acting as obstacles, shows inferior efficiencies with respect to its crystalline counterpart. Additionally, the grain/domain boundaries in a material also act as obstacles to the motion of the electrons and holes, which indicates the beneficial aspect of using either single crystalline materials (technically the most

TABLE 11.1

A Summary of Various Aspects Concerning Materials Used for "Photovoltaic Technology", Including Key Research Opportunities for Such Materials, as in 2016, is presented here.

Material	Cell Efficiency (as %)	Module Efficiency (as %)	Advantages and Options Offered for the Technology	Further Research/Development Opportunities
Matured technologies; deployed at large scale				
Single-crystalline Si	25.6	22.4	Earth-abundant material; >25-year track record; relatively low band-gap of ~1.1 eV allows conversion of a greater fraction of the incident solar photons, but at the same time restricts cell voltage.	Primarily, reduction in recombination losses by more efficient minimization/passivation of defects; improvement in overall absorptivity via reduction in optical losses (i.e., part of "light management").
Poly-crystalline Si	21.3	18.5	Earth-abundant material; >25-year track record; lesser conversion efficiency due to charge carrier recombination at grain/domain boundaries; easier to develop than single-crystalline Si.	
CIGS	21.7	17.5	Tuneable band-gap; allows usage of flexible substrates.	Improvements in "light management"; increase in efficiency via development of tandem cells (for large band gaps); reduction in recombination losses; exploration and further optimization of solution processing routes.
CdTe	21.5	18.6	Allows usage of flexible substrates; has a near-ideal band-gap (of ~1.43 eV) for optimized power output; short energy payback time (i.e., ratio of total energy consumed during fabrication and installation to annual energy saving due to electricity generation from solar).	Reduction in recombination losses; enhanced "light management" to develop thinner cells.

(Continued)

TABLE 11.1 *(Continued)*

A Summary of Various Aspects Concerning Materials Used for "Photovoltaic Technology", Including Key Research Opportunities for Such Materials, as in 2016, is presented here.

Material	Cell Efficiency (as %)	Module Efficiency (as %)	Advantages and Options Offered for the Technology	Further Research/Development Opportunities
Matured technologies; but presently deployed at smaller scale				
Si thin film	11.4	12.2	Allows usage of flexible substrates and development of modules; the module efficiency noted here is for microcrystalline Si/a-Si tandem geometry.	Reduction in recombination losses and improvements in light management; further optimization and developments of the promising tandem cells.
Dye-sensitized TiO_2	11.9	10	Tuneable colors; here absorber is a molecular dye, coated on a porous nanostructured electrode (usually, TiO_2), which upon being "photo-excited" injects electrons into conduction band of TiO_2, and accepts electrons from a redox couple (usually, I^-/I^{3-}) in a non-aqueous electrolyte; band-gaps on lower side.	Improvements and optimization of the redox couple (e.g., Co-based redox couple offer greater voltage, as compared to I-based redox couples); reduction in recombination losses and enhancement in stability upon continued exposure.
Organics	11.5	9.5	Flexible, inexpensive and semi-transparent module development possible via roll-to-roll fabrication on flexible substrates; offer variable colors; bottleneck is the relatively poor cell efficiency and also band-gaps being on the lower side.	Reduction in recombination losses; improvements in "light management" and stability upon continued exposure.
GaAs	28.8	24.1	Has a near-ideal band-gap of ~1.42 eV; very high efficiency; flexible modules; high optical absorptivity allow reduction in cell thickness; presently, at manufacturing level.	Improvement in "light management".
Technologies under development				
Perovskite	21.0	—	They offer possibilities for very facile/cost-effective solution processing and also development of flexible modules; they also offer tuneable band-gaps depending on the exact composition and structure.	Improvements in light management and cell stability; reductions in recombination losses and Pb usage (for the perovskite and quantum dots).
CZTS	12.6	—		
Quantum dots	9.9	—		

Source: Polman A. et al., *Science*, 2016, **352**(6283), 307.

Note: The record cell/module efficiencies have been reported, based on certified measurements. The materials have been further grouped based on the extent of the engineering/technological development using the respective materials.

preferred) or large grain sized (preferably of columnar/elongated shapes) polycrystalline materials, as compared to amorphous or finer grain sized materials. Accordingly, post-processing heat treatments are often designed to increase the grain size.

Moving ahead from the materials aspects, minimization of the optical losses is also another very important criterion for achieving good energy conversion efficiency. The optical losses arise primarily due to the sunlight received on the solar panel getting reflected back before reaching the actual "absorber" or "photovoltaic material." This is one of the major issues with using Si since it has a very reflective back surface and, conventionally, forms the top surface of a solar cell, which is directly exposed to the sunlight through an attached conducting grid structure for current collection. The grid, in turn, is carefully designed to minimize the shading effect on the Si top surface, which, otherwise is another source of optical losses. In order to minimize this problem associated with the reflective surface of Si, the surface is often coated with a very thin layer of oxide (usually a monolayer of Si–O, which reduces reflection by ∼20%), called antireflection coating. However, due to the insulating nature, such antireflection coating leads to a compromise over the efficiency of current collection, due to "electrical losses." More recently, texturing of the top surface of most photovoltaic materials is performed to minimize the reflection losses by virtue of having the originally reflected light strike a "second surface" or protrusions of the texture for getting further absorbed and so on. This leads to enhanced absorption of a light beam before it can finally escape.

> Overall, in scientific and engineering perspective, the efficiency of a solar cell depends on its ability to capture and trap most of the incident light photons (viz., "light management"), ability to convert the same to majority/minority electrical carriers (viz., "conversion management"), and the further ability to get the as-generated electrical carriers efficiently collected for production of current (viz., "carrier management").

In addition the to developing and engineering of single solar cells, integration to form a larger "module" is again very important for the technology because a single solar cell typically produces a voltage of only ∼0.5 V. Accordingly, an usual "12 volt solar panel" or "module," which has dimensions of ∼25 × ∼54 inches, contains about 36 cells connected in series, which if not integrated properly, will lead to substantial electrical losses due to the resistances involved.

11.1.2 Harvesting the Wind Energy via Mechanical Turbines

In the case of wind energy, the natural wind at the earth's atmosphere is utilized to generate electrical energy via mechanical (kinetic) energy. Here it may also be mentioned that one of the most important sources of wind generation is solar heating or, in more precise terms, uneven solar heating across the earth's surface that causes temperature and air pressure gradients leading to wind flow. Due to this, wind energy is sometimes considered to be one of the important manifestations of solar energy itself. Nevertheless, conversion of wind energy to the useful form (i.e., electrical energy) is done with the help of wind turbines (or wind mills, in more popular terminology), which convert the kinetic energy associated with wind to mechanical energy associated with the rotation of turbine blades; which is then further converted into electrical energy via an electrical generator powered by the rotating turbine blades.

In very simple terms, the kinetic energy (E) of wind having a mass of m and average velocity of $<V>$ is $\frac{1}{2}m<V>^2$. Now, the available power (P) associated with the same is a derivative with respect to time (i.e., dE/dt) and, thus, may be written as $\frac{1}{2}m_{fr}<V>^2$, where m_{fr} is the air mass flow rate. The m_{fr} is, in turn, given by $\rho A<V>$, where ρ is the density of the air and A is the area swept by the turbine blades (i.e., proportional to the area of the blades themselves). Accordingly, the available power (or P) is given by $\frac{1}{2}\rho A<V>^3$. For more detailed understanding on these aspects, the reader may refer to References [3,5].

Overall, according to the above relationship, the power available at any instant from a wind mill not only depends on the length of the turbine blades (or the rotors) for a greater sweeping area, but is very sensitive to the wind speed, varying directly as cube of the same. The actual power that can be harvested also depends on the height of the wind mills, that is, the height of the tower supporting the rotor. This is because winds having greater average velocity can be utilized at greater heights. For example, under normal weather conditions, for a 10 m high tower, the highest wind speed that is available can be up to $\sim$7 m/s or slightly higher, which can lead to a wind power density of $<$400 W/m^2. By contrast, a 50 m tower at the same location can allow access to winds of velocities $\sim$9 m/s, leading to a wind power density of $>$400 W/m^2. It is thus no wonder as to why the wind mills that can be seen beside highways passing through deserted areas appear fairly huge in size when close to them. Overall, a single utility scale wind turbine (i.e., those typically used for large-scale electricity generation for powering village, wind power plants, distributed power, etc.) can range from $\sim$100 kW to $\sim$1 MW power rating. Still greater capacity wind turbines (i.e., of capacity beyond 10 MW) have been investigated and presently are under systematic installation at various parts of the world (especially in the United States).

In a generic sense, the basic science or physics involved in a wind mill (as above) is relatively more straightforward compared to that in the photovoltaic technology. However, the engineering aspect and maintenance of all the parts and the interconnections are more stringent. This is primarily because, unlike a photovoltaic system, the wind turbine (or wind mill) has large moving parts in the form of turbine blades (or rotor), which can be kilometers in length (in a bid to obtain a greater sweeping area). Additionally, the drive train (which includes a gearbox and generator), the huge supporting tower for the rotor and drive train, and other equipment (such as controls, electrical cables, interconnections, etc.) are prone to maintenance. All the parts/components in this case have to be structurally very strong. The moving rotors (or blades) are typically made of fiber-reinforced composites to reduce the weight (as compared to that for metallic blades), while at the same time retain good strength (via the fiber reinforcement). However, the damage caused to such structures, not just due to continuous motion, but even due to impact of impurity and foreign particles in the air, is always an area of concern.

11.2 Reducing Negative Environmental Impacts

Over the last decade or so, there has been significant progress worldwide toward the development/growth of energy "harvested" from these renewable energy sources (especially from the solar and wind energy). For instance, in the Indian scenario, the government (via the Ministry of New and Renewable Energy or MNRE) is presently targeting generation of 175 GW power from the energy associated with the renewable sources by 2022; with 100 GW from solar, 60 GW from wind, 10 GW from biomass, and 5 GW from hydro (source: A Report of the Expert Group on 175 GW RE by 2022; published by MNRE in 2015). As per an article published in *The Economic Times* on April 10, 2017, India could add the solar power generation (which include both the roof top solar, as well as larger dedicated solar power stations) by greater than 5.5 GW just in one year's duration. Overall, worldwide, presently the net installed capacity for solar energy (i.e., photovoltaic) is greater than even 300 GW (which is expected to account for nearly 2% of global electricity demand), with Germany, followed by China, Japan, and the United States being the major producers and consumers of the same. Furthermore, in some countries, like Australia, powering of privately owned houses by solar roof-top technology has become quite common. With respect to wind energy, the worldwide capacity presently is ~500 GW, with just the United States itself having a capacity of producing >80 GW of power from the wind. For more details, see Table 11.2 (source: REN21 (2017); *Renewables 2017 Global Status Report*; ISBN 978-3-9818107-6-9),[6] where the data clearly suggest the special impetus toward the development of photovoltaic technology worldwide at least in the last one year. In more specific terms, the % increment in capacity for photovoltaic technology over the last one year (i.e., from 2015–2016) was ~33%, followed by ~13% for wind energy; with the increments for the rest being just ~2%–6%.

In the context of the environmental benefits of using the "green" renewable sources, as compared to the more "conventional" sources like coal and natural gas, it may be noted that, energy harvesting from coal leads to, on an average, emission of CO_2 equivalent to per kilowatt-hour of ~1.4–3.6 pounds of CO_2e/kWh, with natural gas emitting ~0.6–2 pounds of CO_2e/kWh. By contrast, utilization of both solar and

TABLE 11.2

Total Power Harvested from Various Renewable Energy Sources (Not Including Hydro Power), World wide, in 2015 and 2016

Renewable Energy Sources	Net Power Capacity, World-wide (in GW)			% Capacity Increment from 2014 to 2015	% Capacity Increment from 2015 to 2016
	in 2014	in 2015	in 2016		
Bio-power	101	106	112	~4.9%	~5.7%
Geothermal power	12.9	13.2	13.5	~2.3%	~2.2%
Solar photovoltaic	177	228	303	**~28.8%**	**~32.9%**
Solar thermal power	4.3	4.7	4.8	~9.3%	~2.1%
Wind power	370	433	487	~17%	~12.5%
Total renewable power capacity (not including hydroelectricity)	665.2	785	921	~18%	~17.3%

Source: REN21 (2016) and (2017); *Renewables 2016 and 2017 Global Status Reports*; ISBNs 78-3-9818107-0-7 and 978-3-9818107-6-9.

Note: Major increments, as noted for solar photovoltaic, have been highlighted with "bold text".

wind energies leads to CO_2 emissions of only ~0.07–0.2 and ~0.02–0.04 pounds of CO_2e/kWh, respectively (source: Intergovernmental Panel on Climate Change (IPCC) (2011); *Special Report on Renewable Energy Sources and Climate Change Mitigation*; prepared by working group III of IPCC; Cambridge University Press, Cambridge, UK and New York, USA, pp. 1075).

Accordingly, the increasing carbon footprint (or the C-containing greenhouse gas emissions, expressed as CO_2 equivalent or CO_2e) of the environment would get suppressed considerably by greater utilization of the green renewable energy sources.

In addition to directly causing environmental pollution, the carbon footprint also leads to continual warming of the environment, referred to as global warming. In this regard, as an example, just for the United States, electricity production accounts for a significant fraction (i.e., greater than even one-third) of the net carbon footprint or global warming emissions. It is not surprising that the majority of these are caused by coal-based electric power plants, which produce approximately one-fourth of even the total greenhouse emissions of the United States (source: Environmental Protection Agency (2012); *Inventory of U.S. Greenhouse Gas Emissions and Sinks*: 1990–2010). Nevertheless, further advancements in science and engineering is needed, not only to make the "harvesting" technologies more efficient in technical terms, but also to make the process less expensive (i.e., reduction in kWh/$ is a very important aspect) and sustainable for more widespread and long-term usages.

Similar to the case of electricity generation, burning gasoline while driving a car not only emits greenhouse gases like CO_2 and CO, but also gives off other very harmful gases/vapors, such as nitrogen oxides, and also particulate matter and unburnt hydrocarbons. These additionally contribute to air pollution and global warming; and at an alarmingly increasing rate with the improvements in global economy and thus car manufacture/sale. Driving hybrid (having combination of engines powered by gasoline and batteries) and full electric (having engine powered totally by battery) vehicles leads to much suppressed and nearly no emission, respectively, of such gases/vapors/particles. This might lead to a notion that electric vehicles are "zero emission" vehicles. Well, during usage probably yes, but not during manufacturing, primarily since the battery component itself leads to considerable emissions during development; an aspect which is still under debate for quantification purposes (for more details, see a more recent study by the IVL Swedish Environmental Research Institute (Report no.: C43, ISBN 978-91-88319-60-9 (2007); Authors: M. Romare, L. Dahllöf.).

Nevertheless, it is believed that over a full lifetime, electric vehicle would lead to about 50% lesser greenhouse emissions, as compared to a corresponding gasoline vehicle (source: Union of Concerned Scientists (2015); *Gasoline vs Electric – Who Wins on Lifetime Global Warming Emissions?*).

11.3 Need for Efficient Storage of Energy "Harvested" from Renewable Sources

Despite the significant developments and progress made in the science, technology, and efficient implementation of the energy "harvesting" from "green" renewable sources, efficient utilization of them necessitates the development of suitable advanced energy storage technologies and proper integration of the "harvesting" technologies with the "storage" techniques. Most of the "green" renewable energy sources are intermittent in nature. For instance, the solar energy is directly available during the day time ("on hours"), but not during the night time ("off hours"); the durations of which vary at each location on the earth and with the time of the year. Furthermore, places having overcast conditions and excessive particulate pollutants in the atmosphere cause diffused, rather than direct, sunlight to reach the solar cells. These lead to much inferior electric power generation with respect to the rated/expected capacity. On a similar note, the wind speed varies at each location and with time at any given location. As noted in the previous section, the wind power, being very sensitive to the wind speed, fluctuates considerably; thus even rendering a lot of places not suited for wind-mill installation at all (due to fluctuating or very low wind speeds).

These issues immediately stress upon the importance of storage of the electrical energy generated from the renewable sources, which is not only needed during the "off hours" for the solar or wind energy, but also needed during the "on hours," as well, due to the fluctuating nature of both the sources. Overall, the electrical energy harvested from the renewables cannot be directly transferred to a grid, but has to go via an inverter-cum-stabilizer and also preferably via storage technology.

Furthermore, even in the case of conventional electrical energy generation from non-renewable sources, for transfer to the consumers in household, factories, and so on, via power grid, energy storage is essential to store energies at times when production exceeds consumption and use the stored energy during the reverse scenario. This improves the grid stability by leading to improved load levelling (i.e., utilizing surplus energy generated during "on hours" and during the "off hours" of the sources), peak shaving (i.e., accommodating "peaks" and "valleys" in the daily demand curve), and system regulation (i.e., addressing short-term/random fluctuations in demand without the need for regulation by the main or harvesting/supplying plant). In a way, the large-scale grid energy storage leads to overall improvement in the quality and reliability of the power used in household, as well as in industry; irrespective of the fact whether the electricity gets produced by a renewable/green or non-renewable source.

One of the best examples for such development is the huge grid-scale Li-ion battery-based energy storage proposed and getting developed by Elon Musk in the southern regions of Australia. This is expected to have a capacity of 100 MW of power stored over >100 MW hours. Such a development is expected to be capable of powering nearly 30,000 domestic houses, primarily from renewable sources; thus improving the effectiveness of (and also lowering the cost associated with) energy "harvested" from the renewable sources. In the context of reducing the environmental pollution by replacing the gasoline based vehicles by battery run electric vehicles (EVs), the production of electric vehicles has been rising near exponentially in the last few years. For supporting such EVs, the increase in production of Li-ion batteries is also showing a very similar trend.[7] The importance notwithstanding, the successful development and integration of Li-ion batteries of such scale (possibly the only battery chemistry capable of meeting these demands, as of now) on a very regular basis still necessitates considerable scientific-cum-engineering improvements and modifications of the battery technology itself (see Chapter 13 for more details on the same).

11.4 Closure

Preliminary aspects concerning the science behind "harvesting" energy from the renewable/green sources, namely, the sun and the wind, directly in the form of electrical energy have been discussed in this chapter. The brief discussion on the scientific principles is followed by discussion on the key engineering challenges toward efficient and long-term utilization. This is followed by a brief survey of the needs for developing efficient technologies based on the above, especially in the contexts of providing a reliable/quality power supply to consumers and protecting the environment. Some data, based on published survey concerning the increased usage of renewable sources of energy and associated government policies, have also been included here. They emphasize upon the present impetus toward the development and usage of technologies based on the above. Toward the end of the chapter, the needs for proper integration with energy storage systems toward efficient utilization of the energy harvested from the renewable sources (and also non-renewable sources) have been brought to the fore.

REFERENCES

1. Reinders A., P. Verlinden, W. van Sark, A. Freundlich, *Photovoltaic Solar Energy: From Fundamentals to Applications*, John Wiley & Sons, Inc., 2017, ISBN: 978-1-118-92746-5.
2. Kalogirou S. A., *Solar Energy Engineering: Processes and Systems*, 2nd edn, Academic Press, 2013, ISBN-13: 978-0123972705.
3. da Rosa A., *Fundamentals of Renewable Energy Processes*, 1st edn, Academic Press, 2005, eBook ISBN: 9780080477954.
4. Green M. A., S. P. Bremner, Energy conversion approaches and materials for high-efficiency photovoltaics, *Nat. Mater.*, 2017, 16, 23–24.
5. Manwell J. F., J. G. McGowan, A. L. Rogers, *Wind Energy Explained: Theory, Design and Application*, 2nd edn, John Wiley & Sons, Inc., 2004, ISBN: 978-0-470-68628-7.
6. Polman A., M. Knight, E. C. Garnett, B. Ehrler, W. C. Sinke, Photovoltaic materials: Present efficiencies and future challenges, *Science*, 2016, 352 (6283), 307.
7. Feng X., M. Ouyanga, X. Liu, L. Lu, Y. Xia, X. He, Thermal runaway mechanism of lithium ion battery for electric vehicles: A review, *Energy Storage Mater.*, 2018, 10, 246–267.

12

Spectrally Selective Solar Absorbers and Optical Reflectors

Over the last few decades, significant research has been pursued to develop new technology and novel materials to utilize abundant solar energy for many different applications. Concentrating solar power (CSP) technology is also perceived as one of the most promising and economical solar energy harnessing technologies, wherein solar radiation is converted into thermal energy, using different mirror configurations and a receiver.[1,2] The interest in deploying CSP plants has increased significantly in recent years as there is a trend toward maximum utilization of renewable energy to decrease the usage of conventional source of energy for reducing greenhouse gas emission. In order to make CSP plants technologically widespread, significant improvements are implemented for its scalability by introducing solar reflector materials with excellent optical properties and long-term stability.[3] To this end, this chapter emphasizes the current challenges in this area, while highlighting the need for new generation solar reflector materials. More details of some of the new solar absorber coatings and reflector materials are presented in some of the papers published by the research group of one of the authors.[4–13]

12.1 Importance of Solar Mirrors

In the field of solar energy harvesting, photovoltaics (PV) has been widely investigated and deployed in the society at large. However, there is an urgent need to harness solar energy as the fossil fuels are depleting rapidly. Besides, the excessive usage of fossil fuels causing air pollution has now become a threat to the environment. As discussed in Chapter 10, CSP technology is undoubtedly an interesting and promising way of usage of solar energy in the form of thermal energy. One of the major components of CSP systems is spectrally selective absorbers, which provide outstanding performance by absorbing maximum incoming solar radiation ($\alpha \geq 0.95$) and preventing thermal loss of the absorbed radiation in the infra-red wavelength region ($\varepsilon \leq 0.05$). In recent years, researchers have paid attention to develop spectrally selective absorbers using the concept of surface plasma polaritons (SPPs). The SPPs generated in a metal–dielectric interface, possess interesting properties to confine solar energy by generating an additional electromagnetic field at the metal–dielectric interface. The optical constants of metal and dielectric play a crucial role in achieving maximum solar absorption with the help of SPPs. In this chapter, various solar absorber coatings are discussed with a special focus on dielectric–metal–dielectric (DMD)-based absorber coatings.

Solar reflectors play a crucial role by focusing the solar radiation onto the receiver, thereby increasing the efficiency of solar thermal and photovoltaic applications. A reflector material is chosen over lenses in the solar thermal system, since the solar radiation is scattered and lenses are not very effective to focus these scattered light. Reflectance[14] is determined by the wavelength (λ) of the incident light and the angle of incidence (θ) to the mirror surface. "Reflectance" can be expressed as the ratio of intensity of the light reflected from a surface to the intensity of incident radiation.

Basically, a reflector material should satisfy a certain number of requirements.[15] First, they should reflect almost all the incident solar radiation with a minimum amount of absorption onto the receiver tube. Second, the high reflectance property should not be degraded throughout the entire lifetime of the solar collector, i.e., 20–25 years. Essentially, the specular reflectance of any material is dependent on its phase composition and its corresponding plasma frequency. The amount of radiation reflected from a surface depends on the properties of the material and its microscopic nature. For a concentrator to be of high quality, the reflector material should reflect back the maximum amount of incident sunlight.

12.2 Existing Issues with Solar Reflectors

The efficiency of a CSP system is directly related to the performance of solar reflectors employed in the system.[4] The reflector determines the amount of solar radiation that can be transformed into thermal energy. The net solar flux absorbed by the receiver can also be influenced by the shape and position of the sun (the extent of the incident rays from the sun are imperfectly collimated), the quality of the concentrator, accuracy of the tracking system, and most importantly the geographical location of the CSP plant. Major effort is therefore extremely needed to develop reflector materials with all the aforementioned properties, and this has been the major motivation to discuss about new reflector materials in this chapter.

Currently, metallic aluminum[16,17] and silver[18,19] films are considered as superior reflecting materials, which are employed in the CSP plants. Also, significant effort has been made for developing front surface aluminum and silver mirrors. Since a CSP plant is typically placed in an open field, reflectors should possess a long life under harsh environmental conditions and the mirrors must be resistant to abrasion and corrosion. Also, there should be very less degradation. The regular maintenance of the solar reflectors would make the CSP system expensive, since there are challenges related to aluminum and silver. Aluminum degrades rapidly due to abrasion and oxidation under atmospheric conditions,[20] whereas silver is prone to photon-induced corrosion, soiling, and tunneling.[21] Furthermore, the poor durability and self-cleaning capability of the reflectors are serious drawbacks of the CSP plants, and this drives significant research activity in this frontier area.

12.3 Need for Development of New Reflector Materials

The current state-of-the-art reflectors include typical metallic[22] thin films formed on glass substrates. Good reflector materials mainly consist of silver, aluminum, or stainless steel. These materials have received greater attention in the scientific community working on solar energy materials for possessing high reflectance in the solar spectrum. It is well known that the optical reflectivity is strongly dependent on the material and the quality of films that cover the solar reflectors.[23] Metallic thin films, when exposed to atmosphere, undergo oxidation and corrosion, thereby reducing the efficiency of the reflectors.

One of the major requirements of solar reflectors is to protect these materials against all kinds of degradation over time under adverse environments, particularly against humidity and dust. In order to avoid loss of efficiency, metallic thin films are covered with a transparent layer, for example, lacquers, plastics, or anodizing the surface. To protect films from human/nature made and atmospheric corrosion, many of the reflector designs include clear protective overcoats to the reflecting surface. Any such protective overcoat will act as a source of light absorption and/or they may interfere with the optical design, all of which results in diminishing the overall reflectance of the reflectors. Such designs, however, compromise the reflectivity against durability. Nowadays, there is an approach to prepare flexible mirrors, which are bendable and lightweight, however, there should be a balance between the mechanical properties of the mirrors and their optical performance. In some cases, it has been reported that the reflectance properties of such mirrors deteriorate severely and degrade within a couple of months, if the reflecting layer is not protected[24,25] and the coatings delaminate from the substrate due to the lack of proper adhesion.

An extensive literature survey suggests that there are a number of reflector thin films prepared using various materials. However, there is a scarcity of practically applicable materials, which have a combination of high reflectance, endurance under harsh environmental conditions, and low fabrication cost. In order to avoid such issues, it is an interesting idea to fabricate reflector coatings using various alloys[26,27] or adherent promoting layers,[28] instead of using pure metallic layers. These films would not only serve the purpose of using reflector coatings in CSP systems, but also can be a source for performing significant scientific studies.

12.4 CSP Technology

CSP, also referred to as concentrating solar thermal power, represents a powerful, clean, endless, environment friendly, and reliable source of energy with the potential to satisfy electricity requirements in the coming years.[29] One of the most attractive properties of CSP plants is that there is no scope of carbon dioxide (CO_2) emission. A typical CSP system uses an array of mirrors to concentrate sunlight onto the receiver containing a heat-transfer fluid (HTF), which eventually gets heated up by absorbing solar radiation and produces steam. A conventional steam generator is used then to produce electricity. A schematic of the CSP plant is shown in Figure 12.1.

There are two types of solar concentration systems. The first one is concentrated solar thermal (CST), which supplies the required heat for industries, whereas the latter uses solar collectors produces power. There are three categories of solar thermal collectors including low-, medium-, and high-temperature collectors. Both low and mid- temperature collectors are basically flat plates, but applied for different purposes. Low-temperature collectors are often utilized in swimming pools to heat up the water, whereas for commercial and residential usage to supply the warm water, medium temperature collectors are a practical option. For flat plat collectors, no concentration of solar energy takes place. Hence, the parallel solar beam creates a maximum collector temperature of about 80–100 °C. Hence, low- and mid-temperature collectors

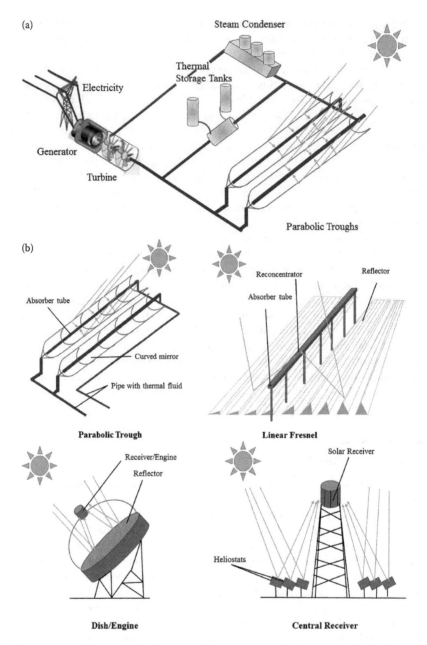

FIGURE 12.1 (a) Concentrated solar power (CSP) systems. (Reproduced with permission from Ref. 48). (b) Different types of solar concentrators. (Reproduced with permission from Ref. 47).

are suitable for domestic applications. Being efficient to reach elevated temperatures (300°C or more), high-temperature collectors have attracted attention for converging diffracted sunlight into receiver tubes using mirrors. The deployment of such collectors can be used for collectors in industries and electric power production. The use of concentrated solar power is not only limited on the earth's surface but also is an efficient way to utilize the solar energy in the space. Space power applications, known as "Solar dynamic" energy systems, have also been identified as cost-effective and energy-saving technology for various spacecraft applications, including solar power satellites, where a reflector serves the purpose of focusing sunlight on to a heat engine (e.g., Brayton cycle type).

For a applications such as cooking, or supplying a heat engine, or turbine–electrical generator, the concentration of solar energy is extremely important. Briefly, CSP technology includes parabolic troughs, linear Fresnel, solar towers, and parabolic dishes to concentrate the solar energy using mirrors.[30]

Parabolic trough systems utilize a large, U-shaped parabolic reflective surface, which focuses solar radiation over an evacuated metal tube located at the focal point of the parabola. The reflector mirrors can be tracked using a sun tracker so that they can be tilted toward the sun, and the solar energy directly gets concentrated on the receiver tube containing HTFs such as water, oil, or molten salts inside so that a temperature of 400°C can be reached. The HTFs transfer the heat through the receiver toward a thermal energy storage facility or to a water boiling facility. The heated fluid can then be converted into steam using which steam turbines are driven and electricity is generated.

In linear Fresnel collectors (LFC), the mobility in tracking the sun can be achieved by placing only one receiver above several mirrors. In case of LFC, the sunlight focused by the reflectors gets focused on the receiver tube carrying the liquid. The major advantages of using LFCs are the easiness of the production and installation of the plants due to the simplistic nature of the supporting structures.

Solar towers are also called as central receivers, where a tall tower is circumferenced by a huge number of mirrors known as heliostats. In this system, the heliostats track the sun in two axes and concentrate the solar radiation onto a receiver. The receiver is placed on the top of the tall tower in which focused solar radiation helps the HTF to reach a temperature up to 565 °C. The heated fluid is generally utilized to produce steam for power generation and supply or that can also be stored for using in cloudy days or winter season. Due to the high heat capacity of molten salt, it can be used as an HTF and stored for days, thereby enhancing electricity production during peak hours in cloudy days or even several hours after sunset.

A Stirling Dish is a parabolic dish-shaped concentrator similar to the shape of a satellite, which concentrates the solar rays exactly onto the focal point of the parabolic dish where a receiver is placed. In order to reduce the fabrication cost, the large dish is usually made of a larger number of smaller mirrors, which forms a shape of parabola. Generally, a two axes solar tracker is used to focus the solar radiation throughout the day. This system concentrates the entire sunlight into a point and hence the temperature rises at the focal point drastically and eventually the conduction of heat increases the temperature of the gas flowing through the receiver. As a result, at high temperature, the gas expands, and this expansion is used to drive a piston inside a cylinder. The movement of the piston serves as the source of energy to produce electricity. The combination of receiver, engine, and generator is considered as a single unit, which is placed at the focal point of the parabolic dish. The size of the receiver for the parabolic dish is small.[31] The major advantage of using an assembly of mirrors[32,33] is to reflect and to focus the solar radiation onto desired position, which enables to heat up the HTF, such as molten salt, steam, oil, and compressed air.[34,35]

As mentioned earlier, a solar reflector plays a pivotal role in concentrating the solar radiation. The scientific community has given major attention to develop different novel reflector materials.[36,37] The mirrors with reflectors[38] having a high reflectance in the solar wavelength range, resistance against environmental degradation, and prolonged life time are the best choice for solar thermal systems. While developing such materials, one needs to pay special attention to humidity and dust resistance properties to ensure the uninterrupted electricity generation using same reflectors for a prolonged duration.[39] There is a trend of using glass reflectors due to their significantly high reflectance. However, there are various drawbacks of using these due to their low resistance against breakage related to wind, high fabrication cost, and inconvenience in transportation. In recent years, scientific community is motivated to develop different types of reflector materials, which are easy to handle and have minimum fabrication cost. In this context, the area of research intended to prepare multifunctional reflectors with an optimum combination of specular reflectance, self-cleaning capabilities, thermal stability, and resistance to corrosion and abrasion has attained a greater significance.[1,40,41,43] Some of the reflector systems developed in recent years are described below, and two examples of such development are also shown in Figures 12.2 and 12.3.

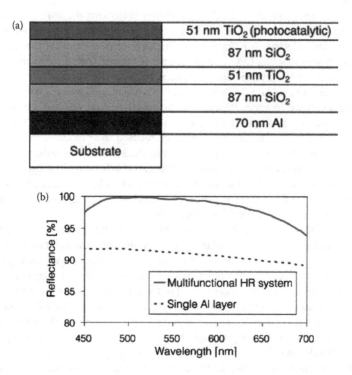

FIGURE 12.2 (a) Layer design of optimized multifunctional high reflective stacks. (b) Measured reflectance of the multifunctional HR system and of a sputtered aluminum single layer. (From Glöß D. et al., *Thin Solid Films*, 2008, 516, 4487–4489.)

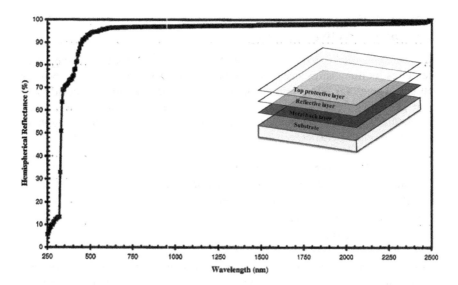

FIGURE 12.3 Typical hemispherical reflectance spectrum for an alumina-protected silver reflector. Inset shows the schematic of advanced solar reflector material structure consisting of silvered mirror with alumina protective coating. (From Kennedy C. E. et al., *Thin Solid Films*, 1997, 304, 303–309.)

12.4.1 Multifunctional High Reflective System

Multifunctional high reflective (HR) layer systems are reported to be fabricated on glass and polyethylene terephthalate (PET) substrates, keeping the deposition temperature constant at 150°C. The design of a multifunctional high reflective layer stack[44] is depicted in Figure 12.2a. The reflectance of the mirror (Figure 12.2b) is significantly higher than that of a single aluminum layer. The reflectance of the HR system is above 98% in the visible spectrum of the electromagnetic spectrum (500–630 nm).

12.4.2 Silver Mirror with Alumina Protective Layer

An ion-beam-assisted physical vapor deposition (IBAD) technique was used to deposit an optically transparent alumina coating[28] for protecting a promising low-cost reflector of a silvered polymer layer.

The structure consists of an alumina protective layer, silver reflective layer and a copper back-protective layer on the PET substrate. The solar-weighted hemispherical reflectance was measured at air mass 1.5 to be 95%. Fringes occurred on spectra as shown in Figure 12.3, depending on the alumina thickness.

This reflector material has exhibited an initial solar-weighted hemispherical reflectance of 95% and has maintained high optical performance in accelerated testing for more than 3000 hours. Additional issues to be addressed for practical applications include the determination of the minimum coating thickness for ensuring optical durability and long-term mechanical stability.

12.5 Solar Selective Absorbers

The development of spectrally selective absorber coatings has been a subject of intense research in energy materials community. Solar selectivity of the absorber coatings depends on selecting the material, design and deposition processes, and so on. However, the primary challenge lies in achieving high efficiency at elevated temperature. Therefore, together with the solar selectivity, the absorber coatings have the following characteristics,

- a. High-temperature stability ($\sim$400°C)
- b. Stability under corrosive and humid environments
- c. Oxidation resistance at operational temperature
- d. High structural and chemical stability
- e. Mechanical stability, etc.

Among all these characteristics, elevated temperature stability is a key criterion for the potential of a solar selective absorber coating. To create spectrally selective solar absorbing surfaces, a number of distinct design principles and physical processes can be used. The solar selective surfaces can be classified into a number of distinct kinds, depending on the design principle, including intrinsic absorber, semiconductor coatings, multi-layer coatings, metal-dielectric composite coatings, textured surfaces and, most importantly, DMD-based coatings. The design of these absorber coatings has been provided with schematic diagrams in Figure 12.4. Some of the coating types and their principles of operation are discussed briefly.

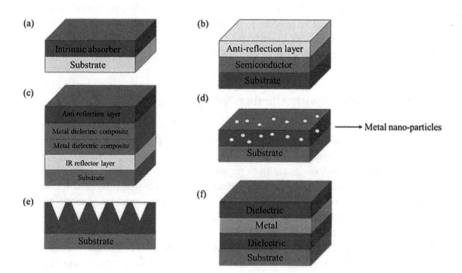

FIGURE 12.4 Different types of solar selective absorber coatings; (a) intrinsic absorber, (b) semiconductor absorber, (c) multilayer absorber, (d) cermet absorber, (e) textured surface, and (f) dielectric–metal–dielectric-based absorber. (From Granqvist C. G., *Adv. Mater.*, 2003, 15, 1789–1803.[42])

12.6 Multilayer Absorbers

For high-temperature applications, the most elementary systems, such as intrinsic or semiconductor absorber coatings, are not suitable. The coatings with multilayer structures have therefore been regarded reliable in elevated temperature in recent days. Several alternative layers of dielectric and semi-transparent metallic layers build a multilayer coating. To enhance transmission to the absorbing layers, anti-reflection coatings are generally placed on top. The multilayer stack can achieve high absorption up to 0.95 and low emission of 0.05 even under high operating temperature.

In a multilayer, various light wave superpositions introduce interference, which depends on each light wave's relative phase. Destructive interference reduces light reflection and improves absorption in a solar absorber. When two interfering waves with a 180° phase difference overlap, destructive interference happens. However, this phase difference for destructive interference correlates with a sinusoidal wave change, which is best accomplished by changing the layer's optical thickness. The above discussion highlights the need to tailor the thickness of the individual layer in multilayer coatings to obtain desired optical characteristics and thermal stability, simultaneously.

12.7 Dielectric/Metal/Dielectric (DMD) Absorbers

The light trapping ray-optics theory using multiple reflections and interference effects (in the case of multilayer films or surface texture) is becoming traditional. In addition, owing to specular reflection from the back surface, the long wavelength gets reflected back from the film. The optical path length of weakly absorbed light must be improved to decrease the loss of long wavelength radiation. The DMD structure, another form of selective layer, can be used to enhance absorption with the help of surface plasmon polaritons (SPPs). SPPs provide a high scattering cross-section at larger angles and increase the length of the optical path. The major challenge of developing DMD coating is to design tunable structures that are able to connect the difference between metal and dielectric optical properties and boost energy absorption without increasing the thermal heat loss.

Figure 12.5 demonstrates the basic distinction between localized surface plasmons and surface plasmon polaritons. The electron cloud oscillates locally around the metallic nanoparticles for LSPs, while the SPPs move through the metal and dielectric interface in the x and y directions with an evanescent decay in the z-direction. Hence, both LSPs and SPPs are manifesting themselves in local field enhancement.

12.7.1 MgF$_2$/Mo/MgF$_2$

Sergeant[46] modeled a DMD multilayer stack in a very different manner using the conventional transfer matrix technique where MgF$_2$ and TiO$_2$ with refractive indices of 1.37 and 2.75 at 1 μm, respectively, fulfilled the purpose of dielectrics in the same coating, while Mo and W were used as metallic layers, sandwiched between dielectrics. The combination of MgF$_2$ and TiO$_2$ was chosen as the dielectric owing to their difference in their refractive indices ($\Delta n = 1.38$; $\lambda = 1$ μm). Two sets of four

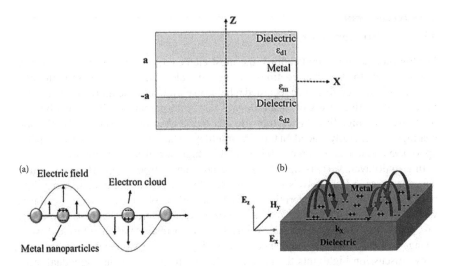

FIGURE 12.5 Schematic of the dielectric–metal–dielectric structures for SPP propagation. (a) Formation of localized surface plasmons (LSPs); (b) The propagation of surface plasmon polaritons (SPPs) in metal–dielectric interface in response to electro-magnetic field. (Adapted from Ref [13]).

layer-free DMD coatings made of Mo, MgF$_2$, TiO$_2$ and W, and MgF$_2$ and TiO$_2$ with number of layers 5, 7, 9, and 11 were optimized at 720 K.

The design of coatings is shown in Figure 12.6. The metal substrate W or Mo is considered as layer 1 ($L = 1$). The optimized coating showed high solar absorptance (>94%) in wavelengths below 2.24 µm and low absorption (<7%) in wavelengths beyond 2.24 µm. The optical characteristics are shown in Figure 12.6. They also showed that the merit function of the absorber layer increases while increasing the number of layers as shown in Figure 12.6. The efficiency, however, saturates as the number of layers exceeds 11. If the number of layers is increased by more than 11, owing to the thickness increase, a significant increase in emittance was noted. Both the DMD absorbers were explored for studying the angular spectral selectivity and the coatings have very high solar absorption. Together wide-angle absorption, solar electivity, and high-temperature stability shows the coating as a potential candidate for CSP systems.

12.7.2 W/WAlN/WAlON/Al$_2$O$_3$

Figure 12.7 shows the development of W/WAlN/WAlON/Al$_2$O$_3$ films on stainless-steel substrates using DC and RF reactive magnetron sputtering.[6,9,10] The W layer acts as an infrared reflecting layer in the tandem absorber coating. The WAlN layer serves the purpose of a primary absorber layer, whereas the WAlON layer acts as a semi-absorber layer. The Al$_2$O$_3$ layer on top works as an anti-reflection layer, which decreases the reflection from the coating. The WAlN, WAlON, and Al$_2$O$_3$ layers have a total thickness of ∼140 nm. In the solar wavelength spectrum, the thin film also showed wide-angle adaptability even when the angle of incidence increases to 58°. More importantly, the short-term thermal stability of the film in air was up to 500°C for 2 hours.

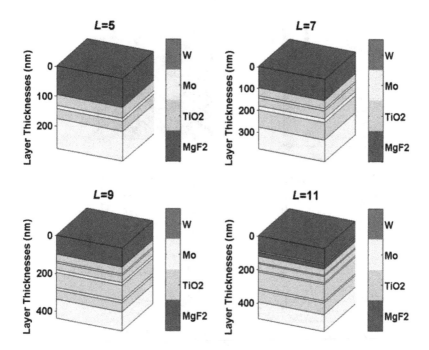

FIGURE 12.6   Schematic of DMD coatings with Mo, MgF$_2$, and TiO$_2$ layers. Optimized stacks with layers $L = 5, 7, 9$, and 11 are shown. (Adapted from Ref [13]).

Figure 12.8 presents an interesting finding, while evaluating the optical constants of the individual layer.[7] Using Bruggeman's effective medium approximation, the origin of high absorption in W/WAlN/WAlON/Al$_2$O$_3$ was described. An excellent arrangement between simulated and experimental reflectance spectra supports the existence of two intermediate layers (26% WAlN–74% WAlON and 60% WAlON–40% Al$_2$O$_3$) at the WAlN/WAlON and WAlON/Al$_2$O$_3$ interfaces. It was found that solar absorption was dominated by a gradation of optical constants from the surface to substrate, facilitating destructive interference and total internal reflections. In order to prove the accuracy of the optical constants and the presence of additional layers, a simulation process to generate the reflectance spectra was carried out with and without intermediate layers. A good correlation between simulated and experimental data confirms the predictability of the optical constants and existence of intermediate layers.

The durability of W/WAlN/WAlON/Al$_2$O$_3$ was examined under harsh environmental conditions and the findings are shown in Figure 12.9. Exposure to high humid environments (95% humidity at 37°C) did not affect the spectral properties of hydrophobic W/WAlN/ WAlON/Al$_2$O$_3$. The thin film even showed susceptibility to degradation in corrosive medium. High scratch resistance and appreciable hardness of the deposited film were established by performing mechanical tests. Selective surfaces without unchanged reflectance spectra validate their long-term thermal stability after a heat treatment at 350°C in air for 550 h (more discussion in References [4,47]).

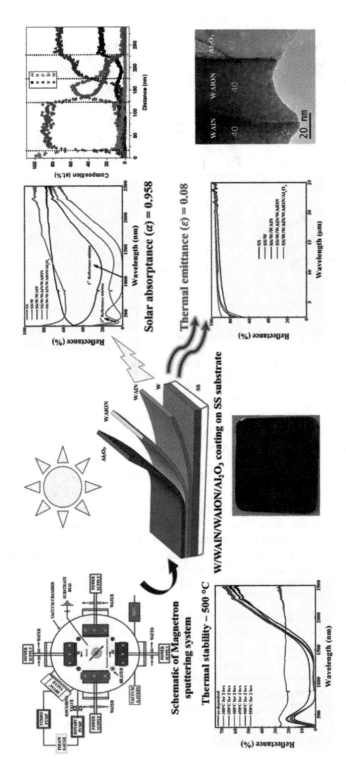

FIGURE 12.7 Optical properties of magnetron sputtered W/WAlN/WAlON/Al$_2$O$_3$.[9,47]

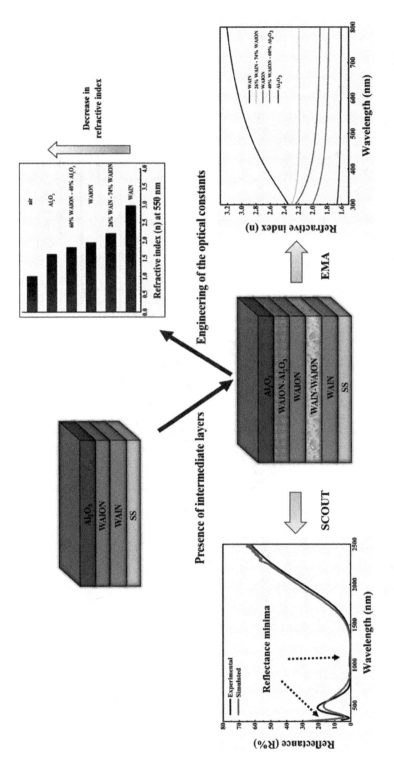

FIGURE 12.8 Schematic illustration of the concepts involved and experimental measurements of the optical properties of W/WAlN/WAlON/Al$_2$O$_3$ coating.[6,47]

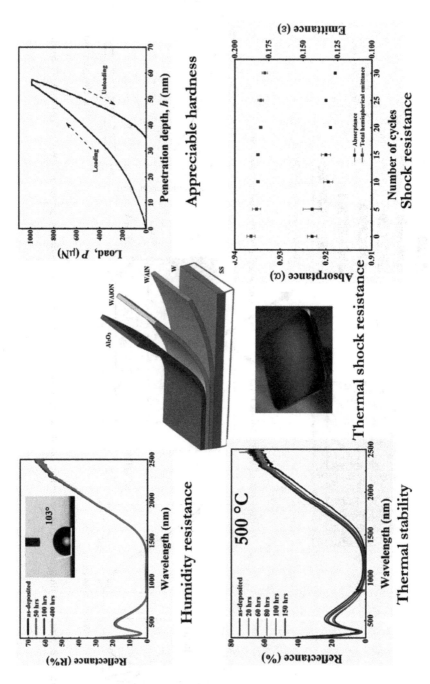

FIGURE 12.9 Summary of the published experimental results to confirm the operational stability of W/WAlN/WAlON/Al$_2$O$_3$, results reported in References [4,47].

12.7.3 TiB$_2$/TiB(N)/Si$_3$N$_4$

The author's research group was involved to probe the influence of deposition parameters on spectral selectivity of magnetron sputtered TiB$_2$/TiB(N)/Si$_3$N$_4$ films.[12] The study shows that thin film selectivity depends on the position of the absorption edge in the solar spectrum. The experimental findings also indicate that selectivity could be influenced by the deposition conditions such as target power, deposition time, and reactive gas flow rate. Figure 12.10 summarizes the key results while optimizing the selectivity of TiB$_2$/TiB(N)/Si$_3$N$_4$. The film exhibited a solar absorptance and thermal emittance of 0.964 and 0.18, respectively. Such characteristics were comprehended by finding out the optical constants of the individual layer.

12.8 Closure

The development of a solar selective reflector and absorber coatings has been discussed in this chapter. In the context of solar reflector materials, the metallic coating exhibits a specular reflectance of more than 85% in the solar spectrum. Two significant design parameters are to be critically considered. The first is the rationale for getting consecutive layers of various compositions, and the second is the thickness of the various layers. While optimizing solar reflectance, both these parameters must be tailored over a small window of the process parameters (magnetron sputtering or vacuum evaporation) to achieve the desired layer thicknesses. In short, the suitability of advanced low-cost reflector components for outdoor solar thermal applications will require a series of trials.

It is apparent from this chapter that electromagnetic wave confinement using a plasmonic geometry DMD stack offers many benefits for converting solar energy into thermal energy. Such physical confinement also allows one to realize the thermal stability of the coating. The ellipsometry measurements are to be conducted to understand the metallic or dielectric nature of DMD-based solar selective absorber coatings in individual layers. However, theoretical and experimental efforts should be made to explore the underlying physical mechanisms, which play an important role in achieving high selectivity.

A dielectric layer with a very high refractive index ($n_d > 2$) and a comparatively dense metal layer has been reported to be capable of satisfying a "zero reflective condition," i.e. maximum absorption. It is therefore suggested that the choice of the appropriate mixture of metals and dielectrics be investigated to increase the absorption characteristics. Despite numerous DMD-based absorber coatings, it should be essential to focus on their corrosion resistance, scratch resistance, hardness, abrasion, hydrophobic nature, and humid environment efficiency. To demonstrate the reliability and weather resistance properties of the absorbers, a year-round performance evaluation under varying climatic circumstances is needed. While assessing the annual output and developing the operational strategy, the variation of solar radiation must be taken into account throughout the year in distinct times.

It is essential to consider optical constants of metals and dielectric of the multilayer DMD films over a broad spectrum of wavelengths, when developing tandem coating. In addition, it is necessary to investigate the theoretical absorption and emittance before beginning experimental processes.

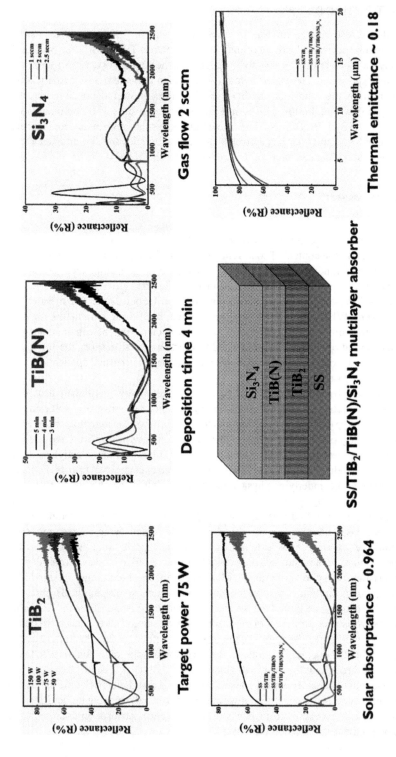

FIGURE 12.10 Process parameter optimization and optical properties of TiB$_2$/TiB(N)/Si$_3$N$_4$.[12,47]

It is also important to investigate how the incident angle of solar radiation can influence the optical properties of the absorber coating in order to develop a highly efficient CSP system. The slight shift in the angle of the incident may result in a significant variation in the optical thickness as well as a phase difference between incident and reflected rays, resulting in a noticeable difference in the optical characteristics of absorbers. Therefore, before practical implementation in CSP systems, the angular dependency of the absorber layer should be unrevealed.

Another problem in the development of solar selective materials is the elevated temperature stability, which is of extreme importance. The extended exposure under concentrated solar radiation is probable to cause inter-diffusion of the atoms from the substrate to the absorber layer, thus modifying interfaces, compositions, and optical properties of the layer. Another approach is to deposit a barrier of metal diffusion layer between the absorber and the substrate. This layer restricts the metal atoms from entering the absorber layer at elevated temperatures and protects the absorber from changing its optical properties. The diffusion of the substrate atom can also be avoided by pre-heating stainless steel in air, before the deposition.

The single dielectric layer on top of the coating is vulnerable in terms of long-term durability under high-temperature conditions. The inclusion on top of the dielectric of another thick anti-reflection layer will protect the film by restricting oxidation from the environment. The top layer should have appreciable hardness and scratch resistance characteristics in addition to the anti-reflection characteristics to make the film suitable for practical applications.

Studying which deposition method or which absorber structure will finally succeed commercially is also extremely intriguing. It would, however, undoubtedly be dictated by the simplicity of manufacturing process and energy expenditure per watt. In a competent thermo-photovoltaic system, an integrated system combining a solar photovoltaic and solar thermal panel, DMD-based solar selective coatings can also be used. In such devices, the DMD-based absorber will absorb the solar energy and get heated up and emit a narrow-banded spectrum above the band gap of the solar cell.

In summary, it will be interesting to see how the applications of DMD-based absorber coatings evolve in terms of material reliability, testing specifications, market strength, and cost. Most importantly, the steady demand for renewable energy along with the CSP scheme would be competitive with standard power generation. The strategic efforts to further advance manufacturing processes, structure and material optimization, energy conversion and efficiency, and performance testing can be expected to have an important effect in the development of superior DMD-based absorber coatings.

REFERENCES

1. Barlev D., R. Vidu, P. Stroeve, Innovation in concentrated solar power, *Sol. Energy Mater. Sol. Cells*, 2011, 95, 2703–2725.
2. Enger R. C., H. Weichel, Solar electric generating system resource requirements, *Sol. Energy*, 1979, 23, 255–261.
3. Kennedy C. E., K. Terwilliger, Optical durability of candidate solar reflectors, *Journal of Solar Energy Engineering, Trans. ASME*, 2005, 127, 262–269.
4. Dan A., A. Soum-Glaude, A. Carling-Plaza, C. Ho, K. Chattopadhyay, H. Barshilia, B. Basu, Photothermal conversion efficiency, temperature and angle dependent emissivity and thermal shock resistance of W/WAlN/WAlON/Al_2O_3-based spectrally selective absorber, *ACS Appl. Energy Mater*, 2019, 2(8), 5557–5567.

5. Alex S., R. Kumar, K. Chattopadhyay, H. C. Barshilia, B. Basu, Thermally evaporated Cu–Al thin film coated flexible glass mirror for concentrated solar power applications, *Mater. Chem. Phys.*, 2019, 232, 221–228.

6. Dan A., B. Basu, T. Echániz, I. González de Arrieta, G. A. López, H. C. Barshilia, Effects of environmental and operational variability on the spectrally selective properties of W/ WAlN/WAlON/Al$_2$O$_3$-based solar absorber coating, *Sol. Energy Mater. Sol. Cells*, 2018, 176, 157–166.

7. Dan A., A. Biswas, P. Sarkar, S. Kashyap, K. Chattopadhyay, H. C. Barshilia, B. Basu, Enhancing spectrally selective response of WWAlN/WAlON/Al$_2$O$_3$-based nanostructured multilayer absorber coating through graded optical constants, *Sol. Energy Mater. Sol. Cells*, 2018, 176, 157–166.

8. Alex S., K. Chattopadhyay, B. Basu, Tailored specular reflectance of Cu- based novel intermetallic alloys, *Sol. Energy Mater. Sol. Cells*, 2016, 149, 66–74.

9. Dan A., J. Jyothi, H. C. Barshilia, K. Chattopadhyay, B. Basu, Spectrally selective absorber coating of WAlN/WAlON/Al$_2$O$_3$ for solar thermal applications, *Sol. Energy Mater. Sol. Cells*, 2016, 157, 716–726.

10. Dan A., H. C. Barshilia, K. Chattopadhyay, B. Basu, Angular solar absorptance and thermal stability of WAlN/WAlON/Al$_2$O$_3$-based solar selective absorber coating, *Appl. Therm. Eng.*, 2016, 109 B, 997–1002.

11. Alex S., S. Sengupta, U. K. Pandey, B. Basu, K. Chattopadhyay, Electrodeposition of δ-phase based Cu–Sn mirror alloy from sulfate-aqueous electrolyte for solar reflector application, *Appl. Therm. Eng.*, 2016, 109 B, 1003–1010.

12. Dan A., K. Chattopadhyay, H. C. Barshilia, B. Basu, Colored selective absorber coating with excellent durability, *Thin Solid Films*, 2016, 620, 17–22.

13. Dan A., H. C. Barshilia, K. Chattopadhyay, B. Basu, Solar energy absorption mediated by surface plasma polaritons in spectrally selective dielectric-metal-dielectric coatings: A critical review, *Renew. Sustain. Energy Rev.*, 2017, 79, 1050–1077.

14. Guidelines, Measurement of solar weighted reflectance of mirror materials for concentrating solar power technology with commercially available instruments, Solar PACES interim Version 1.1, May 2011.

15. Nostell P., A. Ross, B. Karlsson, Ageing of solar booster reflector materials, *Sol. Energy Mater. Sol. Cells*, 1998, 54, 235–246.

16. Almanza R., P. Hernandez, I. Martinez, M. Mazari, Development and mean life of aluminium first-surface mirrors for solar energy applications, *Sol. Energy Mater. Sol. Cells*, 2009, 93, 1647–1651.

17. Benson B. A., Acrylic solar energy reflecting film 3M brand ECP-244 (formerly FEK-244), *Proc. Annu. Meet. - Am. Sect. Int. Sol. Energy Soc.*, 6(1983).

18. Adams R. O., C. W. Nordin, F. J. Fraikor, Co-sputtering of aluminium-silver alloy mirrors for use as solar reflectors, *Thin Solid Films*, 1979, 63, 151–154.

19. Adams R. O., C. W. Nordin, Al-Ag alloy films for solar reflectors, *Thin Solid Films*, 1980, 72, 335–339.

20. Fend T., B. Hoffschmidt, G. Jorgensen, H. Kuster, D. Kruger, R. Pitz-Paal, P. Rietbrock, K. Riffelmann, Comparative assessment of solar concentrator materials, *Sol. Energy*, 2003, 74, 149–155.

21. Schissel P., G. Jorgensen, C. Kennedy, R. Goggin, Silvered-PMAA reflectors, *Sol. Energy Mater. Sol. Cells*, 1994, 33, 183–197.

22. Liao Y., B. Cao, W. C. Wang, L. Q. Zhang, D. Z. Wu, R. G. Jin, A facile method for preparing highly conductive and reflective surface-silvered polyimide films, *Appl. Surf. Sci.*, 2009, 255, 8207–8212.

23. Brogren M., A. Helgesson, B. Karlsson, J. Nilsson, A. Roos, Optical properties, durability, and system aspects of a new aluminium-polymer-laminated steel reflector for solar concentrators, *Sol. Energy Mater. Sol. Cells*, 2004, 82, 387–412.

24. Czanderna A. W., Stability of interfaces in solar-energy materials, *Sol. Energy Mater.*, 1981, 5, 349–377.

25. Roos A., C. G. Ribbing, B. Karlsson, Stainless-steel solar mirrors—A material feasibility study, *Sol. Energy Mater.*, 1989, 18, 233–240.

26. Lin J. C., H. L. Liao, W. D. Jehng, C. A. Tseng, J. J. Shen, S. L. Lee, Effect of annealing on morphology, optical reflectivity, and stress state of Al-(0.19–0.53) wt% Sc thin films prepared by magnetron sputtering, *Surf. Coat. Technol.*, 2010, 205, S146–S151.

27. Hummel R. E., Reflectivity of silver-based and aluminum alloys for solar reflectors, *Sol. Energy*, 1981, 27, 449–455.

28. Kennedy C. E., R. V. Smilgys, D. A. Kirkpatrick, J. S. Ross, Optical performance and durability of solar reflectors protected by an alumina coating, *Thin Solid Films*, 1997, 304, 303–309.

29. http://sycomoreen.free.fr/docs_multimedia/greeneconomy_Ch4_ConcentratingSolarPower.pdf.

30. Sarver T., A. Al-Qaraghuli, L. L. Kazmerski, A comprehensive review of the impact of dust on the use of solar energy: History, investigations, results, literature, and mitigation approaches, *Renew. Sustain. Energy Rev.*, 2013, 22, 698–733.

31. Lior N., Mirrors in the sky: Status, sustainability, and some supporting materials experiments, *Renew. Sustain. Energy Rev.*, 2013, 18, 401–415.

32. Wang M., R. Vandal, S. Thomsen *Durable Concentrating Solar Power*, Guardian Industries Corp, Carleton, USA, pp. 1–5

33. Pettit R. B., E. P. Roth Solar mirror materials: Their properties and uses in solar concentrating collectors, in *Solar Materials Science*, ed. L. E. Murr, Academic Press, New York, 1980, pp. 171–197.

34. Li Z., Z.-G. Wu, Analysis of HTFs, PCMs and fins effects on the thermal performance of shell–tube thermal energy storage units, *Sol. Energy*, 2015, 122, 382–395.

35. Montecchi M., Approximated method for modelling hemispherical reflectance and evaluating near-specular reflectance of CSP mirrors, *Sol. Energy*, 2013, 92, 280–287.

36. DiGrazia M., R. Gee, G. Jorgensen, ReflecTech Mirror Film Attributes and Durability for CSP Applications, *ASME 2009 3rd International Conference on Energy Sustainability*, San Francisco, California, USA, 2, 2009, pp. 677–682.

37. Kutscher C. T., R. Davenport, *Preliminary Results of the Operational Industrial Process Heat Field Tests, SERI/TR-632-385R*, Solar Energy Research Institute, Golden, CO, June, 1981.

38. Delord C., C. Bouquet, R. Couturier, O. Raccurt, Characterizations of the durability of glass mirrors for CSP Development of a methodology, *Solar PACES*, 2012.

39. Bergeron K. D., J. M. Freese, *Cleaning Strategies for Parabolic Trough Collector Fields; Guidelines for Decisions, SAND81-0385*, Sandia National Laboratories, Albuquerque, 1981.

40. Xu Y. J., J. X. Liao, Q. W. Cai, X. X. Yang, Preparation of a highly-reflective TiO_2/SiO_2/Ag thin film with self-cleaning properties by magnetron sputtering for solar front reflectors, *Sol. Energy Mater. Sol. Cells*, 2013, 113, 7–12.

41. Xu Y. J., Q. W. Cai, X. X. Yang, Y. Z. Zuo, H. Song, Z. M. Liu, Y. P. Hang, Preparation of novel SiO_2 protected Ag thin films with high reflectivity by magnetron sputtering for solar front reflectors, *Sol. Energy Mater. Sol. Cells*, 2012, 107, 316–321.

42. Granqvist C. G., Solar energy materials, *Adv. Mater.*, 2003, 15, 1789–1803.

43. Barlev D., R. Vidu, P. Stroeve, Innovation in concentrated solar power, *University of California Davis Solar Energy Collaborative Workshop*, Davis, CA, May 11, 2010.
44. Glöß D., P. Frach, C. Gottfried, S. Klinkenberg, J. S. Liebig, W. Hentsch, H. Liepack, M. Krug, Multifunctional high-reflective and antireflective layer systems with easy-to-clean properties, *Thin Solid Films*, 2008, 516, 4487–4489.
45. http://abcofsolar.com/a-brief-on-types-of-concentrated-solar-power-csp/
46. Sergeant N. P., O. Pincon, M. Agrawal, P. Peumans, Design of wide-angle solar-selective absorbers using aperiodic metal-dielectric stacks, *Opt. Express*, 2009, 17, 22800–22812.
47. Dan A. Spectrally selective tandem absorbers for Photothermal conversion in high temperature solar thermal systems, PhD thesis, Indian Institute of Science, Bangalore, India, January, 2019.
48. Alex S., Development of Cu-based intermetallic reflector materials for Concentrated Solar Power Application, PhD thesis, Indian Institute of Science, Bangalore, 2017.

13

Advanced Electrochemical Energy Storage Technologies and Integration

The need for developing advanced energy storage technologies was introduced briefly in Chapter 11. The overall characteristics of such technologies are usually defined in terms of two parameters, namely, energy density and power density; where the former is the amount of energy that can be stored and released per unit mass or volume and the latter indicates the rate at which the energy can be stored or withdrawn. They are also compared with respect to their position on a plot/map of energy density versus power density, which is known as the Ragone plot (see Figure 13.1). Accordingly, the particular energy storage technology (among different electrical, electrochemical, flywheel, pumped-hydro-based technologies, etc.) to be used is governed by the concerned application, with respect to the energy and power density needed. Nevertheless, it is widely perceived that electrochemical energy storage technologies (viz., batteries and supercapacitors) are suitable for a lot of modern day technological applications. Accordingly, the various technological aspects, advancements, and scopes for further improvements of electrochemical energy storage technologies are discussed in this chapter.

13.1 Historical Perspectives of Electrochemical Energy Storage Technologies

Electrochemical energy storage (EES) is based on basic electrochemical principles, leading to the storage of energy in the form of electrical charge either on the surface or within the bulk of the electrode material in the charged condition, which can be "harnessed" during discharging. The principles governing these aspects were discovered in the late 1700s and the early 1800s. It is believed that the pioneering works of Alessandro Volta leading to the development of the "Volta Pile" is the beginning of "battery" technology, the primary form of the EES systems. Subsequently, various types of EES have been developed, such as different batteries (primary and secondary/rechargeable) and super capacitors, depending on the charge storage mechanism, electrolyte-type and also operating temperature. History says that the first practical EES device (a primary or non-rechargeable battery) was developed in 1836 by a British chemist, John Frederic Daniel, as a modification of the Volta Pile. The first generation Daniel cell consisted of a copper pot (acting as the cathode) having copper sulfate solution (electrolyte for the Cu-containing half-cell or catholyte), which was immersed in an earthenware container having sulfuric acid and a zinc electrode (the anode). A simple schematic of the later modified Daniel cell, along with the cell reactions and attainment of the cell voltage, is presented in Figure 13.2. It is interesting to note that the Daniel cell and its modified versions (like the one in Figure 13.2) have been

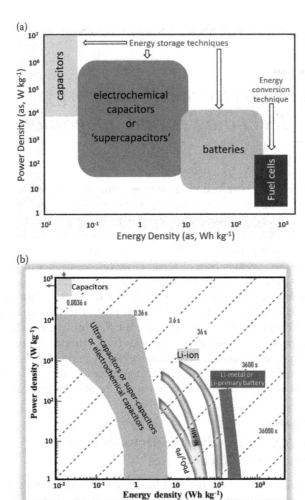

FIGURE 13.1 (a) and (b) Ragone plots, comparing the different energy storage technologies in terms of their ranges of power and energy densities (Adapted from Kotz R., M. Carlen; *Electrochim. Acta*, 2000, 45, 2483), with (b) providing more details of the important battery chemistries presently in use. (Adapted from Simon P., Y. Gogotsi, *Nat. Mater.*, 2008, 7, 845.)

conveniently used for teaching some of the basic electrochemical concepts. After the development of the Daniel cell, few other primary batteries were developed (which could not be recharged after one-time usage).

However, the major thrust toward practical and heavy duty usage of batteries as one of the primary systems/technologies for storage of electrical energy (in the form of electro-chemical energy) came with the development of secondary or rechargeable systems, starting with the lead-acid battery (initiated in 1859). The cell is typically made of a Pb anode and PbO_2 cathode, immersed in H_2SO_4 (acting as the electrolyte). Lead sulfate is produced during the cell reaction, with the release of electrons at the Pb anode. More importantly, the chemical reactions could be reversed by passing a reverse

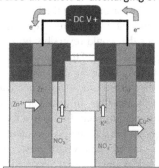

Galvanic cell type	Electrolytic cell type
Galvanic cell type	**Electrolytic cell type**
Normal direction => charging of cell	Reverse direction or discharging of cell
If the concentrations of the ions in the electrolyte is at 1 M, then standard conditions prevail and the cell voltage will be +1.1 V.	Upon application of external voltage of opposing polarity and magnitude > +1.1 V, the reverse or cell charging takes place. In this case, Cu 'dissolves out' electrode and Zn gets 'plated' on the corresponding electrodes.

FIGURE 13.2 A modified version of the basic Daniel cell (having the same basic cell chemistry as invented in 1836), showing both the modes (viz., the spontaneous discharge mode, as well as the reverse or charging mode) and a few associated details. (Adapted from 'Fundamentals of Electrochemistry' online notes hosted by Chemistry, University of Guelph, Canada).

current, thereby recharging it. Even though this system has many disadvantages, including weight, sustainability and use of Pb, it is still used in many heavy duty applications, including in automobiles. It is to be noted here that India is at the moment one of the largest producers of Pb-acid batteries. The development of Pb-acid battery was followed by alkaline cells, the most significant of them being the Ni-Cd and Ni-MH (MH: metal hydride) batteries, with relatively enhanced energy densities and sustainability. It is to be noted that energy density is given by

$$\text{Energy density (Wh/kg)} = \text{electrode capacity (Ah/kg)} \times \text{cell voltage (V)} \quad (13.1)$$

Accordingly, an increase in the charge storage capability of the electrodes and cell voltage, while reductions in the mass of the electrodes lead to enhanced energy densities.

Even though the Pb-acid, Ni-Cd, Ni-MH battery technologies were (and still are) being successfully used for EES applications (such as in toys, portable electronic/communication devices, remote monitoring stations, emergency power/lighting [especially for ensuring un-interrupted power supplies], medical devices [such as pacemakers, hearing aids], computers, vehicles [for aiding in ignition], industrial controls, etc.), the quest for further improvements in the energy density, power density, cycle life (sustainability), and also safety aspects led to the development of Li-ion systems. It started with Li-primary cells, having Li metal as the anode and layered compounds like TiS_2 as cathodes, which possesses very high energy density due to Li being the lightest (resulting in lower cell mass) and one of the more electropositive metals (leading to high voltage). This was developed and demonstrated for the first time by M. Stanley Whittingham while working for ExxonMobil in the 1970s. However, Li-dendrite formation on the Li metal foil and, concomitantly, short circuiting related issues upon attempts to recharge the cell led to the investigation of alternative anode materials that could reversibly host Li, without too much sacrifice of the overall cell voltage. To the best of the authors' knowledge, this was potentially addressed by Rachid Yazami in the early 1980s, when he discovered that graphitic carbon could reversibly intercalate (host) Li at electrochemical potentials not too high compared to the Li/Li$^+$ potentials and thus, in principle, could function as an "intercalation" or "insertion" anode (replacing metallic Li) in a Li-ion cell. At about the same time (in fact, a couple of years earlier), John B. Goodenough (at the University of Oxford, UK) discovered that Li-ions can be reversibly extracted from (and re-inserted into) layered transition metal oxides like LiCoO$_2$ at considerably higher potentials (vs. Li/Li$^+$), which would render them promising for use as cathode material for Li-ion cell. Following the above discoveries, Akira Yoshino, working in Asahi Kasei Corporation company (in Japan), used polyacetylene, as well as some special form of graphitic carbon, to develop for the first time working prototype of Li-ion cell in the mid to late 1980s. This eventually paved the way towards commercialization of Li-ion battery technology by Sony in 1991, which, in turn, led to revolutionizing the EES technology and applications. It is important to note that the some of the pioneering efforts that led to the discovery and development of the Li-ion battery science and technology, as above, were recognized by the "Nobel committee," which awarded Prof. John B. Goodenough, Prof. M. Stanley Whittingham and Dr. Akira Yoshino with the 2019 Nobel Prize in Chemistry for the development of lithium-ion batteries. On a different note, as may be inferred from the above, the very sequence of events that led to the development of Li-ion battery technology highlights the importance of fundamental research conducted at academic and research laboratories, followed by translation of the same to prototype and product development by industrial R&D centres and (later by) production units.

Even though the basic chemistry of Li-ion batteries still remains the same, further understanding of various associated scientific aspects and modifications in the compositions of the electrode materials and electrolyte have been able to increase the energy density from the erstwhile $\sim$150 Wh/kg to $\sim$300 Wh/kg.

> The on-going research promises to lead to still further enhancement in the energy density of Li-ion batteries, without sacrificing on the stability and safety aspects, such that a practically achievable energy density of $\sim$500 Wh/kg is expected in a few years' time.

Nevertheless, present day research is also focused on advancement of the very chemistry (known as "beyond Li-ion chemistry"), with focus on systems such as Li$-$S and Li$-$air. In a bid to protect the Li resources in the world, research groups are also focusing on the development of Na-ion, K-ion, Mg-ion batteries, etc. especially for applications not necessitating too much reduction in the weight/volume of the EES. The rechargeable battery technologies, in order of their evolution, along with the basic components, chemistries, and characteristics are listed in Table 13.1.

Simultaneously, along with the evolution of the battery systems, where charge (or the concerned metal ion) is stored within the bulk/lattice of the electrode materials,

TABLE 13.1

Summary of the Various Features, Including Basic Components, Chemistries, and Advantages/Disadvantages of the Presently Used Rechargeable Battery Systems, in Order of Their Evolution

Battery System	Typical Electrodes and Electrolyte	Usual Cell Voltage (V)	Energy Density (Wh/kg)	Power Density (W/kg)	Usual Cycle Life (Cycles)	Self-Discharge	Other Aspects
Pb-acid	Anode: Pb; cathode: PbO_2; electrolyte: H_2SO_4 (active ions in electrolyte: HSO_4^- and H^+)	2.1	35–45	50–100	<500	Low	One of the oldest, being invented in 1859; is an environmental hazard due to the presence of Pb; used in gasoline vehicles just for ignition
Ni–Cd	Anode: Cd; cathode: NiOOH; electrolyte: KOH (active ion in electrolyte: OH^-)	1.2	40–60	140–180	500–2000	Low	Invented in 1899 (one of the first to use alkali electrolyte); used in a few portable electronics, toys, power tools, etc.
Ni–MH	Anode: ANi5 (an intermetallic, with A = La, Ce, Nd, Pr, etc.); cathode: NiOOH; electrolyte: KOH (active ion in electrolyte: OH^-)	1.2	60–120	220–500	<3000	High	Invented in 1967, replaced NiCd in most portable electronic applications; also tried in some electric vehicles
Ni–Zn	Anode: Zn; cathode: NiO; electrolyte: NaOH or KOH (active ion in electrolyte: OH^-)	1.6	100	>3000	50–200	High	Developed in 1901, but significantly stabilized in the year 2000; good for power tools avoiding storage under charged conditions
Ag–Zn	Anode: Zn; cathode: Ag_2O; electrolyte: NaOH or KOH (active ion in electrolyte: OH^-)	1.5–1.9	130	–	<30	–	Used for special applications like missile or space shuttle launch and other military usages
Na–S	Anode: molten Na; cathode: molten S; electrolyte: β-Al_2O_3 solid (active ion in electrolyte: Na^+)	2	130–150	–	1500	High	Has disadvantage of molten electrodes; used for stationary large-scale applications like grid storage
Li-ion	Anode: graphite, Si, $Li_4Ti_5O_{12}$ (or LTO), etc.; cathode: $LiCoO_2$ (or LCO), $LiNi_{1/3}Mn_{1/3}Co_{1/3}O_2$ (or Li-NMC), $LiFePO_4$ (or LFP); electrolyte: Li-salt in EC/DMC/PC (i.e., aprotic/organic solvents) (active ion in electrolyte: Li^+)	3–4 V	200–250	350–450	1000–2000	Medium	Commercialized since 1991; dominates the market as power source for portable/heavy electronics and power tools; presently is the most suited battery for electric vehicles and storage of energy "harvested" from renewables

development of electrochemical capacitors was also envisaged which would allow faster storage and release of charges from the surfaces. The advantage of electrochemical capacitors (or supercapacitors) over conventional electrostatic capacitors is that the capacities are considerably greater for the former. The very basic relationship for capacitance is as follows:

$$\text{Capacitance}(C) = \varepsilon_r \varepsilon_0 A/d \qquad (13.2)$$

where ε_0 is the permittivity of free space, ε_r is the relative permittivity, A is the area of the surfaces hosting the charges, and d is the separation between the charged layers. Accordingly, enhancement in capacitance in the case of electrochemical capacitors was achieved using charge storage in electrochemical double layers (in the presence of liquid electrolytes), thus by reducing the distance (d) between the charged layers significantly (from mm to a few angstroms!) (as depicted in Figure 13.3). Further improvements were achieved via a considerable increase in the specific surface areas of the electrode materials by developing nano-porous materials like activated carbon (and now even nanoporous metals like Au, etc.). This is an area of on-going research, with the as-developed technology being known as either supercapacitor or ultra-capacitor or electrochemical capacitor. Along with the same, further improvements in the surface charge storage capacities are presently being attempted via adopting surface faradic reactions (popularly known as "pseudocapacitance"). Again, with respect to the energy stored, similar to the case of batteries, it is the product of the charge stored at the surfaces of the electrodes and the voltage between them. For the case of a "perfect" capacitor (having only non-faradic surface charge storage), where voltage (V) increment is linear with charge (q) build-up or time, the energy stored is $1/2\ qV$; but not strictly valid for pseudocapacitors (which show non-linear voltage–charge profile due to interfering faradic reactions).

Overall, the last decade has witnessed considerable research in trying to bridge the gap between batteries and capacitors (as depicted in the Ragone plot in Figure 13.1) and also sometimes use them together (like in the present and upcoming electric vehicles) to store a greater amount of charge (with battery technology), while releasing/accumulating some of them very fast (with supercapacitor technology). Overall, the present generation EES are being tuned to render them suitable for still advanced and heavier duty applications, such as in hybrid/full electric vehicles, decentralized peak power-shaving, load leveling, and in integration with intermittent renewable energy sources, such as solar cells. Nevertheless, for these applications, the cycle life, specific power, specific energy, and the price per kilowatt-hour need to be still improved considerably (thus forming the motivation behind continued research and developments).

13.2 Looking Inside the Electrochemical Energy Storage Technologies

By now we are aware of the importance of technologies concerning advanced batteries and supercapacitors. These technologies, when in use for heavy duty applications, like in automobiles or in association with solar cells, are composed of relatively huge "packs" for attaining the as-required overall energy storage of say ~25 kWh, with a voltage of ~360 V (for electric vehicles like Nissan leaf). The overall battery pack consists of multiple actual cell modules, which in turn consist of individual cells (4 of them,

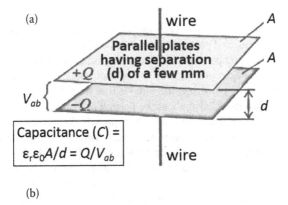

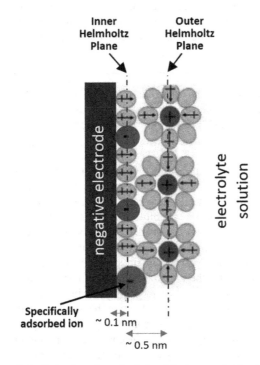

Helmholtz model of electrochemical double layer at electrode/electrolyte interface; crossed arrows represent dipoles

FIGURE 13.3 Schematic representations of (a) basic parallel plate capacitor, having separation between the charged plates (d) of the order of millimeters. (Adapted from; Addison Wesley Longman Inc.), and (b) basic model of the interfacial double layer formed at the interface between a charged "electrode" and solution containing solvated ions. Note that in this case (i.e., for EDLC), each electrode/electrolyte interface has two capacitors in series; namely, one between the charged electrode surface and IHP, while the other between IHP and OHP, both having d of the order of a few Angstroms. (Adapted from; 'Electrochemistry – Electrochemical Energy Conversion and Storage - Batteries, Fuel Cells and Electrochemical. Capacitors' - P. A. Christensen [EOLSS]).

each weighing $\sim$4 kg in Nissan leaf battery pack) connected in series and parallel (2 in parallel and 2 in series for the Nissan leaf cell module), along with sensors, controllers, battery management systems, and packaging items (to be discussed in more detail later in Section 13.5). Each of the components of a battery pack is based on some basic science and engineering aspect. For example, the sensors and controllers are based on concepts of electrical and electronics engineering, the same is true for the battery management system; but with a detailed knowledge about the desired cell operational features and cut-off voltages. The functional unit of the battery pack is the cell modules, which in turn consist of individual Li-ion batteries, with each battery being made of one or more elementary cells with a suitable cathode, anode, electrolyte, and other associated components like current collector, separator, etc. A flow diagram depicting the hierarchy in the construction of the battery pack technology, including the relationships with basic materials and electrochemical sciences, has been presented in Figure 13.4.

As per the more general perspective, there is no denying that the core to the performance of an elementary cell in a battery pack is electrochemical concepts (which got partly discussed in the earlier Chapters 6 and 7 and will get further discussed in more specific terms in the following Section 13.3).

Furthermore, the performance depends totally on the characteristics of the materials of which the anodes, cathodes, and electrolytes are made of, along with the compatibility and integration of these basic components.

For instance, the more "conventional" Li-ion cells work perfectly well with graphitic carbon as anode and $LiCoO_2$ as cathode materials, when the application is limited to consumer electronic devices (where such cells presently have monopoly). However, for usage in electric vehicles, the capabilities of getting charged very fast and also delivering energy at a rapid rate are needed. For this application, battery getting charged very fast (to a significant extent in about 10 min) is needed because nobody would like to wait in a battery charging station for hour(s) to get the car battery recharged while on a ride. Furthermore, capabilities of fast charging and discharging would allow recharging the battery to a good extent during regenerative breaking and also provide energy needed for rapid acceleration, respectively, without the need for resorting to an additional supercapacitor (connected to the battery) to serve the above purposes. Additionally, the Li-ion chemistry needs to be safer for usage in such mobile applications, since the very possibility of a car battery catching fire and exploding is extremely dangerous and a failure as a technology.

To achieve these, or in other words to develop suitable "advanced" Li-ion battery for usage in electric vehicles, one has to either play with the electrochemistry or the materials used in the basic components.

If looked into these issues in scientific terms since, there is no problem with the basic electrochemical concepts involved in the Li-ion cells per se (a schematic representation of the inside of a single Li-ion cell is shown in Figure 13.5), the problem must be lying with the materials; and it is indeed so. The electrochemical

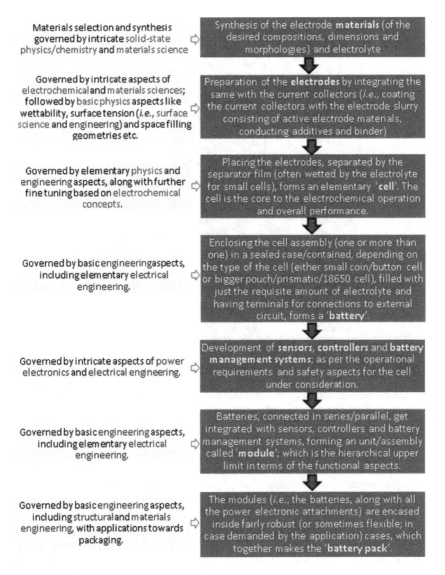

FIGURE 13.4 Flow diagram showing the steps (in a hierarchical sequence) involved in developing a "battery pack," along with the disciplines pertaining to which the concepts/knowledge are needed at each step.

potential at which the presently used graphitic carbon based anodes intercalate or store Li is very close to the electrochemical potential for Li-plating and formation of Li dendrites. The Li dendrites, if formed (especially at higher current densities) can lead to short circuiting by contacting the cathode through the separator, with the heat generated capable of causing fire in the presence of the flammable organic liquid electrolyte. Thus graphitic carbon does not totally address the problem which led to the rejection of Li metal *per se* as the anode material, especially for the high power applications. The $LiCoO_2$ cathode also operates at high electrochemical

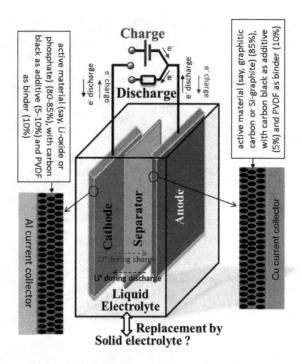

FIGURE 13.5 Schematic representation of the inside of a Li-ion cell, along with depiction of the directions of the flow of the Li-ion (through the electrolyte and across the separator) and electron (through the external circuit) during discharge and charge half cycles. The cathode and anode compositions mentioned are merely representative and may differ.

potentials, thus capable of resulting in a potential difference with the anode or cell voltage of ~4 V. However, this may lead to oxidation of the electrolyte and also oxygen evolution from the cathode itself (at the higher end of the potential limit or during accidental "overcharging"); which also makes the battery unsafe and prone to catching fire. Also, the conventional graphite electrode is not capable of allowing Li-intercalation and de-intercalation at a rapid rate, which limits the power density of the battery and leads to long charging times. It must also be mentioned here that the capacity or energy density of a Li-ion cell gets primarily limited by the present cathode material used (i.e., $LiCoO_2$); which is good enough for use in consumer electronics, but needs improvements for the heavy duty applications like in electric vehicles (to reduce the mass and volume of the battery pack).

In the above context, it can now be appreciated that looking beyond the presently used anode, cathode, and electrolyte materials is mandatory. In more specific terms, the anode materials should allow faster electrochemical insertion/removal of Li-ions and at a slightly higher potential compared to the conventional bulk graphitic carbon, but at the same time not compromise on the cycling stability (or cycle life). The candidate materials being metallic materials (such as Sn, Si, etc.) and nanoscaled graphenic carbons (such as reduced graphene oxides); but all of which have other issues, such as cycling stability for the former and sloping potential profile (i.e., negligible energy delivered at constant cell voltage) for the latter. The materials under investigation and considered as good replacements for $LiCoO_2$ are primarily the Li-NMCs (i.e., $LiNi_xMn_yCo_zO_2$; being contributed significantly by Professor Jeff Dahn's group at Dalhousie University, Canada, Professor Arumugam Manthiram's group at The University of Texas at Austin, USA, and other researchers), Li-excess-NMCs (i.e., $Li_{1+b}Ni_xMn_yCo_zO_2$), and polyanion-based Li compounds (originating with $LiFePO_4$ [possibly the safest cathode material operating at ~3.5 V vs. Li/Li$^+$, and again pioneered by Professor Goodenough's group, this time at The University of Texas at Austin], and with other polyanionic groups such as sulfates, pyrophosphate, etc. being presently investigated); but all of them still need further understanding and tuning in scientific terms for regular practical usage. For improving the safety of Li-ion cells, replacement of liquid electrolyte by inorganic solid electrolyte, which presumably would not allow Li-dendrites to pass through and are, of course, not flammable, is theoretically the best solution. However, most of the solid electrolytes, as investigated to-date possess considerably lower Li-ion conductivities (between 10^{-3} and 10^{-5} S/cm), as compared to the conventional liquid counterpart (~10^{-2} S/cm), along with issues related to incompatibility with the electrode materials at the electrode–electrolyte interfaces; the latter being a more significant issue.

13.3 Correlations between Chemical Sciences, Materials Science, Electrochemical Science, and Battery/Supercapacitor Technology

The basic operating principles of a battery or supercapacitor are based on some of the concepts of electrochemistry. Overall, during discharge of the concerned electrochemical cell, electrons get transferred from the anode to the cathode under the influence of electrochemical potential difference, leading to current, as desired, through an external circuit. This is balanced by the flow of positive ions, again from the anode to the cathode (and/or negative ions in the opposite direction), but through the electrolyte (see Figures 13.2 and 13.5). The reverse takes place during charging of the cell (in the case of a rechargeable system).

13.3.1 Charge Carrying Capacity of Electrodes

The duration over which the current flow can be achieved (or the time that the cell lasts during one discharge half cycle at a constant current) depends on the net charge (in the form of charge carrier ions; for example, Li-ions in the case of a Li-ion battery) that is hosted and successfully released by the electrodes constituting the cell. In quantitative terms, this property of the electrodes is known as the specific capacity (i.e., charge stored per unit mass or per unit volume of the electrode).

It may be mentioned here that the theoretical specific capacity of an insertion-based electrode material (in mAh/g), as for alkali metal-ion batteries, is defined in terms of the net charge (~ no. of Li-ions; since charge of 1 e$^-$ = charge on one Li$^+$) it can theoretically store per unit mass; with the net charge being defined in terms of current (in mA) passed over one hour. With respect to this unit, the theoretical specific capacity (Sp. Cap.), in terms of the theoretical number of Li$^+$ ions (i.e., the amount of charge) stored per unit mass of electrode, can be estimated as

$$Sp.\ Cap.\ (in\ mAh/g) = nF/3.6M \tag{13.3}$$

where n is the number of moles of Li-ions/number of moles of the host atom(s), M is the molar mass of the host atom/molecule, and F is Faraday's constant. Another way of inferring the capacity is, say, capacity of an electrode is $\sim$1000 mAh/g, which implies that if a current of 1000 mA is passed for 1 h, the electrode is "expected to" get fully filled with Li to its theoretical capacity. It may be noted that, in this case, the current is commonly referred to as the C-rate. Similarly, C/20 implies that the corresponding current (in this case, 1000/20 = 50 mA/g of the active electrode material) is "supposed" to "fill" the active electrode material fully (and also empty it fully) with Li in 20 h. Of course, the actual capacity achieved with an electrode corresponding to any applied current will be limited, among other factors, by the kinetics of the charge transfer reaction at the electrode—electrolyte interface, and also the kinetics of Li-transport (i.e., Li-ion diffusivity) within the bulk/lattice of the concerned electrode material; which, in turn, also may vary at different states of charges and with the occurrence of phase transformations during Li-insertion/removal.

While the charge carrier ions are typically hosted within the bulk structure in the case of battery electrodes, they are hosted just along the surface (usually in the form of electrochemical double layer) in the case of supercapacitor electrodes. Hence, in that case, the capacity of the electrodes (typically denoted in terms of Farads per unit mass or per unit volume) depends primarily on the ease of forming an electrochemical double layer and the specific active surface area of the electrodes exposed to the electrolyte. The overall energy density depends upon the combination of the specific capacity of the electrode and the potential difference between the cathode and the anode (i.e., E [per unit mass; say Wh/kg] = electrode capacity [per unit mass; say in terms of Ah/kg or mAh/g] × cell voltage, V; as also in Equation 13.1).

13.3.2 The Cell Voltage, Dependence on Electrodes and Electrochemical Parameters

With regard to the cell voltage, it is supposed to be the difference between the electrochemical potential of the cathode and the anode at any point of time (or state-of-charge) with respect to any common reference electrode. In very basic terms, the potential difference or the cell voltage depends on the difference between the energy levels corresponding to the highest energy electrons (such as "fermi energy level") or the energy levels corresponding to the actual elementary redox reaction where the electron enters/leaves (such as the energy corresponding to the transition metal redox under consideration) (for more details, see Refs. [1,2]). In the case of both a supercapacitor and battery, the basic electrochemical redox reaction takes place at the interface between the electrode and electrolytes, which is what needs careful control and optimization; with the kinetics of the same and the overpotential needed for realizing a certain magnitude of current. The above aspects eventually influence not only the power density, but also the cell voltage and, hence, energy density.

From a more materials perspective in the case of insertion-based electrodes (as in the case of batteries), for a given charge carrier ion (such as Li-ion in a Li-ion cell), the electrode materials where the charge carrier ions are more strongly bonded with the host (i.e., located at lower energy sites) usually necessitate greater expenditure of energy for removing them from the structure of the host and release of more energy when the ion is hosted back during the reverse reaction. Accordingly, stronger is the bonding between the host material and charge carrier ion, the greater is the electrochemical potential (against any reference electrode) for ion-insertion/removal. Recalling Chapter 6, as per the basic Nernst equation, $\Delta\Phi$ (or voltage) $= -\Delta G/z\mathcal{F}$ (see Equation 6.15). Accordingly, such materials are more suited as cathode materials (e.g., Li–T_M–oxides; TM: transition metal), as compared to materials where the corresponding bonding is weaker (such as Li–graphite intercalation compound or Li–Sn/Li–Si alloys). In the case of Li–T^M–oxides, the overall change in energy during Li-insertion (or removal), in turn, depends on the associated crystallographic site of Li+ (say, tetragonal/octahedral), the molecular orbital level of the TM-O bonds (see chapter 3) to which the associated electron goes into (or gets removed from) and the overall bonding-cum-structure of the material. Greater the energy released (or consumed) while inserting (or removing) Li from the structure and greater the difference in energy between the energy level associated with the concerned electron in the material and that in the reference (say, Li metal), the greater will be the associated voltage.

Nevertheless, as also mentioned above, the actual cell voltage at a concerned state-of-charge and current (or rate of charge transfer) depends considerably on the influence of the electrochemical polarization or overpotential (refer to Chapter 7) at the two electrodes needed to support the current; with a greater polarization loss implying a lesser cell voltage. A part of the polarization also depends on the resistance (i.e., electrochemical impedance) toward the electrochemical charge transfer reaction (primarily, for the basic redox reaction, such as $Li^+ + e^- \leftrightarrow Li$) at the electrode–electrolyte interface. Furthermore, the overall resistance or impedance of the cell also influences the cell voltage, which also depends on the resistance toward the flow of ions through the electrolyte, especially when organic electrolytes are used.

13.3.3 The Cycle Life

The cycle life of an electrochemical cell is defined in terms of the number of discharge–charge cycles the cell can undergo without having the overall capacity and voltage get lowered to levels that would render it unusable for the concerned purpose. In the case of commercial Li-ion cells, the reduction in cell capacity to below 80% of the rated capacity upon repeated cycling is usually considered to be the end of "cycle life" of the cell. In general, the cycle life depends on the integrity of the electrodes (including the inherent crystallographic structure) and other components, as also the build-up of impedance at the electrode–electrolyte interface, over multiple discharge–charge cycles. Usually, the integrity of the electrode material upon repeated discharge–charge cycles is better for those which involve minimal changes in the structure/phase and dimensions during electrochemical Li-insertion/removal. Additionally, reducing the size scale to nanosized levels often leads to enhancement in the integrity and, thus, the cycling stability of Li-ion batteries. However, accrued specific surface area in the case of nanoscaled electrode materials causes excessive deleterious side reactions at the electrode/electrolyte interface, unless the interface is suitably engineered. In general, supercapacitors possess considerably greater cycle life (up to even $\sim$10,000 cycles), as compared to batteries (e.g., cycle life of Li-ion batteries is $\sim$2000 cycles). This is because their functioning does not involve any bulk structural/dimensional change of the electrodes.

Overall, it must be emphasized here that right from basic chemistry to materials engineering, all have significant roles in the actual electrochemical performance in the above context, as well as other features of electrochemical energy storage systems.

13.3.4 Various Interrelated Aspects and Challenges

As indicated by the preceding sub-sections, the first step toward the development of a battery and supercapacitor is the identification of suitable electrode materials. In the case of supercapacitor, especially for the case of electrochemical double-layer capacitor (EDLC), it is much simpler in the sense that the material to be used should possess high specific surface area in the form of meso- or nano-porosities, while at the same time allow access to the solvated ions in the electrolyte (i.e., those forming the EDL) to the interior of such pores. It goes without saying that the electrode material needs to have good electronic conductivity and also be stable under the electrochemical environment. One of the most popular and low-cost electrode material for EDLC, viz., activated carbon (having specific surface area of $\sim$3000 m^2/g), uses various optimized chemical synthesis routes (including etching by KOH or by certain vapors of carbon derived from coal, coke, pitch, etc.) for the development of porosities in the desired range (typically between 10 and 2 nm; for the best performances, as of date). This immediately highlights the contributions of chemical synthesis (as discussed in basic terms in Chapter 5) toward electrochemical energy storage. A bit more complication arises in the case of pseudo-capacitive electrode materials, choices for which include metal-oxides and conducting polymers; synthesis of which necessitates the attainment of the desired structural features and oxidation states, but without compromising on the specific surface area.

The aspects concerning structural features, compositional aspects, and oxidation states, in addition to dimensional scales, become even more stringent in the case of developing materials for application in batteries. For instance, most of the cathode materials used in alkali metal-ion batteries (like Li-ion systems) are based on either oxides or poly-anionic compounds (such as phosphates, sulfates, silicates, etc.) of transition metals or T_M (such as Co, Ni, Mn, Fe, etc.). In these materials, the transition metals are entrusted with the "job" of maintaining charge neutrality by changing their oxidation states to more positive (i.e., oxidized state) or less positive values (i.e., reduced state, or in most cases the pristine) during Li-removal or Li-insertion, respectively. As for example, during reversible electrochemical Li-extraction from $LiCoO_2$, the following change in oxidation state of Co takes place: $Li^{(+1)}Co^{(+3)}O_2^{(-2)} \leftrightarrow Li_{1-x}^{(+1)}Co_{1-x}^{(-3)}Co_x^{(+4)}O_2^{(-2)})$. Similarly, for $LiFePO_4$, the oxidation state change taking place over the entire reversible Li-extraction or charging half cycle can be denoted as $Li^{(+1)}Fe^{(+2)}P^{(+5)}O_4^{(-2)} \leftrightarrow Fe^{(+3)}P^{(+5)}O_4^{(-2)}$.

It is the energy level corresponding to the concerned redox couple (relative to the energy level for the Li/Li$^+$ redox couple) in the concerned structure, which is one of the factors deciding the electrochemical potential of the cathode material, in addition to the energy needed/released for Li-removal/insertion from/to the specific sites.

The oxide or poly-anion network provides the structural framework (like layered oxide network for $LiCoO_2$ or cubic oxide network for $LiMn_2O_4$, etc.) that may allow facile insertion and bulk (lattice) transport of the charge carrier after the surface redox reaction (e.g., $Li^+ + e^- \leftrightarrow Li$), with the strength of the (transition) metal-oxide bond also influencing the energy level corresponding to the redox couple. Accordingly, based on the above discussion and example, achieving the right combination of composition and structure in terms of the right transition metal cation (as needed), in the desired oxidation state, and the desired oxide or poly-anion framework is a must (for more details, see Refs. [2,3]).

Talking about the anion/polyanion network, the electrochemical potential for removal/insertion of Li-ions from the cathode materials (and in turn, the cell voltage) depends on the co-ordination of the Li-ions in the structure. The lower are the energy levels of Li-ions and the associated electrons in the hybridized T_M $3d$–O $2p$ hybridized antibonding orbitals in the cathode structure, the greater is the energy needed/obtained to remove/insert the Li-ion and the concerned electron; resulting in greater electrochemical potential (w.r.t. anode). In this regard, for the same anion and transition metal, the presence of Li-ions with tetrahedral co-ordination of anion leads to greater cell voltage, as compared to Li-ions residing with the octahedral co-ordination. $Li_xMn_2O_4$ (the spinel structured cathode material) is a classic example to demonstrate this because insertion of Li till $x = 1$ takes place in the tetrahedral voids in the spinel structure at potential of ~4 V against Li/Li$^+$ (the charge compensating transition metal redox being Mn$^{4+/3+}$). However, for $x > 1$, despite the transition metal redox still being Mn$^{4+/3+}$, the filling-up of Li takes place at a reduced potential of ~3 V corresponding to Li now entering the octahedral voids, as opposed to the tetrahedral voids (see Figure 13.6). Further complication happens when x is close to 2 because the average oxidation state of Mn decreases to ~+3; the "Jahn–Teller" active state (see Section 3.4). Accordingly, the "Jahn–Teller" distortion leads to tetrahedral distortion of the erstwhile cubic spinel structure due to the oxygen anions being "pushed" along the z-axis as a result of the repulsion between the electrons in the p_z orbitals of the co-ordinating oxygen anion and the now singly occupied d_{z2} orbital of Mn^{3+} (as discussed in fundamental terms earlier in Section 3.4 of Chapter 3). Such structural distortion generates considerable internal stresses during electrochemical cycling and leads to lowering of integrity of the cathode material; hence drastic fade in electrochemical Li-storage capacity with continued cycling.

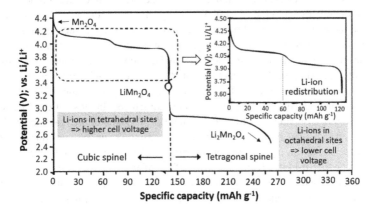

FIGURE 13.6 A typical discharge profile (against Li) of a LiMn$_2$O$_4$-based cathode material for Li-ion batteries, starting with Mn$_2$O$_4$ (the fully delithiated state or $x = 0$ in Li$_x$Mn$_2$O$_4$) showing the different electrochemical potentials corresponding to x (in Li$_x$Mn$_2$O$_4$) < 1 and > 1. (Adapted from Liu C. et al., *Materials Today*, 2016, 19, 109.)

> Accordingly, the electrochemical performance of electrode materials depends significantly on, not only the composition and crystal structure, but also on the associated cation co-ordination and basic bonding characteristics, including the crystal field stabilization/splitting energy (again, as discussed earlier in Chapter 3).

From a more technological view-point, this is also an example which indicates that structural distortions and changes during charge carrier alkali metal ion (viz., Li^+ in the previous case) insertion/extraction in the host cathode/anode materials should be minimized (better avoided) to lead to an enhanced cycle life of the concerned electrode and the cell as a whole.

However, the complications do not stop at that. Synthesis conditions often lead to the formation of impurity phases (due to the simultaneous presence of quite a few elements) and also loss of volatile components such as Li or Na (as the case may be). The syntheses are carried out via a variety of routes, such as solid-state synthesis, hydrothermal synthesis, sol–gel, mechano–thermal synthesis, and so on. It is imperative that the conditions are optimized w.r.t. the synthesis process adopted, since a minor change in the condition can lead to instability in composition, phase assemblage, and structural aspects; all being fairly interlinked to each other.

Last, but not the least, optimization of the dimensional aspects (in particular the particle size) is also equally important. This is because, in the case of battery electrodes, it is not just the surface reaction, but also involves insertion and diffusive transport of the charge carrier (say, Li-ion or Na-ion, as in Li-ion or Na-ion batteries) into the bulk post the surface redox reaction. Accordingly, with a reduction in particle size, the greater specific surface area results in more available surface sites for the surface redox reaction and more lattice "channels" exposed to the electrolyte at the electrode–electrolyte interface for the subsequent (or prior) bulk transport of the charge carrier ions. Furthermore, finer dimension reduces the bulk transport distance, thus leading to "filling-up" of all available lattice sites by the Li- or Na-ions in a lesser time (via solid–state diffusion). Both these, in addition to the intrinsic structural features influencing the facileness of bulk transport (i.e., overall diffusivity), determine the overall kinetics of the electrochemical alkali-ion insertion/removal and accordingly the rate capability of the electrode; or in a more general perspective, the power density of the battery, as a whole.

On the other hand, a very high specific surface area can also lead to excess of undesirable side (i.e., surface) reactions of the electrodes with the solute–solvent species in the electrolyte. Such surface reactions, that take place at the "bare" electrode surface primarily during the first cycle, lead to an irreversible loss of the Li-content from the system by formation of Li-containing salts as some of the reaction products, which is very harmful if such reactions keep continuing for a longer duration (i.e., if no intact passivation layer forms of the surface of the electrodes). However, if the surface layer that initially forms on the surface of the electrode due to electrolyte decompositions, which commonly known as the solid electrolyte interface (or SEI) layer, is stable over repeated lithiation/delithiation, it prevents further direct contact between the electrode and electrolyte (*i.e.*, passivating) post the first (or first few) cycles and thus suppresses the irreversible Li-loss in the subsequent cycles. These phenomena occur primarily at the anode side (for more details, see Refs. [4–7] where the negative potentials are capable of reducing the electrolyte in contact with the electrode surface). Additionally, surface reactions can also lead to irreversible structural changes at the surface of the electrode particles, especially at the cathode side (for more details, see Refs. [8,9]); thus rendering the surface not suited for facile transport of the charge carrier alkali metal ion (i.e., Li or Na).

In order to minimize such undesirable surface reactions, thin coatings on the electrode particle surfaces are often applied, ensuring that such coatings (such as graphene-based or even conducting polymer based) are as thin (thickness typically in the nano-regime), but uniform, as possible. It is also important that the coatings do not have any negative impact on the surface impedance toward the electrochemical reaction and transfer of the charge carrier ions between the electrode and the electrolyte. On a different note, during operation or discharge−charge of a cell, most of the electrode materials undergo dimensional changes in response to the electrochemical insertion/removal alkali metal-ion (i.e., Li-ion, Na-ion, etc., as the case may be) into/from the electrode material and associated structural modifications. Repeated occurrences of such dimensional changes and concomitant stress developments (either due to overall constraining effects of the electrode particles/films or co-existence of a local structural mismatch in the very lattice) often lead to mechanical fracture/disintegration of the electrode materials. This, in turn, causes loss in usable capacity and also repeated breakdown of the surface films, rendering them less passivating and leading to continued irreversible loss of the charge carrier ion from the system (for more details, see Ref. [10]). These phenomena have been observed to be more severe for higher capacity electrode materials, such as "alloying-reaction" based anode materials (viz., Si, Sn, etc.), in comparison to lower capacity intercalation-based graphitic carbon for Li-ion system. Hence, the above also leads to a compromise between the achievable capacity or energy density and cycle life.

With regard to the electrolytes themselves, as another core component of a battery or supercapacitor, the materials used (whether liquid or solid) need to have stability against the electrode materials and over the entire potential range applicable toward the operation. Furthermore, despite maintaining the good ionic conductivity, the electrolytes used should not become electronic conductors under any condition or cell voltage; which, otherwise will lead to a situation similar to short circuiting of the cell. In fact, at present the instability of the liquid electrolyte beyond a potential window of ∼4.4 V is the primary cause for the voltage restriction of the present generation Li-ion batteries to typically less than 4 V, even though electrode materials have been developed which can increase the cell voltage to even greater than 5 V. To address the same, and also the safety concerns associated with the flammable liquid electrolytes, solid electrolytes are investigated which are expected to be safer and also be stable over a wider voltage window. However, due to solid-state diffusion being slower than diffusion via liquid phase, the ionic conductivity of the charge carrier alkali metal-ion is lesser in the case of a solid electrolyte; which limits the power density of a cell. Accordingly, a lot of chemistry and materials science is involved toward the development of a solid electrolyte, starting with the right composition, achieving a nearly fully dense solid body free of impurity phases (to ensure greater ionic conductivity) and a crystal structure again having facile transport paths through the lattice for Li and Na-ions (as the case may be). The materials engineering should also ensure integrity and good interfacial contact between the solid electrolyte and the solid electrode (i.e., a solid/solid interface in this case) during cell development and also during discharge−charge cycles. Overall, the above discussion describes in very general terms a small part of the influences that chemical sciences and materials sciences have over the development and performance of electrochemical energy storage; in fact, being sort of the "core" aspect of the overall technology. Figure 13.4

summarizes all these correlations between the various disciplines of science and engineering, leading to the build-up of the battery technology, in very precise terms.

13.4 Advancement in Supercapacitor and Battery Science and Technology

The concept of supercapacitors came from the science involving spontaneous formation and charge storage in electrochemical double layers at the interface between the electrode (a piece of solid) and the electrolyte (composed of ions having some polarity). The distance between the layers of charges in the electrochemical double layer (EDLC) being of order of a few angstroms, as compared to the separation of a few millimeters between two physical charged plates (as in basic capacitors), led to the enhancement of the capacitance by nearly 7–8 orders of magnitude, from $\sim 10^{-12}$ to $\sim 10^{-5}$ F/cm^2. This was followed by tremendous effort to enhance the specific surface areas of the electrode materials to further increase the capacitance; with the development of nanomaterials such as, activated carbons (having nanosized porosities between ~ 2 and 50 nm; i.e., mesoporosity), resulting in achievement of specific capacitances of the order of $\sim 10^2$ F/g, just based on the EDLC mechanism. Devices based on such mechanisms and materials are possibly the best with respect to power density (see Figure 13.1), i.e., applications requiring withdrawing of very high currents (or nearly "instantaneous" discharge–charge), but without necessitating retention of the same over even a few minutes (i.e., sans the need for high energy density).

However, to go toward the right-hand side in the Ragone plot (as in Figure 13.1), further enhancement of the energy density or charge storage capacity was needed; but without lowering the power densities to levels typical of batteries. Accordingly, in addition to increasing the specific surface area by nanostructuring, there have been extensive developments toward further improving the specific capacitance of the materials used for the electrodes. Accordingly, a class of materials has been realized that can allow very fast faradic reactions restricted to just the surface, with no interference/ contribution from the bulk (i.e., no diffusive solid-state transport post or preceding the redox reactions). Some of the examples include conducting polymers such as polyaniline (or PANI), poly(3,4-ethylenedioxythiophene) (or PEDOT) and transition metal oxides such as RuO_2, MnO_2, etc. With this concept, the specific capacitance has been further enhanced to even $\sim 10^3$ F/g, albeit with some compromise on the power density (or the achievable kinetics), as indicated in the Ragone plot (see Figure 13.1). One of the bottlenecks toward further enhancement in the energy density of the supercapacitor is the electrolyte type. Supercapacitors primarily use aqueous electrolytes (i.e., some salt dissolved in water) which reduce the operational voltage window to ~ 1 V due to the instability of the electrolyte itself beyond that. Even though the voltage window can easily be increased to 3 V with the use of organic electrolytes (i.e., solvent based on organic materials), the electrolyte resistance (or resistance to ionic movements in the electrolyte) increases drastically upon the use of such electrolytes, rendering the power density not good enough for application as supercapacitors. For more details on recent developments on supercapacitor materials and associated aspects, see Ref. [11].

The rechargeable batteries, used for advanced technological applications, have seen extensive evolution of even the core chemistry from lead-acid, via nickel metal-hydride to Li-ion systems (and now further evolving toward other alkali metal-ion systems like

Na-ion or Mg-ion). Of course, the most commonly used battery system as of now is the Li-ion system. Undoubtedly a majority of the portable electronic devices used in our day-to-day life, starting from basic cell phone, smart phone, laptop, digital camera, camcorder, tablet, power tools, and so on, use Li-ion batteries as the power source. Li-ion battery is also finding more heavy duty applications, like in electric vehicles, storage of energy harvested from renewables, and grid energy storage. As a matter of fact, the present day Li-ion battery technology is totally apt for the consumer electronic segment. However, the present generation Li-ion batteries are yet not capable of running electric vehicles to the best of the necessities and customer satisfaction. The prohibiting aspects include charging durations (recollect: nobody would like to wait for an hour to recharge car batteries during a drive), the safety concerns (a car cannot have a battery which might just "explode" sometime!), the "mileage" upon full charge, and the cost (especially considering that the present generation batteries do not usually last for more than a couple of years).

> In more technological terminologies, the present day Li-ion batteries may be considered to still lack in terms of power density, energy density, cycling stability, and safety aspects; as needed for application in electric vehicles.

Accordingly, more recently finer scaled electrode materials, such as those based on well-ordered graphitic carbon, active nanoparticles (such as $LiFePO_4$, $LiNiMnCoO_2$, SiO_2, Si, etc.), "decorating" conducting reinforcements (such as carbon nanotubes or graphene), electrodes based on nanowire/nanorod morphologies, and so on, have been developed. Such developments, which have and are leading to novel/improved electrodes having considerably greater Li-storage capacities (for Li-ion batteries, energy density has increased from $\sim$150 Wh/kg in 1995 to $\sim$250 Wh/kg in 2017), rate capabilities (i.e., ability to get charged and deliver energy very fast), cycling stability (or cycle life), and also safety aspects, have necessitated careful optimization/control of the processing/ synthesis conditions and structural aspects. In the context of nanostructured electrode materials, it may be noted that, the finer dimensional scale enhances the cycling stability due to reduced overall dimensional changes (thus, improved mechanical integrity upon repeated lithiation/delithiation, especially for electrode materials like Si), allowing for faster charging–discharging of a Li-ion cell due to the reduced Li-transport distance within the bulk of the electrode and a greater electrode–electrolyte interfacial area for the electrochemical redox reaction to take place. These lead to considerable operational advantages at higher current densities or greater discharge–charge rates. However, the greater electrode–electrolyte interfacial area also leads to accrued occurrences/ kinetics of undesirable side reactions at the electrode–electrolyte interface; viz., a problem which has been preventing the usage of nanoscaled electrode materials. Such undesirable side reactions, if not controlled, consume Li-ions irreversibly, which are then lost to the cell, and also enhance the overall impedance of the cell, thus negatively affecting the rate capability of the electrode (or power density of the cell) and also the cycling stability of the electrode (thus, cycle life of the cell).

The above, once again, highlights the influence of the in-depth understanding and interplay between various basic scientific principles, leading to advancements in technology. For example, of late (i.e., in mid-2018), charging of battery packs of full electric vehicles in a few hours has become fairly common with most makes; with Tesla claiming that one of its models can get nearly 80% charged in just half an hour. This is already a significant step toward realization of a good fleet of full electric vehicles, as alternative to gasoline-run vehicles, and accordingly a clean environment for the future. Nevertheless, still further technological developments are needed, which will again have their roots toward basic science/engineering right from the level of identifying suitable combination of elements for developing new electrode – electrolyte materials that are still superior and also less expensive, for realizing the dream of 100% electric vehicles on road.

For instance, considering that solid-state Li-ion batteries have the potential to be one of the "safest" energy storage systems (but presently having various issues at scientific/engineering levels); more recently focus has also been directed toward the understanding and development of Li-ion conducting solid-state electrolytes, with the aim of eventually replacing nearly all the liquid electrolyte containing Li-ion batteries by solid-state batteries for applications involving mobility.

Looking further ahead and considering possible depletion of Li-based resources (which are anyway very localized) in the future, focus has also been directed toward understanding and development of electrode materials for the upcoming Na-ion battery technology since Na-based resources are relatively more abundant and widespread (a research and development area, which may be very important in the Indian context; especially for grid energy storage and supporting energy harvested from renewables).

In fact, due to so many interconnected aspects governing the operation and usage of Li-ion batteries, combinatorial investigations pertaining to electrode materials and different aspects of such electrochemical energy storage systems is needed, as discussed by Fleischauer, Hatchard, Bonakdarpour, and Dahn.[12]

13.5 Efficient Integration and Usage of Such Technologies

Even though developments and advancements of the concerned devices themselves (viz., the supercapacitors and batteries) are core to their efficient usage as per requirements, they are not sufficient. For example, when we are charging a cell phone battery or even a car battery, we are just relying on the indicator to tell us the percentage of charge. This seems too trivial to have a mention. In practice, it is not so. Neither is the actual voltage in a cell, ever constant for a given duration, nor is the incremental charge state ever so smooth. There can be an initial steep rise in voltage during charge, followed by a slow rise, which can be made up of small "steps," followed by another steep rise toward the end (see Figure 13.6 as an example). However, we do not get any information on those and all we know is that the cell, as purchased, is a 3.3 V Li-ion cell and believe that the entire charge–discharge takes place either at a

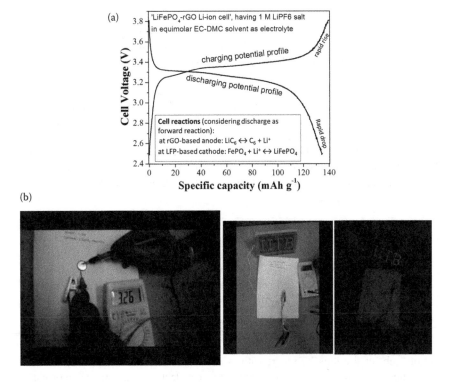

FIGURE 13.7 (a) A typical discharge–charge profile recorded at a constant current density (equivalent to the C/5 rate or current theoretically expected to discharge/charge the cell fully in 5 h) for a Li-ion coin (or button) cell having reduced graphene oxide (rGO) as the anode material and LiFePO$_4$ as the cathode material (with the liquid electrolyte having LiPF$_6$ salt dissolved in an equimolar mixture of ethylene carbonate and dimethyl carbonate solvents). Note the rapid increase/decrease of cell voltage toward the end of the charge/discharge half cycles. The actual photograph of the concerned CR2032 coin cell used for lighting a set of light-emitting diodes (LEDs), along with the as-measured cell voltage, is also shown in (b).

constant voltage or due to a very gradual and smooth rise/drop in voltage (as shown in Figure 13.7 for a laboratory scale LiFePO$_4$/rGO "full" Li-ion coin cell).

Of course, as a mere user, we may not need to know the details. However, the "additional component," which by itself is another technology, ensures that the output voltage or current is at a fairly constant level, as needed for the application, so that it can provide the output to the external "load" in the desired form. This "additional component/technology" also ensures that the battery does not get "overcharged" or "undercharged," which implies that the voltage gets cut-off before it goes beyond the safety limits on the either side. This would have been trivial had the voltage been always constant or risen/fallen at a constant rate, which is not the case. The voltage also depends on the current level and a sudden surge in the current withdrawn or "inserted" can lead to a sudden fluctuation in the voltage so much as to raise/lower the same above/below the safety limits, unless the "additional component/technology" is capable of detecting the same beforehand and "cutting" it off. The other important aspect controlled by the "additional component/technology" is the internal temperature of a cell, which again has to be maintained within a specific limit for efficient and safe operation, but which may fluctuate significantly depending on the current load and cell conditions. Charge–voltage balancing among cells in series/parallel is yet another important function of this "additional component/technology," which is known as the "battery management system" (or BMS). BMS is a power electronic set-up and an integral part of any "battery pack" (as briefly introduced toward the beginning of Section 13.2) used for "advanced applications," right from portable electronic devices to electric vehicles or grid energy storage. In addition to controlling the functionalities of a battery pack, the BMS also plays an important role with respect to the safety aspects (and prevention of overheating or thermal runaway of battery packs), which also includes prevention of "over-charge" and "over-discharge" of the constituent cells. The latter function becomes more challenging when there are imbalances between cells constituting a pack.

Accordingly, as a first step, successful development of a good battery or even supercapacitor "pack" also necessitates the development of an additional very efficient electronic device (based on expertise with electrical/electronic engineering and knowledge about the associated elementary electrochemical cell) and efficient integration of the same with the concerned cell(s).

Now what truly is a "battery pack" in itself? It is itself a full device based on a lot of "sub-integrations." To start with, a basic cell (say, a Li-ion battery cell or a supercapacitor cell) consists of a cathode (positive electrode) and an anode (negative electrode), which are integrated with an electrolyte containing lithium ions and separated by a micro/meso-porous polymer membrane acting as the separator. Arrangement and integration of a number of such elementary cells in a "sealed case" forms a battery or supercapacitor. The physical form of the final battery/supercapacitor is governed by the geometry and structural features of the "sealed case," which may be flat and flexible for a "pouch cell," or flat but hard and bulky for a "prismatic cell," or even "cylindrical cell" like 18,650 (see Figure 13.8 for schematics and their specific advantages/disadvantages). Now, as per requirement and desired output voltage–current–capacity, multiple batteries/supercapacitors are integrated into a module, either in parallel (to increase current) or in series (to increase voltage) or combination of both, which typically house them in a metallic casing with terminals. Subsequently, multiple modules are integrated into what is known as the "battery pack," which not only house the modules inside a casing but also houses the interconnected controllers and sensors (refer back to Figure 13.4). Just for example, the present generation 90 kWh "battery pack" in a Tesla electric vehicle (viz., Tesla S) is made up of ~7600 Li-ion cells (of 18,650 form in this case), each of capacity of ~3.0–3.4 Ah and cell voltage of ~3.5–4.0 V. As another similar example, the 30 kWh "battery pack" for Nissan Leaf (another electric vehicle) is made of 48 modules in series, with each module having four batteries (pouch cells, in this case), but each having ~30–33 Ah capacity and operating voltage of ~3.5–4 V. Of course, the Li-ion cells in the two "battery packs" described above have two different chemistries (i.e., different cathode materials; refer to Section 13.2); thus differing at the very core, but tuned to satisfy the respective requirements of the two makes. As per the requirement, the battery packs may also be connected to supercapacitor packs for providing surge currents, if and when needed.

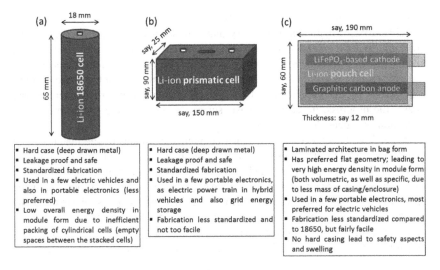

FIGURE 13.8   Schematic representations and typical features of the different battery geometries (of Li-ion cells) used for the heavy duty applications (i.e., (a) 18,650, (b) prismatic, and (c) pouch cells). A photograph of a typical coin cell (viz., the cell geometry used for very low power applications, such as in miniature electronic, healthcare devices, and basic laboratory-scale experiments), was presented in the previous Figure 13.7b.

> It is now fairly straightforward to realize that the development of just a working and safe battery/supercapacitor pack suited to the desired application gets initiated at atomic level research/understanding, tuning and scaling-up based on chemical/materials science, followed by efficient electrical or power electronic engineering for the various integration steps.

Of course, there are other challenges, including design aspects, involved with integrating the energy storage systems with the actual application, whether with electric vehicles or with grid/photovoltaic. Such integrations and design aspects are equally important for achieving the desired performance, just like the core science and engineering associated with the concerned device/technology itself.

13.6 Closure

In a way, this chapter provides insights into "communications" and "co-ordinations" between the basic knowledge pertaining to chemistry, electrochemistry, and materials science, for translating the same into technologies corresponding to electrochemical energy storage (i.e., supercapacitors and batteries) and associated usages of the same at the various levels. Accordingly, starting by providing a brief historical perspective, followed by discussion on the correlations between the various scientific aspects inherent to such technologies and advancements made in recent times, the chapter eventually describes aspects concerning efficient integration toward the successful usage of such technologies. It may be appreciated here that it is the very basic aspects of solid-state chemistry, electrochemistry, materials science and materials engineering which influence the behavior and performance of individual (electrochemical) cells; which, when efficiently interconnected based on concepts of electrical engineering or power electronics leads to the desired ratings and performances of the entire "battery pack," which is put to application. Hence, in the context of electrochemical energy storage, the progress, in terms of "basic science → engineering → technology → application" has been described here.

Nevertheless, beyond the focus of this chapter, the choices of the materials and technologies to be used are also partly governed by the economic and geo-political aspects of the concerned place and time. For example, economic aspects govern whether cathode materials to be used in Li-ion batteries should or should not contain the expensive Co as the transition metal. Furthermore, in many countries, like in India, the focus has been toward developing technologies beyond the Li-ion battery system (such as the upcoming Na-ion battery) due to scarcity of suitable Li-containing precursors, as compared to the widespread availability and distribution of Na-containing precursors.

REFERENCES

1. Goodenough J. B., K. -S. Park, The Li-ion rechargeable battery: A perspective, *J. Am. Chem. Soc.*, 2013, 135, 1167–1176.
2. Liu C., Z. G. Neale, G. Cao, Understanding electrochemical potentials of cathode materials in rechargeable batteries, *Mater. Today*, 2016, 19, 109–123.
3. Gutierrez A., N. A. Benedek, A. Manthiram, Crystal-chemical guide for understanding redox energy variations of $M^{2+/3+}$ couples in polyanion cathodes for lithium-ion batteries, *Chem. Mater.*, 2013, 25, 4010–4016.
4. An S. J., J. Li, C. Daniel, D. Mohanty, S. Nagpure, D. L. Wood III, The state of understanding of the lithium-ion-battery graphite solid electrolyte interphase (SEI) and its relationship to formation cycling, *Carbon*, 2016, 105, 52–76.
5. Peled E., S. Menkin, Review-SEI: past, present and future, *J. Electrochem. Soc.*, 2017, 164, A1703–A1719.
6. Aurbach D., Y. Ein-Eli, O. Chusid (Youngman), Y. Carmeli, M. Babai, H. Yamin, The correlation between the surface chemistry and the performance of Li-carbon intercalation anodes for rechargeable 'Rocking-Chair' type batteries, *J. Electrochem. Soc.*, 1994, 141, 603–611.
7. Mukhopadhyay A., A. Tokranov, X. Xiao, B. W. Sheldon, Stress development due to surface processes in graphite electrodes for Li-ion batteries: A first report, *Electrochim. Acta.*, 2012, 66, 28–37.
8. Rozier P., J. M. Tarascon, A. Manthiram, Review-Li-Rich layered oxide cathodes for next-generation Li-ion batteries: Chances and challenges, *J. Electrochem. Soc.*, 2015, 162, A2490–A2499.
9. Matsui M., K. Dokko, K. Kanamura, Dynamic behavior of surface film on $LiCoO_2$ thin film electrode, *J. Power Sources*, 2008, 177, 184–193.
10. Mukhopadhyay A., B. W. Sheldon, Deformation and stress in electrode materials for Li-ion batteries, *Prog. Mater. Sci.*, 2014, 63, 58–116.
11. González A., E. Goikolea, J. A. Barrena, R. Mysyk, Review on supercapacitors: technologies and materials, *Renew. Sustain. Energy Rev.*, 2016, 58, 1189–1206.
12. Fleischauer M. D., T. D. Hatchard, A. Bonakdarpour, J. R. Dahn, Combinatorial investigations of advanced Li-ion rechargeable battery electrode materials, *Meas. Sci. Technol.*, 2005, 16, 212.

14

Ceramics for Armor Applications

The application of advanced materials spans across various strategic sectors. One such application is body armor for soldiers in the battlefield. Quite rationally, there has been a thrust to develop light-weight body armor materials. This chapter discusses the challenges in the intelligent design of armor materials in general as well as various critical aspects to be considered for high-performance armor materials. In contrast to monolithic materials with homogeneous microstructures, functionally graded materials with gradient in microstructure and properties have been widely developed for biomedical, defense, and energy-related applications. In simplistic terms, functionally graded materials are usually made by stacking multiple layers of different compositions, while adapting the suitable processing approach. Depending on the microstructural characteristics (phase assemblage and microstructure), a gradient in properties can be achieved. This chapter will provide the summary of the published results from the author's group on the development of functionally graded materials for armor applications. While discussing the results, it will be emphasized as to how the materials science-based processing approach can be integrated with physics and mechanics-based measurement and analysis to develop a holistic understanding of the property gradient in functionally graded materials. This will emphasize the need for interdisciplinary measurements and understanding. More details of the results presented in this chapter can be found elsewhere.[1-3]

14.1 Development of Ceramic Armors

Armor materials do not only protect the army vehicles but also the army personnel at the warfront. The primary purpose of armor materials is to deflect or to diffuse impact forces from projectiles or weapons. Various materials are used for body armor applications, which include ceramics, laminated composites, and ballistic fabrics. Among the engineering ceramics, B_4C, SiC, Al_2O_3, TiB_2, AlN, and Syndie (synthetic diamond composite) are widely regarded as armor applications for both personnel and vehicle protection. Such usage is possible as they have a spectrum of properties, including low density, superior compressive strength, hardness, and greater energy absorption capacity.[4-7] These properties allow them to sustain a high impact velocity for dwell/penetration transition and to undergo deformation induced hardening.[8,9] For ballistic applications, the armor performance is largely dependent on the material system (geometry, structure, and properties) and energy absorption mechanisms. The material systems must have the capability to withstand multiple hits.

In the context of armor applications, if the ceramic gets shattered upon impact of a projectile, there exists further abrading of the projectile by pulverized particles. Also, when a sharp projectile penetrates a ceramic, the entrance channel becomes ragged, as compared to that while penetrating a ductile metal. The presence of a ragged channel results in asymmetric pressures and affects the projectile geometry.[10] Therefore, tougher ceramic composites are in demand, which can effectively arrest the projectile as well as minimize the penetration of projection by undergoing shattering, bending, or change of path.

In recent years, the advancements in materials science and technology have enabled the development of metal matrix composites (MMCs). Enhanced toughness in MMCs arises due to toughening through crack bridging, wherein the crack propagation is stopped/hindered by the metal-rich phase. By incorporating a graded interface layer in the region between metal and ceramic, relaxation in thermal stresses can be achieved in metal–ceramic layered composites. The following sections summarizes the development of body armors based on ceramic materials for defeating small-calibre armor piercing projectiles. An emphasize will be given to the properties as well as the design requirements and their evaluation techniques.

14.1.1 Classification

After World War II, there has been resurgence in the development of armor materials owing to rise in threat, wherein the armors as flak jackets, kevlar, and ceramics came into existence. In general, there are two classes of armor systems, namely, body armors and vehicle armors. As the name suggests, body armors are used as personal protection equipment which consists of combat helmets, bullet proof vests, ballistic face masks, and bomb suits. On the other hand, vehicle armors shield vital war equipment such as tanks, frigates, fighter planes, and armored fighting vehicles. Based on material classes, the armor materials can be categorized into three groups as, natural, synthetic, and man-made materials, which are listed in Table 14.1.

14.1.2 Property Requirements for an Armor System

Functional aspects of armor materials require properties such as lower density, high hardness, compressive strength, fracture toughness, Hugoniot elastic limit (HEL), elastic modulus, and ballistic efficiency (Figure 14.1). Low density enables personnel comfort as low weights are convenient for the armor wearer. The combination of high hardness and compressive strength allows effective erosion and defeat of the projectiles, while high fracture toughness leads to improved damage tolerance/failure resistance; a higher HEL results in better performance under dynamic loading. Also, high elastic modulus increases the resistance to contact damage. Ballistic efficiency

TABLE 14.1

Various Classes of Armor Materials[1-3]

Class	Application
Natural materials	Leather, bone, linen, laminated wool
Metals	Bronze, iron plate, rolled steel
Synthetic material	Kevlar
Composites	MMC, CMC, glass–nylon interwoven fibers and FGM's (as ceramic-foam, ceramic-metal)
Ceramics	TiB_2, alumina, SiC, B_4C, BeO borides, Si_3N_4, TiC, WC

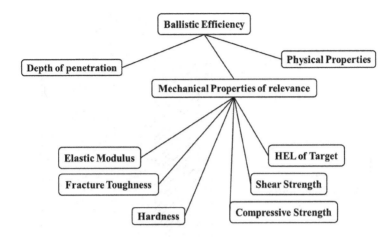

FIGURE 14.1 Summary of various factors affecting the ballistic efficiency of a material.

is defined as the ability to resist projectile penetration, and for armor applications, materials with high ballistic efficiency are preferred. The ballistic efficiency depends on a number of factors, including geometry of projectiles as well as mechanical properties of armor materials and projectile, as depicted in Figure 14.1. The typical values of ballistic efficiency for different armor materials are tabulated in Table 14.2 and shown in Figure 14.2.

The primary factors, which affect the ballistic performance of a ceramic, are density, internal friction, and compression strength. The design for an advanced armor system should consider the desirable properties, such as resistance against large impact load. Second, the materials must provide multi-hit capability. Materials with different advantageous properties can be stacked intelligently so that ballistic properties can be competently achieved. The combination of high hardness and low density in monolithic ceramics aids in resisting maximum shattering. However, complete penetration may not be possible. This requires high fracture toughness, which can be

TABLE 14.2

Ballistic Efficiency of Different Classes of Armor Materials[1-3]

Material	Projectile	DOP (mm)	Ballistic Efficiency	Hardness (GPa)
Aluminum	Pointed	265	–	1.0
Glass	Pointed	200	4.2	5.5
AD85	Pointed	122	8.6	8.8
AD995	Pointed	50	11.7	15
Zirconia	Pointed	68	9.3	11.2
TiB_2 (Ceradyne)		38	11.1	27
Aluminum	Blunt	75	–	1.0
Glass	Blunt	46	1.6	5.5
Zirconia	Blunt	42	1.3	11.2
TiB_2–2.5% $MoSi_2$	Pointed	10–12	5–5.2	29.9
TiB_2–10% $MoSi_2$	Pointed	10–12	5–5.2	22.9

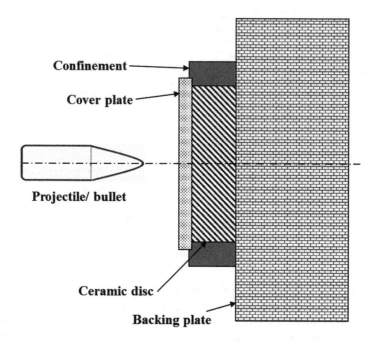

FIGURE 14.2 Typical test configuration to determine the ballistic efficiency using the DOP test. (Adapted from Ref. 3).

achieved, either by a layered material design wherein ceramic is backed by a metal or by implementing functionally graded structure. The top ceramic surface allows the armor to blunt the projectile and also induces shock waves to propagate through the underlying structure, upon the impact. The soft metal backing aids in preventing target penetration by the residual broken fragments, thereby acting as the catcher. Also, during the initial impact, metal backing provides flexible support to the facing material as well as absorbs the projectile energy. Perforation of ceramic resulting from the projectile impacting on the target generally consists of three stages, as explained below and is illustrated in Figures 14.3 and 14.4.[23]

    a. *Shattering*: Upon projectile impact with the armor, the projectile undergoes fracture and disrupts the surface of the ceramic plate, where the projectile shatters.

    b. *Erosion*: After the breaking of the ceramic plate surface, cracking occurs in the ceramic, while projection is still being defeated through erosion.

    c. *Catching*: Here, the ceramic–metal layered composite system reduces the velocity of the projectile by transfer of momentum, which results in prevention of the penetration of projectile.

The kinetic energy of the projectile during the ceramic fragmentation is roughly distributed as ∼0.2% in fracturing the ceramic, 20%–40% in deforming the metallic backing plate, 10%–15% in projectile deformation, and some parts are carried by the ceramic debris. It can be noted that significant energy is utilized in deformation of the backing plate.

The usage of multilayered plates over monolithic plates for armor applications has been well reported. For example, impact experiments by Corran et al.[11] showed that the total thickness exceeding a critical value is required for achieving superior ballistic resistance. The improvement in the ballistic limit against blunt nose projectile has been

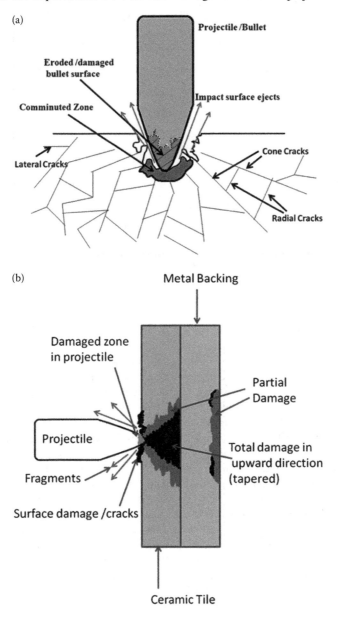

FIGURE 14.3 (a) Typical fracture pattern after impact by a projectile (Adapted from Shockey D. A. et al., *Int. J. Impact Eng.*, 1990, 9(3), 263–275.), (b) schematic illustration of tapered crack zone formation after impact (Adapted from Gupta N., B. Basu, Hot pressing and spark plasma sintering techniques of intermetallic matrix composites. In R. Mitra, *Intermetallic Matrix Composites*, Elsevier, 2018, pp. 243–302).

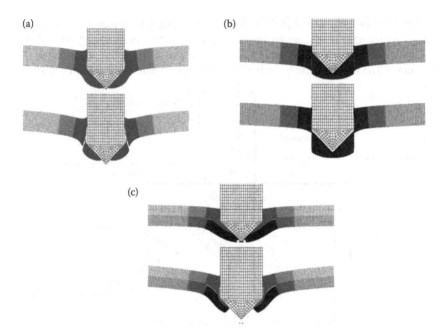

FIGURE 14.4 FEA showing the failure modes under impact by heavy conical nose projectile on (a) monolithic plate of high ductility material, (b) monolithic plate of low ductility material, and (c) double-layered shield.[9]

reported by Teng et al., wherein a double-layered shield has ~25% higher ballistic limit as compared to monolithic shields. Also, the ductility and strength of the material are shown to dictate the perforation resistance of monolithic plates.[12] Using the FEA model, it was demonstrated how the deformation and fracture take place on an armor plate upon the impact of a projectile. In case of a monolithic ductile plate, plastic deformation preceded by deep necking was the principal failure mode, followed by tensile tearing, as shown in Figure 14.4a. In contrast, the failure mode in low ductility plate was shear plugging (Figure 14.4b). Considering penetration, it is clear from the above statement that materials with high ductility certainly are advantageous. Finite-element analysis (FEA)-based optimization studies were performed by Teng et al.[12] on a double-layered design considering the ballistic resistance and maximizing the energy dissipation potential. The double-layered design consisted of a high ductility and low strength material as the top layer and the bottom layer is of material with a low ductility and high strength. The penetration tests on a numerical platform showed that the top layer underwent deep indentation, followed by necking before the failure, while shear plugging resulted in the failure of the bottom layer (Figure 14.4c). Plastic energy dissipation increases during failure mode transition from shear plugging to tensile tearing.

While probing the role of the backing plate, Rosenberg et al.[13] reported lower ballistic performance when using a thick backing configuration, due to the loss of shear strength of the material at high shock pressures. By using ceramic materials with strength greater than that of the projectile strength, low-aspect ratio projectiles can be defeated. However, in case of projectiles with a high-aspect ratio, severe deformation and damage happen leading to penetration.[13] The above discussion signifies that the ballistic performance of an armor system is strongly affected by the geometry and material of the projectile and shield, layering sequence of the material (ceramic/metal), confinement, and so on.

14.1.3 Ballistic Performance

The ballistic performance can be quantitatively analyzed by the depth of penetration (DOP) and ballistic efficiency, using impact experiments. These test procedures are discussed in detail in later sections. When a projectile hits a target, an ideal armor material should be able to completely absorb the projectile energy and shatter the projectile in such a manner that the projectile motion is completely stopped. DOP is a measure to quantify as to how deep a projectile can penetrate inside a target. The experimental test set up is quite different than the conventional set up to measure mechanical properties. A representative set-up is shown in Figure 14.2. Ceramic is located at the front portion of the target and a metal backing is provided at the rear. In a standard test protocol, the residual penetration depth into the reference backing material is compared. X-ray radiographs are used to obtain quantitative information. Ballistic efficiency (η) is estimated as,

$$\eta = \frac{(P_0 - P_r) * D_0}{(t_c * D_c)} \tag{14.1}$$

where,
P_0 = Reference penetration without the ceramic layer
P_r = Residual penetration of the projectile
D_0 = Density of the backing plate
D_c = Density of ceramic
t_c = Thickness of the ceramic layer

Figure 14.3a shows the fracture pattern in ceramic when applied with impact load.[14] The intersection of a variety of cracks, namely, radial cracks, hoop cracks, and cracks in the plane normal to the projectile axis can be noted. Here, the comminuted region is often followed by concentric ring cracks, which run outwards at an angle from the impact site. These cracks propagate in a direction normal to the principal tensile stress. The interaction of these cracks with cone and radial cracks results in fragment formation.

14.2 Overview of Functionally Graded Armor Materials

Functionally graded materials (FGMs) have emerged as a new class of materials, which has a gradient in both microstructure and properties. This aspect has engaged engineers and scientists in the area of materials science for the past three decades, to investigate about FGMs and their use in a diverse engineering application. A typical FGM involves compositional variation from one face to other of the material, as shown in Figure 14.5. Such variations can be discrete or continuous and the material distribution can be designed to various spatial specifications without any discrete

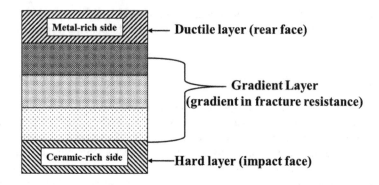

FIGURE 14.5   Schematic of a conceptual design of a functionally graded armor material.

interface boundaries. While designing FGMs, it is vital to consider the variation in coefficient of thermal expansions of the constituents, which can result in cracking or delamination. Even processing stages involved in the fabrication of FGMs (such as sintering) can impart residual stresses due to rapid cooling in sintering cycle, leading to cracking/delamination. Also, the functional gradation results in higher deformation stresses due to asymmetry of the structure, when compared to its homogeneous symmetric counterpart with overall identical material composition. Thus, the design of FGMs must be carried out by carefully considering the selection of the material constituents and the processing cycle.

In the context of processing FGMs, with precise control over the graded material composition and a wide range of material combination, powder metallurgy-based processing routes are one of the reliable candidate choices. Also, the powder metallurgy process, like sintering can restrict grain growth during high-temperature processing, which can aid in the achievement of better mechanical properties. As discussed later, spark plasma sintering (SPS) is an appropriate consolidation route for laminated and functionally graded composites, since SPS makes possible sintering at lower temperatures and in shorter duration. By carefully placing the powder mixtures as layers in the die, the FGM structure can be fabricated via SPS. By shaping the dies, near net shaping can be achieved in SPS for fabricating shapes such as disk, cylinder ring, and so on.

14.3   Summary of Published Results on TiB$_2$-Based FGM

The TiB$_2$-Ti bilayered composites (BLC) with Ti as metallic reinforcement were consolidated using SPS technique. Densification of TiB$_2$-(x wt.% Ti), (x = 10 and 20) homogeneous composites and BLC was achieved by adopting to a two-stage sintering (TSS) scheme in SPS. In TSS, the first stage consisted of holding at temperature of 1273 K for 2 min at pressure of 20 MPa, while in the subsequent second stage, the compact is held at 1823 K for 5 min at 30 MPa. Processing of BLC involves obtaining a green body by carefully stacking powder layer of TiB$_2$-(20 wt.% Ti) over TiB$_2$-(10 wt.% Ti) in a graphite die, before placing in the SPS chamber. Monolithic TiB$_2$ and Ti are consolidated through single stage SPS (SSS). While TiB$_2$ is sintered at

2173 K for 10 min under 60 MPa, Ti is sintered at 1223 K for 2 min under 30 MPa under an argon environment. In order to minimize the residual stress and cracking, the uniaxial pressure is increased gradually from 5 to 20 MPa, while at temperature of 873–1273 K, respectively, and finally to 30 MPa, at 1823 K. Post-final holding stage of sintering, the sample is cooled naturally under a vacuum environment with the power turned off and the sintering pressure maintained.

14.3.1 Microstructure and Mechanical Properties

Transmission electron microscopy (TEM) is used to characterize the finer scale microstructure of the sintered ceramic composites. Figure 14.6 shows the representative TEM images under bright-field conditions and the selected area diffraction patterns (SADP). One of the major processing challenges is to retain the metallic phase in a reactive ceramic system. In the present context, the retention of Ti in spark plasma sintered TiB_2-Ti-based BLC was therefore a challenge. The micrographs show the presence of a faceted grain morphology with the size in the range of 100–600 nm as well as round edged grains with size more than 1 μm. Upon analyzing the SADP from the triple pocket phase, the presence of α-Ti is confirmed (inset in Figure 14.6). Although not shown extensively, TEM analysis also indicates the presence of defects such as dislocations and twins (interlamellar spacing/twin width < ~100 nm) in the grain boundary phase, in addition to few thickness fringes at the triple pocket.

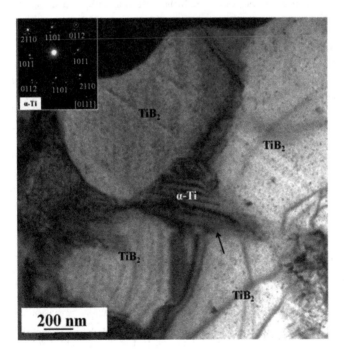

FIGURE 14.6 Bright-field TEM micrograph of TiB_2-Ti composites revealing the presence of α-Ti at the triple point; and dispersion of Ti around the TiB_2 phase, with the inset showing SADP taken from α-Ti. (Reproduced with permission from, Ref. 1).

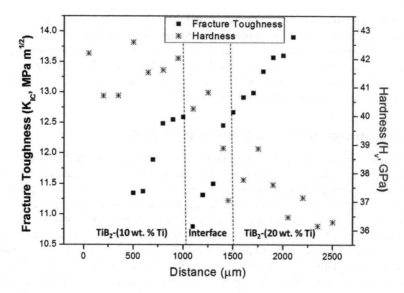

FIGURE 14.7 Experimentally measured variation in hardness and indentation toughness across the thickness of TiB$_2$-Ti-based bilayered ceramic composites. (Reproduced with permission from, Ref. 1).

The mechanical properties of all the SPSed ceramic composites, namely, hardness and indentation toughness were measured across the cross-section at multiple locations. Figure 14.7 summarizes the through thickness distribution of hardness and toughness of BLC. Hardness lies in the range of 39–45, 36–42, and 35–40 GPa for the TiB$_2$-(10 wt.% Ti) layer, interlayer, and TiB$_2$-(20 wt.% Ti), respectively. The measurement of fracture toughness (K_{IC}) is performed by using the instrumented indentation technique, wherein the crack length and the indent diagonals measured by SEM are used, with the elastic modulus determined using nanoindentation experiments. From Figure 14.7 it can be noted that the toughness consistently decreases across the cross-section of BLC, with the variations being 12.5–14 and 11.5–12.5 MPa m$^{1/2}$, respectively, for TiB$_2$-(20 wt.% Ti) and TiB$_2$-(10 wt.% Ti). As compared to monolithic TiB$_2$, with a toughness of 5 MPa m$^{1/2}$, it is clear that the TiB$_2$–Ti system substantially enhances the toughness. In view of the gradual and continuous, variation in properties across the cross-section of the BLC, it can be stated that SPS allows one to develop BLC in the TiB$_2$–Ti system.

14.3.2 Dynamic Compression Properties

For armor applications, one of the performance-limiting properties is the dynamic compression strength, which can be measured at high strain rates. For other engineering applications, the strength property is determined using quasi-static tension/compression tests. Conventionally, split Hopkinson pressure bar (SHPB) is used to evaluate the dynamic compression response of materials. In the experimental set-up to obtain results as shown in Figure 14.8, the cylindrical ceramic sample is sandwiched between the high strength maraging steel-based incident bar and transmitter bar with the diameter being 12.7 mm and length being 1525 and 1350 mm, respectively. To avoid

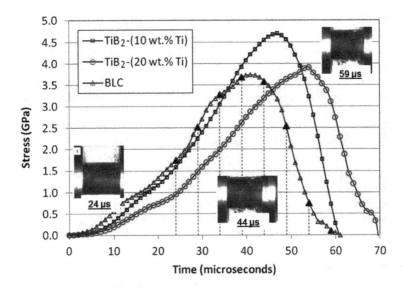

FIGURE 14.8 Dynamic compression response of TiB$_2$-(10 wt.% Ti), TiB$_2$-(20 wt.% Ti) and their bilayered composite, as measured using the SHPB test with the insets showing high-speed images recorded of BLC under impact loading at different timepoints. (Adapted and Reproduced with permission from, Ref. 1).

damage at the bar ends, high strength WC platens having matching impedance are used. To impact the incident bar, a maraging steel-based striker bar, with 12.7 mm diameter and 200 mm length is used. This allows one to generate a compressive stress wave in the incident bar. This compressive wave traverses the incident bar and gets reflected at the specimen interface and a part of it is transmitted to the transmitter bar via the specimen. The occurrence of incidence, reflection, and transmission events are measured using a strain gauge pair located at the mid-length of incident and transmitter bars at diametrically opposite sides of the bars, with the signal sampling rate being 1 MHz. To achieve force equilibrium before failure and for reliable measurements, the slope of the compressive stress wave is controlled using a copper pulse shaper. While obtaining strain rate ($\dot{\varepsilon}_S$), engineering strain (ε_s), and engineering stress (σ_S) in the specimen, incident (ε_I), reflected (ε_R), and transmitted (ε_T) strain signals are recorded.
$\dot{\varepsilon}_S$, ε_S, and σ_S are calculated from the measured signals as,

$$\dot{\varepsilon}_S(t) = -2c_b \frac{\varepsilon_R(t)}{l_s}$$

$$\varepsilon_S(t) = -\frac{2c_b}{l_s} \int_0^t \varepsilon_R(t)dt \qquad (14.2)$$

$$\sigma_S(t) = \frac{E_b A_b}{A_S} \varepsilon_T(t)$$

where, c_b is the wave speed in the bars, E_b is the elastic modulus of the bar material, A_b and A_S are, respectively, the area of cross-section of the bar and the specimen, and l_s is the specimen length.

The representative dynamic compression response of TiB_2-(10 wt.% Ti) and TiB_2-(20 wt.% Ti) composites as well as BLC are shown in Figure 14.8. The average strength of TiB_2-(10 wt.% Ti), TiB_2-(20 wt.% Ti) and BLC are 4.6, 4, and 3.7 GPa, respectively, under dynamic loading at a nominal strain rate of 600 s^{-1}. Thus, the variations in strength among the ceramic composites are minimal, with the maximum difference being 7.5% between BLC and TiB_2-(20 wt.% Ti). To understand the failure mechanisms, the high-speed real-time photographs of the BLC impact, corresponding to different levels of stress, are illustrated in Figure 14.8. In all the images, the left side of the specimen, which is in contact with the incident bar is the TiB_2-(10 wt.% Ti) layer. At 24 μs in the first image of Figure 14.8, no visible damage is seen. Further, at 29 μs, a horizontal crack extending along the specimen length was noted. At this time instant, the stress is ~2.7 GPa, nearly 73% of the BLC dynamic strength. As the crack opened more at the mid-length as compared to the ends, it is possible that the crack initiation could have occurred at the interface of TiB_2-(10 wt.% Ti) and TiB_2-(20 wt.% Ti) layers and subsequently progressed into the other layers.

14.4 Correlation between Theory and Experimental Measurements of Dynamic Strength

One of the important aspects is to establish a theoretical framework to rationalize the experimental results. In this section, it will be demonstrated how simple mechanics-based theory can be adapted to explain dynamic compression results. As the developed BLC shows dynamic strength lower as compared to its homogeneous compositions, the following analysis explains the reasons for such results. In the BLC, the two layers have varying elastic modulus and Poisson's ratio. From nano-indentation tests, the elastic moduli of TiB_2-(10 wt.% Ti) and TiB_2-(20 wt.% Ti) are found to be 288.7 ± 11.4 and 244.1 ± 24.4 GPa, respectively. The interlayer shows an elastic modulus of 280.1 ± 29.6 GPa. The Poisson ratio is calculated using the simple rule of mixtures, from the following expression:

$$\nu_c = \nu_1 V_1 + \nu_2 V_2 \tag{14.3}$$

where V is the volume fraction of the phases present in the composites and the subscripts 1 and 2 are for two individual phases. The Poisson ratio for TiB_2-(10 wt.% Ti) and TiB_2-(20 wt.% Ti) are 0.133 and 0.156, respectively. Under compressive loading, TiB_2-(10 wt.% Ti) will constrain the radial expansion of TiB_2-(20 wt.% Ti), owing to relatively high stiffness of TiB_2-(10 wt.% Ti). This results in radial and circumferential tensile stresses in the TiB_2-(10 wt.% Ti) layer. By approximating BLC as a two-layered material (as the interlayer has elastic modulus close to TiB_2-10 wt.% Ti), a simplified analysis can be performed to obtain an estimate of radial and circumferential stresses, as explained in Figure 14.9. Though same axial stress, σ_{zz} is experienced by individual layers, the mismatch in elastic properties leads to a difference in radius change ($-\Delta r_1$ to Δr_2). By equating (Δr_1 to Δr_2), radial (σ_{rr}) and circumferential ($\sigma_{\theta\theta}$) stresses in each layer are given as,[15]

$$\sigma_{rr} = \sigma_{\theta\theta} = \frac{E_2\nu_1 - E_1\nu_2}{E_2(1-\nu_1) + E_1(1-\nu_2)}\sigma_{zz} = \alpha\sigma_{zz} \tag{14.4}$$

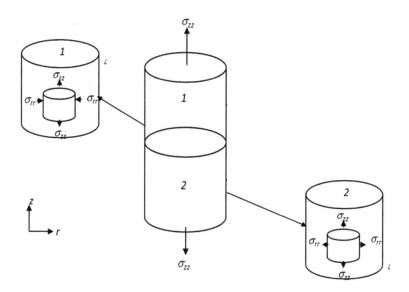

FIGURE 14.9 Schematic illustration of general stress state in bilayered ceramic composites, under applied axial stress (σ_{zz}). (Reproduced and Adapted, with permission from, Ref. 1).

In Equation 14.4, E and ν denote the elastic modulus and Poisson's ratio, respectively, and subscripts 1 and 2 refer to TiB$_2$-(10 wt.% Ti) and TiB$_2$-(20 wt.% Ti), respectively. Using the elastic constants for each layer in Equation 14.4, it is found that σ_{rr} and $\sigma_{\theta\theta}$ are 2.8% of the applied axial stress (σ_{zz}). In layers 1 and 2, these stresses are tensile and compressive, respectively. Thus, the tensile stress in a relatively stiff layer of the BLC results in overall load-carrying capacity lower than that of its homogeneous compositions. Using the Mohr–Coulomb theory, the failure strength of brittle composites with high compressive strength is defined as,[16,17]

$$m\sigma_1 - \sigma_3 = \sigma_c \qquad (14.5)$$

where, m is the ratio of compressive strength to tensile strength and σ_1 and σ_3 are the largest and smallest principal stresses. For brittle ceramics, it is assumed that $m = 8$.[18] For TiB$_2$-(10 wt.% Ti), using Equation 14.5, it can be deduced that $\sigma_1 = 0.028\sigma$ and $\sigma_3 = -\sigma$, where σ is the magnitude of the applied axial stress. Using these values for σ_1 and σ_3, and denoting dynamic compressive strength of the TiB$_2$-(10 wt.% Ti) layer as σ_{s1}, Equation 14.5 can be rewritten as,

$$\sigma = \frac{\sigma_{s1}}{(1 + m\alpha)} \qquad (14.6)$$

Using Equation 14.6, the dynamic strength of BLC is estimated to be 3.76 GPa, which agrees well with the measured value of 3.7 GPa. Thus, it can be conclusively stated that strength of BLC can be predicted reasonably well with a simplistic continuum model.

TABLE 14.3

Physical and Mechanical Properties of Some Armor Materials[22]

Ceramic	Density (g/cm³)	Elastic Modulus (GPa)	Toughness (MPa m^{1/2})	Compressive Strength (MPa)	Hardness (GPa)
AD85	3.43	224	3.2	2175	8.8
AD90	3.58	268	3.3	2345	10.6
AD96	3.74	310	3.7	2660	12.3
AD995	3.90	383	4.7	2785	15.0
TiB$_2$ (Ceradye)	4.52	414	4.1	5700	27.0
TiB$_2$ (Cercom)	4.52	538	5.2	6000	26.1
ZrO$_2$ (Nilcra, MS)	5.72	205	12.0	1900	11.2
Soda lime glass	2.50	69	0.73	966	5.5

Abbreviation: AD, alumina.

The primary motivation for using a BLC design is to mitigate stresses due to property mismatch and associated cracking of the material. The above calculation unravels that the induced tensile stresses due to mismatch, particularly in the 10 wt.% Ti layer could result in reducing the strength of BLC to levels lower than the strength of the 20 wt.% Ti layer. This highlights that the material architecture (layer thickness and its properties) can have a significant effect on the BLC performance. By substituting the strength of the second layer (σ_{s2}) for σ in Equation 14.5 and rearranging other terms, we can obtain the following relationship,

$$\alpha = \frac{1}{m}\left(\frac{\sigma_{s1}}{\sigma_{s2}} - 1\right)$$

(14.7)

For this value of α, the strength of BLC can be expected to be same as the strength of the weaker layer. The values of α greater than that given by Equation 14.6 will result in a reduction in the strength compared to strength of the weaker layer.

At the closure, it can be summarized that SPS with a two-stage scheme allows the fabrication of TiB$_2$-Ti-based homogeneous as well as bi-layered composites with varying Ti contents (10–20 wt.%). The addition of Ti as sinter aid results in tougher and stronger TiB$_2$ without affecting its hardness. Comparing these mechanical properties together with the earlier reported ballistic properties,[20] TiB$_2$-Ti are potential candidate materials for armor applications (see Table 14.3).

Based on the above discussion, it emerges that if TiB$_2$-(10 wt.% Ti) is used as the front face of an armor, one can achieve a better damage tolerance. This would allow the cracks to be arrested from being propagated into the material, due to crack deflection and ductile metal bridging. Moreover, at faster loading rate, flaws may not get enough time to grow subcritically causing the material to transmit a higher load before catastrophic failure.[19–21]

14.5 Closure

In this chapter, various design considerations for armor materials are presented. While such conceptually designed armors can potentially offer significant performance advantages over their conventional monolithic counterpart, the major challenge is to fabricate such materials. In particular, the residual stresses that would be developed at various layers in case of multilayered functionally gradient armors can be a major issue in obtaining the uncracked dense armors. The property requirements of the ceramic armor as well as the dynamic mechanics involved are also discussed in this chapter. The results from the author's group are presented to demonstrate how to develop with enhanced indentation toughness in the range of 11.5–13.3 MPa m$^{1/2}$ along with nanoscale hardness in the range of 33.9–39.0 GPa. The bi-layered composites had a characteristic interlayer with gradation in hardness and toughness.

Also, this chapter presents the measurements of some unusual properties, that is, dynamic strength under high strain rate, which are not determined for conventional materials. The earlier published results suggested that an average dynamic strength of 4 GPa under compression could be achieved in TiB$_2$/Ti BLC, which is comparable to many armor ceramics. The correlation between experimental results and theoretical prediction is always regarded as an important aspect in new materials development. To illustrate this, the predictions based on Mohr–Coulomb theory are shown to rationalize the dynamic strength properties.

REFERENCES

1. Gupta N., V. Parameswaran, B. Basu, Microstructure development, nanomechanical and dynamic compression properties of spark plasma sintered TiB2-Ti based homogenous and Bi-layered composite, *Metall. Mater. Trans. A.*, 2015, 45(10), 4646–4664.
2. Gupta N., V. V. Bhanu Prasad, V. Madhu, B. Basu, Ballistic studies on TiB2-Ti functionally graded armor ceramics, *Def. Sci. J.*, 2012, 62(6), 382–389.
3. Gupta N., B. Basu, Hot pressing and spark plasma sintering techniques of intermetallic matrix composites, Intermetallic Matrix Composites. In R. Mitra, *Intermetallic Matrix Composites*, Elsevier, 2018, pp. 243–302.
4. Tarry C. A., United States Patent [19], Patent no. 5443917, 22 Aug 1995.
5. Holmquist T. J., A. M. Rajendran, D. W. Templeton, K. D. Bishnoi, A ceramic armor database, TARDEC Technical Report, Jan 1999.
6. Kumar K. S., M. S. DiPietro, Ballistic penetration response of intermetallic matrix composites, *Scr. Metall. Mater.*, 1995, 32(5), 793–798.
7. Madhu V., K. Ramanjaneyulu, T. Balakrishna Bhat, N. K. Gupta, An experimental study of penetration resistance of ceramic armour subjected to projectile impact, *Int. J. Impact Eng.*, 2005, 32(1–4), 337–350.
8. Lundberg P., R. Renstrom, B. Lundberg, Impact of metallic projectiles on ceramic targets: Transition between interface defeat and penetration, *Int. J. Impact Eng.*, 2000, 24, 259–275.
9. Qiao P., M. Yang, F. Bobaru, Impact mechanics and high-energy absorbing materials: Review, *J Aerospace Eng.*, 2008, 21(4), 235–248.
10. http://www.martinfrost.ws/htmlfiles/june2008/chobham_armour.html
11. Corran R. S. J., P. J. Shadbolt, C. Ruiz, Impact Loading of plates-an experimental investigation, *Int. J. Impact Eng.*, 1983, 1(1), 3–22.
12. Teng X., T. Wierzbicki, M. Huang, Ballistic resistance of double-layered armor plates, *Int. J. Impact Eng.*, 2008, 35, 870–884.
13. Rosenberg Z., S. J. Bless, N. S. Brar, On the influence of the loss of shear strength on the ballistic performance of brittle solids, *Int. J. Impact Eng.*, 1990, 9(1), 45–49.

14. Huang F. L., L. S. Zhang, Investigation on ballistic performance of armor ceramics against long-rod penetration, *Metall. Mater. Trans. A.*, 2007, 38, 2891–2895.
15. Sadd M. H., *Elasticity: Theory, Applications and Numerics*, Elsevier Academic Press, New York, 2005.
16. Khoei A. R., *Computational Plasticity in Powder Forming Process*, Elsevier, Amsterdam, 2005, p. 449.
17. Yu M., Advances in strength theories for materials under complex stress state in the 20th Century, *Appl. Mech. Rev.*, 2002, 55(5), 169–129.
18. Basu B., M. Kalin, *Tribology of Ceramics and Composites*, John Wiley & Sons Inc., New Jersey, 2011.
19. Faber K. T., A. G. Evans, Crack deflection processes—I. Theory *Acta Metall. Mater.*, 1983, 31(4), 565–576.
20. Munro. R. G., Material properties of titanium diboride, *J. Res. Natl. Inst. Stand. Technol.*, 2000, 105(5), 709–720.
21. Li Q., Y. B. Xu, M. N. Bassim, Dynamic mechanical behavior of pure titanium, *J. Mater. Process. Technol.*, 2004, 155–156, 1889–1892.
22. Woodward R. L., W. A. Gooch Jr., R. G. O'Donnell, W. J. Perciballi, B. J. Baxter, S. D. Pattie, A study of fragmentation in the ballistic impact of ceramics, *Int. J. Impact Eng.*, 1994, 15(5), 605–618.
23. Shockey D. A., A. H. Marchand, S. R. Skaggs, G. E. Cort, M. W. Burkett, R. Parker, Failure phenomenology of confined ceramic targets and impacting rods, *Int. J. Impact Eng.*, 1990, 9(3), 263–275.

15

Functionally Graded Materials for Bone Tissue Engineering Applications

Human bone is a classic example of natural hybrid composites with functional gradation and characteristic piezoelectric properties, which have significant relevance towards governing its various metabolic activities. In order to mimic its electrical transport properties, considerable efforts are invested to develop hydroxyapatite (HA)-based composites with piezoelectric phases. However, in such electro-active composites, the surface chemistry of excellent biocomposite HA is compromised. Against this backdrop, this chapter discusses the published results of functionally graded materials such as, HA–BaTiO$_3$–HA, HA–CaTiO$_3$–HA, and HA–(Na, K)$_{0.5}$NbO$_3$–HA with a particular focus on their dielectric and electrical properties.[25,26] Such interdisciplinary approach towards the development of functionally graded materials for bone tissue engineering applications has two important consequences: (i) the surface chemistry of the excellent biocompatible materials will not be affected and (ii) the polarizability and/or the electro-active response of the implant can be improved to meet such property requirements of natural living bone. Further, the experimentally measured variation in frequency-/temperature-dependent electrical transport properties is discussed in reference to activation energies of various structural defects. In addition, the values of the dielectric constant, ac conductivity, relaxation behavior, and so on have been compared with those of the natural living bone. Overall, the chapter reveals the importance of interdisciplinary approach in developing the bone-mimicking materials.

15.1 Introduction

Natural bone consists of a spectrum of combination of unique properties such as, mechanical and bioelectrical along with their functional gradation depending on the anatomical location. It is known that the living bone is an electrically active tissue which is noticed in the form of piezoelectricity, streaming potential, pyroelectricity, ferroelectricity, etc. These properties direct a number of metabolic activities in bone.[1–5] For example, the presence of piezoelectricity plays a key role in regulating the bone growth, maintaining its structure, and repairing the bone fractures.[6–9] Bone is also a classical example of a functionally graded material and such gradation assists in various biophysical/chemical/mechanical/electrical activities. For the simplistic case, long bones can be a typical example of how functional gradation occurs in bone. The outer cortical bone which is a compact dense part encases the cancellous bone which is a spongy or trabecular in nature. By analyzing the cross-section of the long bone, structural changes from the cortical to cancellous bone are gradual without the presence of abrupt interfaces which facilitates in the smooth variation of mechanical properties. Similarly, tendon to bone and cartilage to bone junctions can withstand a variety of mechanical loadings.[10] These properties together provide electromechanical couplings and therefore, the bone behaves as a transducer.[11] Mascaren suggested that bone is also a "bioeletret."[12]

Due to the electro-active nature of the bone, polarized implant or external electrical stimulation improves the bone growth and heals fractures.[13–17] In this regard, development of materials having similar electrochemical/electromechanical responses to that of bone are promising substitutes as compared to conventional implants for orthopedic applications. In general, the implants consisting of heterogeneous structures such as composites, are often associated with the critical issues like presence of abrupt interfaces, which sometimes leads to interfacial failure after a certain period. The functionally graded materials (FGMs) are a new class of materials, specifically designed to overcome these drawbacks.

> FGMs are generally composite materials in which properties such as grain size, texturization level, density, microstructure, and composition gradually change in more than one spatial direction, according to the application.

FGMs eliminate the macroscopic boundaries within materials, which results in a continuous change in the mechanical, physical, and biochemical properties.[18]

In general, FGMs possess two classic features which make them an excellent substitute to orthopedic implants. First, the gradual gradation in the composition and microstructure across the whole volume can provide the bone mimicking response and simultaneously minimizes the consequences of stress shielding effect.[19] Second, they possess superior mechanical properties compared to monolithic and composite structures. Previously, FGMs are developed having at least one ingredient as a metallic phase. These FGMs of metal and metal–polymer-based compositions generate debris/microscopic particulates due to friction between articulating surfaces of implants, degradation of the implant, bone resorption, and so on.[20,21] Due to these disadvantages, functionally graded ceramics (FGCs) have gained much prominence because of their endurance to be operative under severe conditions of temperature, corrosive environment, abrasion, large mechanical loadings, and thermal-induced stresses. In addition to these properties, ceramics can potentially mimic the electro-active response of natural bone. Overall, the development of electro-active FGMs can be anticipated to provide better functional response as far as the integration of implant with the host bone tissue and its survival period is concerned.

Figure 15.1a represents the electro-micro environment of living bone where the piezoelectric collagen fiber component in combination with the non-piezoelectric mineralized component provides the functional gradation to the bone. To mimic such a microenvironment of spatial distribution of interdispersed piezoelectric/non-piezoelectric microzones (Figure 15.1b), Yu et al.[22] suggested the calibrated laser treatment of piezoelectric $Na_{0.5}K_{0.5}NbO_3$ (NKN). The laser treatment induces non-piezoelectricity in NKN in that particular microsized zone which has been demonstrated to potentially mimic the microscopic level functional gradation of bone. It has been demonstrated that such a distribution of microscopic piezoelectric and

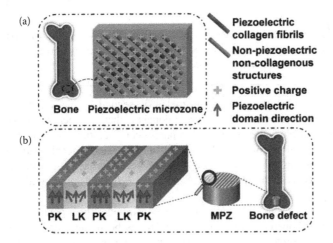

FIGURE 15.1   Schematic illustrating the (a) spatial distribution of microscopic zones of piezoelectric collagen and non-piezoelectric mineral phase in bone and (b) bone mimicking spatial distribution of piezoelectric $Na_{0.5}K_{0.5}NbO_3$. MPZ refers to microscale piezoelectric zones, PK and LK represent the piezoelectric and non-piezoelectric microzones, respectively.[22]

non-piezoelectric zones induces the osteogenic differentiation as bone regeneration as compared to mere piezoelectric or mere non-piezoelectric phases.[22]

Synthetic hydroxyapatite (HA) is among the most appealing materials for orthopedic applications due to its excellent biological response and structural similarity with the mineral component of natural bone.[23,24] However, the poor mechanical response refrains its widespread applications. In addition, as far as the resemblance of electrical properties of living bone with that of HA are concerned, a composite of HA with some electro-active ceramics can mimic the electrical response of the bone. However, with the composite approach, the surface chemistry of HA and consequently, its excellent biocompatibility is sacrificed. To address such a critical issue, the development of FGMs by inserting some electro-active ceramic layers as an intermediary layer between two HA layers is an interesting approach. Such an approach of FGM development increases the electrical activity of HA without any effect on its biological response.

Toward this end, the potentiality of perovskites such as $Na_{0.5}K_{0.5}NbO_3$ (NKN), $BaTiO_3$, and $CaTiO_3$ has been revealed in this chapter.[25,26] For bone tissue engineering applications, the niobate ceramics have been demonstrated to be an appealing alternative.[27,28] NKN possesses quite high value of electromechanical coupling coefficient (0.44), piezoelectric strain coefficient ($d_{33} = 161$ pC/N), and dielectric constant (657).[29] In addition, the ferroelectric NKN ($Na_xK_yNbO_3$; $0 \leq x \leq 0.8$ and $0.2 \leq y \leq 1$) has been suggested as a prospective piezoelectric biocompatible material.[30] The principle ingredient in the NKN composite is niobium which has been depicted to exhibit superior cellular activity for human osteoblast cells when it is coated on stainless-steel substrates as compared to uncoated steel substrates, as indicated by Navarrete et al.[31]

The biocompatibility of $BaTiO_3$ (BT) has been represented by a number of *in-vitro/in-vivo* studies.[14,32–34] The incorporation of $BaTiO_3$ as a secondary phase in a ceramic matrix significantly increases the fracture toughness.[35]

> The presence of the piezoelectric secondary phase in the path of propagating crack dissipates the crack energy by the process of alignment of domains, which consequently toughens the material.

Therefore, the incorporation of piezoelectric secondary phases in any ceramic matrix can be suggested to provide the additional toughening mechanism. A composite of HA with some piezoelectric secondary phases can potentially provide the improved electro-mechanical response.[36,37] It has been demonstrated that cell viability and cell proliferation of human osteoblasts cells are enhanced on the composite of polymers such as poly (vinylidene-trifluoroethylene) with $BaTiO_3$ as compared to the basic polymer matrix.[38]

Perovskite $CaTiO_3$ (CT) has also been demonstrated to be a potential prospective material for orthopedic application.[39] $CaTiO_3$ has been suggested to significantly improve the osseointegration.[40,41] In addition, the electrical conductivity of HA has been reported to increase with the incorporation of $CaTiO_3$ as the secondary phase in the HA matrix.[42] Overall, the perovskites have potentiality to improve the electrical and mechanical response of HA. The development of FGMs using these perovskites as an intermediary layer between HA layers can improve the functional performance of HA without any influence on its surface chemistry.

In view of the fact that polarization plays a very important role in governing various bone metabolic activities, this chapter focuses on the various approaches to improve the polarizability of HA without any change in the surface chemistry. These approaches can be substantiated by the development of functionally graded materials (FGM) of piezoelectric $Na_{0.5}K_{0.5}NbO_3$ (NKN), $BaTiO_3$ (BT), and non-piezoelectric material $CaTiO_3$ (CT) phases.[25,26] The processing of such FGMs offers major challenges due to the difference in the coefficient of thermal expansions between HA, NKN, BT, and CT phases. To compensate such differences in thermal expansion, a buffer layer has been suggested to be introduced between HA and perovskite layers. However, the composition of the buffer layer needs to be optimized. For example, for the FGM containing NKN as the intermediary layer, the buffer layer composition has been optimized as HA:NKN::1:7.[25] In addition, the processing route also plays an important role in developing the integrated layers in FGMs. In this chapter, the efficacy of spark plasma sintering route to develop such FGMs has been discussed.

15.2 Microstructure of HA-Based FGMs

In view of the perceived importance of microstructure on the electrical transport properties, it is important to describe first the key microstructural features. Figure 15.2 demonstrates the scanning electron microscopy (SEM) images around interfacial regions of synthesized functionally graded HA-NKN-HA (Figure 15.2a and b), HA-BT-HA (Figure 15.2c and d), and HA-CT-HA (Figure 15.2e and f) FGMs. The entire layered arrangement is completely integrated. No sign of any kind of delamination or crack between the layers is observed. Rather, few of the interfacial regions contain tiny fraction of residual porosity. Here, the interfacial regions are examined for the fractured samples. The absence of any kind of interfacial crack or

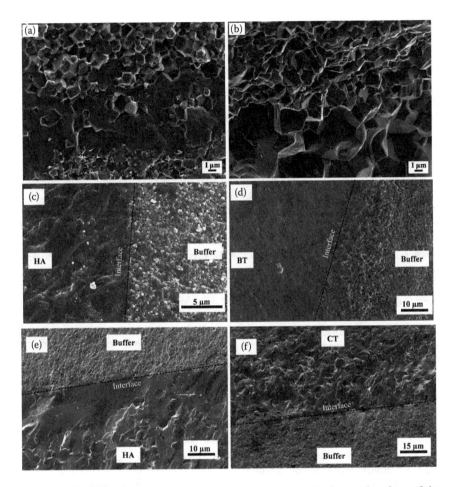

FIGURE 15.2 Scanning electron microscopy (SEM) images of the fractured surfaces of the interfacial regions in functionally graded materials. (a and b) SEM image depicting the interfacial region between HA and buffer (HA–NKN) layers and buffer and NKN layers of the functionally graded HA–NKN–HA sample, (c and d) interfacial region of functionally graded HA–BT–HA and (e and f) interfacial region of HA–CT–HA. (Reproduced with permission from publisher Dubey A. K. et al. *RSC Adv.*, 2014, 4(47), 24601–24611; Saxena A. et al. *Ceramics Int.*, 2019, 45(6), 6673–6683.)

any sign of delamination, even at microscopic level, suggests that the insertion of optimal buffer layers can significantly diminish the thermal stresses between HA and NKN/BT/CT layers.

15.3 Dielectric Response of HA and FGMs

The electrical properties of HA has been of interest because of the two crucial factors. First, its structural and compositional similarity with that of the bone, as mentioned above and second, it is the electro-active nature of bone tissue. The electrical properties of HA can most convincingly be described on the basis of its structure. Generally, HA occurs in two crystalline phases, monoclinic ($P2_1/b$) and hexagonal ($P6_3/m$).[43,44] The calcium ions in HA occupy two different positions in the lattice, named as Ca(I) and Ca(II). Ca(I) occupies a column along the c-axis which are coordinated by the oxygen of the PO_4^{3-} tetrahedra.[45,46] These are also called the interstitial sites. Ca(II) forms an equilateral triangle perpendicular to the c-axis. The dipole, formed by the OH⁻ ions, lies along the c-axis at the center of the equilateral triangle.[46] The difference between the monoclinic and hexagonal phases of HA is the orientation/ordering of the hydroxyl group along the c-axis.[44] In monoclinic HA, all the OH⁻ ions are oriented in the same direction (•O–H•••O–H•••O–H•) in a particular column and their orientation is opposite in the adjacent column.[44,46] This structure is said to be the most ordered, highly stoichiometric with no OH⁻ defects, and thermodynamically the most stable crystal structure of HA.[47,48] While, in hexagonal HA, the OH⁻ ions are oriented in two-fold disorder (•O–H•••H–O•), that is, the adjacent OH⁻ ions are oppositely oriented along the c-axis.[43,46] Hexagonal HA is suggested to be the most stable structure at room temperature.[46,49] This is due to the fact that the OH⁻ ions in the lattice become highly unstable above 200°C.[50] In addition, in the hexagonal HA, the adjacent OH⁻ ions would have steric interference.[44] These processes together produce a small amount of defects at the OH⁻ sites as well as disordered orientation of the OH⁻ ions in the lattice. Therefore, in the process, hexagonal HA becomes the most stable structure at room temperature. The reorientation of the OH⁻ ions in the lattice significantly contributes to the dielectric response of HA.

NKN is one of the alkaline niobate-based ceramics having excellent piezoelectric properties and high Curie temperature.[51] NKN possesses a perovskite structure where the A-site is shared by cations Na⁺ and K⁺, in which each ion occupies half of the lattice sites.[52] In addition, the B-site is occupied by Nb^{5+}. It needs to be mentioned that the alkali metals (Na⁺ and K⁺) are volatile at elevated temperature ($\geq 800°C$) which creates defects in the structure.[53] In the process, oxygen vacancies are also formed. Oxygen vacancies together with alkali metal defects significantly influence the conductivity of the niobate ceramics.

Now, we describe the summary of the published results on dielectric responses with an aim to unravel the influence of BT/CT/NKN phases on dielectric and electrical responses of HA.[25,26] Figure 15.3a and b represents the variation of dielectric constant and loss with respect to temperature at few selected frequencies for HA and functionally graded HA–NKN–HA. The dielectric response of pure HA is suggested to be similar to as reported by Orlovskii et al.[54] It can be observed that as the temperature increases, the dielectric constant increases. As the frequency increases from 1 to 100 kHz, the increase in the dielectric constant becomes more gradual. The dielectric loss curve indicates the peaks, corresponding to each frequency and these peaks shift toward higher temperature, as the frequency increases. These peaks are probably associated with the presence of space charge and dipolar polarizations. With an increase in frequency, these peaks shift toward the higher temperature region to satisfy relationship, $\omega\tau = 1$.[54] As mentioned above, the OH⁻ ions form dipoles in HA, which are oriented along the c-axis and perpendicular to the triangle formed by Ca^{2+} triangles.[55] The migration of protons (H⁺) is also responsible for conduction in HA.[50] The generation of thermal defects at elevated temperatures (>300°C) gives rise to dispersion in the dielectric behavior. In the lower temperature (~200°C) range, HA exhibits a structural phase transition from monoclinic ($P2_1/b$) to hexagonal ($P6_3/m$).[57,58]

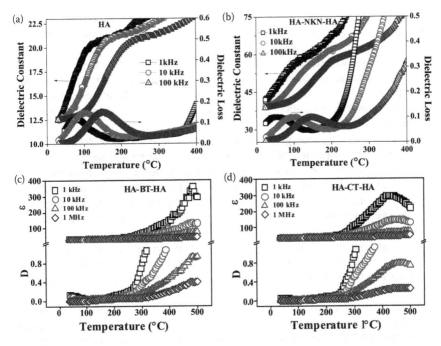

FIGURE 15.3   Variation of dielectric constant and loss with temperature of (a) HA, (b) functionally graded HA–NKN–HA sample, (c) HA–BT–HA, and (d) HA–CT–HA at few selected frequencies. (Reproduced with permission from publisher Dubey A. K. et al. *RSC Adv.*, 2014, 4(47), 24601–24611; Saxena A. et al. *Ceramics Int.*, 2019, 45(6), 6673–6683.)

The dielectric behavior of the functionally graded HA-NKN-HA sample demonstrates similar dependence of dielectric constant and loss with temperature and frequency as that of HA (Figure 15.3b). Above 200°C, the sharp rise in dielectric loss in FGMs is associated with the space charge polarization, which can be resulted due to the number of interfaces of different electrical conductivity layers. At room temperature and 10 kHz of frequency, the dielectric constants of HA, NKN, and HA–NKN–HA are measured as 12, 610, and 38, respectively.

It clearly suggests that such a concept of development of functionally graded materials using a highly piezoelectric NKN phase as an intermediary layer increases the polarizability of HA by more than three times.

As far as the variation of dielectric constant with frequency is concerned, it decreases with an increase in frequency. In addition to the space charge and dipolar polarizations, interfacial polarization plays a prominent role in the dielectric variation. The multi-layered configuration in the functionally graded system results in the additional relaxation phenomenon. The low-frequency (<1 kHz) dispersion of dielectric behavior, especially in the high-temperature region (>200°C), is the signature of presence of dc conductivity.[59,60] The dielectric and electrical response of such materials at lower frequencies ($f \ll 1/2\pi RC$) are dictated by the resistive component as the capacitive component acts as an open circuit. With an increase in frequency, the relaxation of polarization processes is initiated. In this sequence, the space charge polarization relaxes first at lower frequency ($<10^3$ Hz). It is then followed by the relaxation of dipolar polarization, which occurs at relatively higher frequencies. After the relaxation of the lower frequency polarization, the intrinsic value of dielectric constant can be obtained. In general, the spark plasma sintering of such samples, in vacuum, increases the conductivity of the samples.

Let us now draw an analogy between the natural bone and piezoelectric FGM. A similar dielectric dispersion behavior is reported for the bone tissue as well.[61] The biological tissues also demonstrate the dielectric dispersion with frequency such as α (<10 kHz), β ($\sim$MHz), and γ ($\sim$GHz).[62,63] These dispersion phenomena are associated with a number of polarization processes. The relaxation phenomenon in the dry human cortical bone is demonstrated to be associated with the dipolar polarization phenomenon.[64] The living bone is reported to be ferroelectric in nature, that is, it possesses spontaneous polarization, which is reversible with the polarity of the applied field.[5] The dielectric constant of bone is suggested to be about 10.[65] The dielectric and electrical behavior of bone is dependent on bone density, for example, specific capacitance strongly varies with bone density.[66]

Figure 15.3c and d demonstrates the temperature and frequency dependence of dielectric constant (ε) and loss (D) for HA–BT–HA and HA–CT–HA FGM compositions.[26] The insertion of BaTiO$_3$ and CaTiO$_3$ as intermediary layers between HA layers has significantly improved the dielectric response of HA-based FGMs. At 1 kHz of frequency and room temperature (35°C), the dielectric constant values or polarizability of HA–BT–HA (34) and HA–CT–HA (34) FGMs have been increased by two times to that of mere HA (17). In the lower temperature region (<200°C), the dielectric behavior is almost independent of temperature. However, both the dielectric constant and loss increases significantly for the temperature above 200°C (Figure 15.3c and d). Such an increase in dielectric constant/loss is more significant at lower frequencies (<10 kHz) as compared to higher frequencies (≥ 10 kHz). The significant increase in dielectric constant and loss values with temperature can be suggested to be associated with a number of interfacial regions, present in FGMs. The presence of such multiple interfacial regions and phases with different conductivity properties contribute to space charge polarization. As far as the frequency dependence of dielectric constant and loss is concerned, these values decrease with an increase in frequency. However, a temperature-independent behavior is observed for the frequency, above 100 kHz. Due to the variable valency of Ti (Ti^{4+}/Ti^{3+}), it loses a small amount of oxygen during high-temperature processing as,[25,26]

$$O_o^* \leftrightarrow V_o + \frac{1}{2}O_2 \uparrow \tag{15.1}$$

$$V_o \leftrightarrow V_o^{\cdot\cdot} + 2e' \tag{15.2}$$

$$Ti^{4+} + e' \leftrightarrow Ti^{3+} \tag{15.3}$$

The creation of such defects (oxygen vacancy, V_o and free electrons, e') increases the ionic and electronic conduction at elevated temperature.

15.4 AC Conductivity Behavior

In a number of studies from the author's group, the role of electrical conductivity of biomaterials on cell/tissue response is emphasized. Figure 15.4a, c, and d represents the variation of AC conductivity as a function of temperature and frequency (1, 10,100 kHz, and 1 MHz) for functionally graded HA–NKN–HA, HA–BT–HA, and HA–CT–HA, respectively. The temperature-dependent ac conductivity of functionally graded HA–NKN–HA increases with an increase in temperature in an exponential manner. After a maxima, the conduction is observed to vary linearly with temperature. The

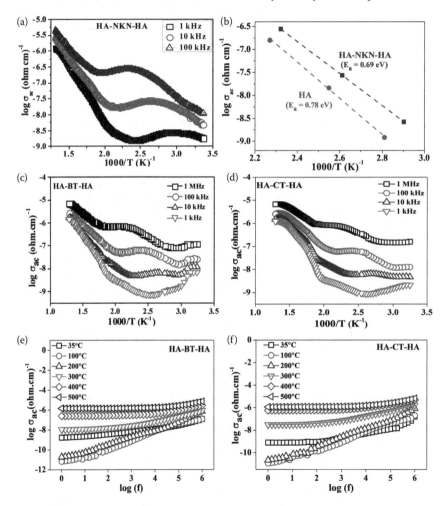

FIGURE 15.4 Temperature-dependent AC conductivity of (a) developed functionally graded HA–NKN–HA, (c) HA–BT–HA, and (d) HA–CT–HA samples. (b) AC conductivity response with temperature at crossover frequencies for HA and HA–NKN–HA. Frequency-dependent AC conductivity of (e) HA–BT–HA, (f) HA–CT–HA. (Reproduced with permission from publisher Dubey A. K. et al. *RSC Adv.*, 2014, 4(47), 24601–24611; Saxena A. et al. *Ceramics Int.*, 2019, 45(6), 6673–6683.)

maximum represents the relaxation process. The peak positions are shifted to the higher temperature, when the frequency is increased from 1 to 100 kHz. The ac conductivity corresponding to the peak maximum has been plotted with that particular temperature (Figure 15.4b). Here, the conductivity is observed to vary linearly with temperature, following the Arrhenius relationship, $\sigma = \sigma_0 \exp(-E_d/kT)$, where σ_0, E_a, and k are the pre-exponential factor, activation energy, and Boltzmann constant, respectively. The activation energy evaluated from the linear fitting of HA is 0.78 eV and functionally graded HA–NKN–HA is 0.69 eV. The activation energy value for HA is suggested to be associated with the proton conduction mechanism.[67,68] It can, therefore, be concluded that the mechanism of conduction does not get altered after incorporating NKN layers between HA layers. It was interesting to notice that the variation of ac conductivity with respect to temperature for functionally graded HA–NKN–HA is similar to that of the pure HA phase.

The value of ac conductivity for the functionally graded HA–NKN–HA sample is 5.7×10^{-9} Ω cm^{-1}, measured at a frequency of 10 kHz at room temperature. The ac conductivity of the living bone has been reported to be order of 10^{-9}–10^{-10} Ω cm^{-1}.[69]

The room temperature dc resistivity (ρ_{dc}) values for HA and functionally graded HA–NKN–HA samples are measured to be 2.8×10^{14} and 2.0×10^{13} Ω cm, respectively. Such a comparison establishes the bone-like electrical transport properties in functionally graded HA–NKN–HA. The electrical transport properties can be explained from the point of view of structural changes.

Based on the described results of conductivity, the structural characteristics of HA, which facilitates the electrical properties to it, has been discussed herewith. In addition, the electrical characteristics of NKN are also discussed. As mentioned, HA occurs in both, hexagonal and monoclinic close-packed structures. In both the structures, hydroxyl ions (OH$^-$) are located at the center of planar Ca^{2+} triangles directed along the c-axis.[43] The electrical characteristics of HA are primarily decided by the presence of OH$^-$ ions in the lattice, followed by H$^+$ protons and O^{2-} ions.[70] The other ions in the lattice such as Ca^{2+} and PO$_4^3$ ions rarely contribute to its conductivity. As mentioned, the protons (H$^+$) associated with the hydroxyl (OH$^-$) group becomes highly unstable at elevated temperature (>200°C).[56] In the process, it may get dissociated from the hydroxyl (OH$^-$) group and conduct along the c-axis. In the lattice, the adjacent OH$^-$ ions along the c-axis are separated by a distance of 0.344 nm while the distance between the adjacent columns of OH$^-$ ions is 0.95 nm.[71] It is suggested through theoretical calculations that the activation energy of migration/hopping of the OH$^-$ ions through adjacent columns (c-axis) is more than 10 eV while along the c-axis columns it is about 2.0 eV.[71] This large difference in the activation energy indicates that the migration/hopping of the hydroxyl (OH$^-$) group along the c-axis is more favorable. Therefore, the hydroxyl group or the OH$^-$ ions has a pseudo one-dimensional character. Based on this, HA is also suggested as a one-dimensional (1D) proton (H$^+$) conductor.[68,72,73] The conduction of protons is mainly proceeded with the formation of vacancy at OH$^-$ ion site in the lattice. The vacancy is mainly formed via two processes. First during synthesis, structural defects of OH$^-$ ions take place due to the sintering atmosphere and various other processing means. Second, at elevated temperatures ($> 800°C$), the process of de-hydroxylation takes place which can be described as,[74]

$$Ca_{10}(PO_4)_6(OH)_2 \rightarrow Ca_{10}(PO_4)_6(OH)_{2-2x}O_x(\square)_x + H_2O(g)\uparrow \qquad (15.4)$$

where, $\square$ indicates vacancy. Such a process results in the formation of OH^- and H^+ ion vacancies on the two adjacent OH^- ion sites.[77] These vacancies lead to the proton (H^+) conduction at elevated temperature as,

$$OH^- + O^{2-}\square + \square + OH^- \rightarrow O^{2-}\square + OH^- + \square + OH^- \qquad (15.5)$$

However, such a mechanism of conduction of protons (H^+) is not supposed to be feasible due to the vacancy of the OH^- ions which consequently hinders the conduction path of the protons (H^+). In addition, the distance between the adjacent OH^- ions is large (0.344 nm) compared to that of the nearest PO_4^{3-} tetrahedra (0.307 nm). Therefore, oxygen of the PO_4^{3-} is likely to form a hydrogen bond with the proton and the conduction of protons can be proceeded as,

$$2OH^- + PO_4^{3-} \rightarrow O^{2-} + HPO_4^{2-} + OH \rightarrow O^{2-} + PO_4^{3-}\text{-}HOH + \square_{OH}^- \qquad (15.6)$$

where, $\square$ represents vacancy at OH^- lattice site and O^{2-} is oxide, dissociated from OH^- ion.[75,76] It is also suggested that the surface of HA is ionic in nature and therefore, water molecules are strongly bounded to it. Consequently, at lower temperatures ($<100°C$), proton (H^+) migration due to adsorbed water in HA is suggested to be a major contributor in its conductivity.[77] Other conduction species such as oxide (O^{2-}) ions and hydroxyl (OH^-) ions also contribute to the conduction mechanism at very high temperatures ($>800°C$).

In the case of NKN, sodium and potassium are the volatile alkali metals and therefore, on conventional sintering these metals are volatized, leaving defects in the perovskite structure of NKN. However, if sintering of NKN is carried out under a vacuum/inert environment, the volatization of alkali metals can be restricted as,

$$2(M)_M^* + \frac{1}{2}O_2 \leftrightarrow (M)_2O(\text{gas}) + 2V_M' + 2h^\bullet \qquad (15.7)$$

where M indicates alkali elements, Na, and K. The vacuum/inert atmosphere sintering also leads to the formation of oxygen vacancies, which can be described as Equations 15.1 and 15.2.

Spark plasma sintering (SPS), a vacuum sintering process has been utilized to develop all such FGMs (HA–NKN–HA, HA–BT–HA and HA–CT–HA). SPS of HA leads to the dehydroxylation and therefore, structural OH^- defects are created. The conductivity behavior of HA is due to the process of proton (H^+) conduction via dehydroxylation of the crystal OH^- ions. Apart from the presence of polar OH^- groups, the space charge contributes significantly to the increase in the conductivity with an increase in the temperature.

For the functionally graded HA–BT–HA and HA–CT–HA materials, at higher temperatures ($>300°C$), the conductivity increases linearly with the temperature. Prior to the linear region, a diffused maxima, representing a relaxation process, is observed.

The maxima shift toward higher temperature with the increase in temperature. At elevated temperatures, the linear variation of ac conductivity follows the Arrhenius equation. The high-temperature downfall in the curve can be due to the presence of defects in (HA–BT/HA–CT) buffer, $BaTiO_3$, $CaTiO_3$, and interfaces which contribute to an increase in conductivity with temperature (Figure 15.4c and d), as compared to that of pure HA.

Figure 15.4e and f represents the variation of ac conductivity as a function of temperature and frequency for functionally graded HA–BT–HA and HA–CT–HA. The conductivity is observed to follow the "universal" power law as,[78]

$$\sigma(\omega) = \sigma(0) + A\omega^n \qquad (15.8)$$

where ω, n, $\sigma(0)$, and A are the angular frequency, a constant ($0 < n < 1$), the low-frequency conductivity and strength of polarizability, respectively. The exponent (n) represents the interaction between mobile ions and the lattice. Such a type of conduction behavior is of R-C type.[79–81]

In HA, the alignment of dipoles, formed due to OH^- ions in the channels of the Ca^{2+} triangles, contribute significantly to the resistive properties in HA. However, the immobile ions such as PO_4^{3-} and Ca^{2+} contribute to the capacitive behavior.[70]

Figure 15.4e and f illustrates the frequency-dependent conductivity at various temperatures. It is observed that with an increase in frequency, conductivity also increases for both the FGMs. In the lower temperature (<100°C) and frequency (upto 10^2 Hz) regions, the ac conductivity is almost independent of frequency. However, at higher frequencies (>10^2 Hz), the ac conductivity increases linearly. In contrast, at higher temperatures (>200°C), a sharp increase in ac conductivity with frequency is observed. Such a behavior can be suggested to follow Jonscher's power law. The electrical conductivity of FGMs appear to be associated with the migration of defect charge carriers in HA, $BaTiO_3$, $CaTiO_3$ and in buffer layers. In contrast, the space charge regions in the proximity of grain boundaries and interfaces give rise to capacitive behavior.

The hopping frequency corresponds to the frequency at which a significant increase in conductivity takes place with an increase in the temperature.[82] With an increase in temperature, the hopping frequency shifts toward the higher frequency region. At higher temperatures (>200°C), the steep behavior of the curve decreases with an increase in the frequency. Such a behavior can be explained by a correlated barrier height (CBH) model.[83] This model suggests that the polarons (quasi particles), created by the electric field induced lattice distortions, are also responsible for conduction in solids. The conduction takes place via the hopping of polarons between defects. The nature of such conduction mechanisms are of translational type, where the hopping of charge carriers such as H^+ and OH^- ions (in HA) and oxygen vacancies (in $BaTiO_3/CaTiO_3/NKN$) are involved.[82,83] Further, the decrease in the steepness in the conductivity curve with an increase in temperature can be suggested to be associated with the decrease in binding energy, required to create the oxygen vacancies and OH^- defects.[84]

15.5 Impedance Spectroscopy

A more comprehensive understanding of the electrical transport properties of piezoelectric materials demands the use of impedance spectroscopy. First, let us explain the principle. The general appearance of the impedance spectrum is semicircular arcs and the shape of the arc changes with an increase in temperature. The length of the vector from the origin to any point on the semicircular arc gives the magnitude of impedance at a particular frequency. The right side of the spectrum is the low-frequency region, while the left side (origin) is of high frequency. The conventional materials, generally utilized, are polycrystalline in nature, that is, having grains and grain boundaries. Therefore, their electrical characteristics are distinct and this can be clearly depicted in the impedance spectrum. The complete semicircular arc in the impedance plot can be represented by the parallel RC circuit. Such a type of impedance response is known as the ideal Debye-type behavior. However, the ideal Debye-type behavior is highly impractical for real materials. To compensate the deviation from the ideal Debye-type behavior, the constant phase element (CPE) has been introduced in place of capacitor in the RC circuit.[85] CPE considers the inhomogeneous response in bulk and grain boundary regions.[86] In addition, CPE explains the frequency dependence of impedance in the high-frequency region, which follow the Jonscher's power law. The CPE value can be calculated by the expression $C = (R^{1-n}C_Q)^{1/n}$, where the parameters C_Q and n are capacitance and a constant, respectively ($n > 0$, for non-ideal case).[87]

> The impedance spectroscopy analyses separate the electrical contributions of grain, grain boundary, and the material–electrode interface from the overall spectrum in terms of resistance and CPE. Therefore, each microstructural constituent in the material can be represented as a parallel combination of resistance and CPE.

In general, the sample–electrode interface contribution is not considered. Under standard conditions, the series combination of two R-CPE circuits is used to represent the contribution of grain and grain boundary. Each of these combinations (R and CPE in parallel) possesses a particular time constant which is a representative of the relaxation behavior of that contribution. The multiple relaxation phenomena depict the dielectric relaxation of each microstructural constituent (grains, grain boundary). Grains which mainly consist of dipoles, small amount of free charge carriers, and defects have a high relaxation frequency and therefore, are depicted in the high-frequency region of the impedance plot. On the other hand, grain boundaries which are mainly comprised of space charge have lower relaxation frequency and therefore, are represented in the lower frequency region. The impedance plot of the polycrystalline material is a combination of multiple semicircular arcs.

> The maxima of a particular semi-circular arc represent the relaxation frequency for that particular contribution (grain/grain boundary). The expression, $\omega\tau_r = (2\pi f_r RC) = 1$, is used to obtain the relaxation frequency, where τ_r is the time constant.[88]

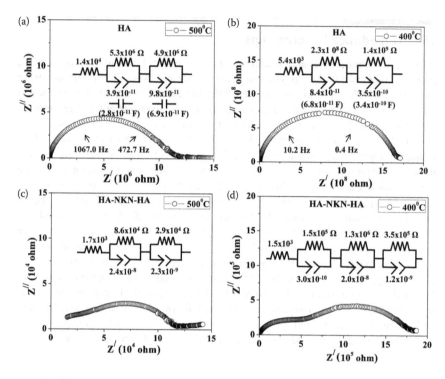

FIGURE 15.5 Complex plane impedance spectroscopy plots for HA (a and b) and functionally graded HA–NKN–HA (c and d) at 500°C and 400°C. In the inset of plots, the electrical equivalent circuit is represented as the parallel combination of resistance and constant phase element (CPE) for each contribution from the sample such as grain, grain boundaries, and sample electrode interface. (Reproduced with permission from publisher Dubey A. K. et al. *RSC Adv.*, 2014, 4(47), 24601–24611; Saxena A. et al. *Ceramics Int.*, 2019, 45(6), 6673–6683.)

Against the backdrop of the scientific principles of impedance spectroscopy, we shall now describe the representative impedance spectroscopy results of HA-based FGM structures. Figure 15.5 represents the complex plane impedance plots for HA and HA–NKN–HA FGMs at a few selected temperatures (400 and 500°C). As can be seen from Figure 15.5, the center of the semi-circular arc, corresponding to each contribution, is observed to be below the real x-axis, which represents the non-Debye type relaxation behavior.[88]

The high-frequency semicircular arc (passing through origin) represents the contribution from grain and the low-frequency arc represents the grain boundary contribution.

The fitting is usually performed using an electrical equivalent circuit with various combinations of R and CPE. The resistance and CPE values, corresponding to each contribution (grain/grain boundary), are provided in the inset of the impedance spectrum (Figure 15.5a to d). These values are obtained by best fitting of the spectra through Z-View software. The decreased value of real (Z') and imaginary (Z'') parts of the impedance with temperature is associated with the decrease in the ac conductivity.[89] In order to understand the entire spectrum of mechanism of conduction, simultaneous analyses of complex impedance (Z^*) and complex modulus (M^*) response are required. More resistive components are highlighted in the impedance plot whereas the least capacitive component is highlighted in the modulus plot.[25] The real and imaginary components of electric modulus and impedance are interrelated as, $M' = 2\pi f C_0 Z''$, $M'' = 2\pi f C_0 Z'$; where f and C_0 are the frequency and open cell capacitance, respectively.[56] The maxima of both, the impedance and modulus spectroscopy curves at different points on the frequency scale, are a representative of involvement of multiple relaxation processes.

Also, the full width at half maxima (FWHM) of 1.14 decades on frequency scale is representative of an ideal Debye type response. A FWHM value higher than 1.14 decades suggests the presence of multiple phases/compositions of different conductivities.

Hodge et al.[88] demonstrated that the broadening of the peaks or peak separation depends on the relative conductivity of the phases.

In the impedance spectroscopy plot, the peak height is proportional to the resistance of that particular contribution (grain/grain boundary). However, in modulus spectroscopy plot, the height of the peak is inversely proportional to the capacitance of that particular contribution. Also, in the spectroscopy plots, the peak shifts toward the lower/higher frequency side with temperature variation can be represented in terms of higher/lower relaxation times.

15.6 Closure

The critical analyses of the published results clearly establish that the electrical polarizability of hydroxyapatite can be enhanced significantly via interdisplinary approach to realize the bone-like electrical properties, using the piezoelectrics and biocompatible sodium potassium niobate, $BaTiO_3$, and non-piezoelectric $CaTiO_3$ as an intermediary layer in the FGM structure. The delamination of the layers or the cracking can be avoided by inserting the buffer interlayers between HA and NKN layers in case of HA–NKN–HA, HA–BT composite layers in case of HA–BT–HA, and HA–CT composite layers for HA–CT–HA. It is important to note that the challenges in processing of dense FGMs can be accomplished by adopting the optimal spark plasma sintering conditions. The interdisciplinary concept of development of functionally graded bioelectroceramics can potentially provide the bone mimicking mechanical and electrical properties combination without affecting the excellent bioactivity of hydroxyapatite. As far as the mechanism of conduction is concerned, the proton migration is the dominant conduction mechanism in HA and FGMs. However, oxygen vacancies are responsible for conduction in NKN, $BaTiO_3$ as well as in $CaTiO_3$.

At the closure, it needs to be highlighted that the FGM structure in HA-based systems represents key advances to closely mimic bone-like functional properties. It, however, remains to be comprehensively explored how these FGMs would modulate bone regeneration in pre-clinical studies involving rabbit models. This should be the focus of future research to realize the biomedical potential of HA-based FGMs.

REFERENCES

1. Fukada E., I. Yasuda, On the piezoelectric effect of bone, *J. Phys. Soc. Jap.*, 1957, 12, 1158–1162.
2. Dreyer C. J., Properties of stressed bone, *Nature*, 1961, 90, 1217.
3. Lang S. B., Pyroelectric effect in bone and tendon, *Nature*, 1966, 212, 704–705.
4. Cerquiglini S., M. Cignitti, M. Marchetti, A. Salleo, On the origin of electrical effects produced by stress in the hard tissues of living organisms, 1967, 6(24), 2651–2660.
5. El Messiery M. A., G. W. Hastings, S. Rakowski, Ferro-electricity of dry cortical bone, *J. Biomed. Eng.*, 1979, 1, 63–65.
6. Marino A. A., R. O. Becker, Piezoelectric Effect and Growth Control in Bone, *Nature*, 1970, 228, 473–474.
7. Ciuchi I. V., L. P. Curecheriu, C. E. Ciomaga, A. V. Sandu, L. Mitoseriu, Impedance spectroscopy characterization of bone tissues, *J. Adv. Res. Phys.*, 2010, 1(1), 011007.
8. Baxter F. R., C. R. Bowen, I. G. Turner, A. C. E. Dent, Electrically active bioceramics: A review of interfacial responses, *Ann. Biomed. Eng.*, 2010, 38(6), 2079.
9. Hastings G. W., F. A. Mahmud, Electrical effects in Bone, *J. Biomed. Eng.*, 1988, 10, 515.
10. Genin G. M., A. Kent, V. Birman et al., Functional grading of mineral and collagen in the attachment of tendon to bone, *Biophys. J.*, 2009, 97, 976–985.
11. Isaacson B. M., R. D. Bloebaum, Bone bioelectricity: What have we learned in the past 160 years? *J. Biomed. Mater. Res. Part A.*, 2010, 95A, 1270.
12. Mascarenhas S., The electret effect in bone and biopolymers and the water bound-water problem, *Ann. N. Y. Acad. Sci.*, 1974, 238, 36–52.
13. Teng N. C., S. Nakamura, Y. Takagi, Y. Yamashita, M. Ohgaki, K. Yamashita, A new approach to enhancement of bone formation by electrically polarized hydroxyapatite, *J. Dent. Res.*, 2001, 80, 1925.
14. Feng J. Q., H. P. Yuan, X. D. Zhang, Promotion of osteogenesis by a piezoelectric biological ceramic, *Biomater.*, 1997, 18, 1531.
15. Pickering S. A. W., B. E. Scammell, Electromagnetic fields for bone healing, *Low. Extremity Wounds*, 2002, 1(3), 152.
16. Ryaby J. T., Clinical effects of electromagnetic and electric fields on fracture healing, *Clin. Orthop.*, 1998, 355, S205.
17. Oishi M., S. T. Onesti, Electrical bone graft stimulation for spinal fusion: A review, *Neurosurg*, 2000, 47, 1041.
18. Besisa D. H. A., E. M. M. Ewais, *Advances in Functionally Graded Materials and Structures*, ed. F. Ebrahimini, InTech., London, UK, 2016, Ch. 1, pp. 1–31.
19. Sola A., D. Bellucci, V. Cannillo, Functionally graded materials for orthopedic applications—An update on design and manufacturing, *Biotech. Adv.*, 2016, 34(5), 504–531.
20. Zhou C., C. Deng, X. Chen, X. Zhao, Y. Chen, Y. Fan, X. Zhang, Mechanical and biological properties of the micro-/nano-grain functionally graded hydroxyapatite bioceramics for bone tissue engineering. *J. Mech. Behav. Biomed. Mater.*, 2015, 48, 1–11.
21. Petit C., L. Montanaro, P. Palmero, Functionally graded ceramics for biomedical application: Concept, manufacturing, and properties, *Int. J. Appl. Ceram. Tec.*, 2018, 15(4), 820–840.

22. Yu P., C. Ning, Y. Zhang, G. Tan, Z. Lin, S. Liu, X. Wang, H. Yang, K. Li, X. Yi, Y. Zhu, C. Mao, Bone-inspired spatially specific piezoelectricity induces bone regeneration, *Theranostics*, 2017 Aug 11, 7(13), 3387–3397.

23. Fathi M. H., V. Mortazavi, S. I. R. Esfahani, Bioactivity evaluation of synthetic nanocrystalline hydroxyapatite, *Dent. Res. J.*, 2008, 5(2), 81–87.

24. Sobczak A., Z. Kowalski, Z. Wzorek, Preparation of hyroxyapatite from animal bones, *ActaBioeng. Biomech.*, 2009, 11(4), 23.

25. Dubey A. K., K. Kakimoto, A. Obata, T. Kasuga, Enhanced polarization of hydroxyapatite using the design concept of functionally graded materials with sodium potassium niobate, *RSC Adv.*, 2014, 4(47), 24601–24611.

26. Saxena A., S. Gupta, B. Singh, A. K. Dubey, Improved functional response of spark plasma sintered hydroxyapatite based functionally graded materials: An impedance spectroscopy perspective, *Ceramics Int.*, 2019, 45(6), 6673–6683.

27. Wang Q., J. Yang, W. Zhang, R. Khoie, Y. Li, J. Zhu, Z. Chen, Manufacture and Cytotoxicity of a Lead-free Piezoelectric Ceramic as a Bone Substitute-Consolidation of Porous Lithium Sodium Potassium Niobate by Cold Isostatic Pressing, *Int. J. Oral Sci.*, 2009, 1(2), 99.

28. Jalalian A., A. M. Grishin, Biocompatible ferroelctric (Na,K)NbO$_3$ nanofibers, *Appl. Phys. Lett.*, 2012, 100, 012904.

29. Kakimoto K., Y. Hayakawa, I. Kagomiya, Low-temperature sintering of dense (Na,K)NbO$_3$ piezoelectric ceramics using the citrate precursor technique, *J. Am. Ceram. Soc.*, 2010, 93, 2423.

30. Nilsson K., J. Lidman, K. Ljungstrom, C. Kjellman, Biocompatible material for implants, U. S. patent 6, 2003, 526, 984 B1.

31. Olivares-Navarrete R., J. J. Olaya, C. Ramirez, S. E. Rodil, Biocompatibility of niobium coatings, *Coatings*, 2011, 1, 72.

32. Park Y. J., Y. H. Jeong, Y. R. Lee, S. R. Noh, H. J. Song, Effect of negatively polarized barium titanate thin films on Ti, on osteoblast cell activity, *J. Dent. Res.*, 2003, 82, B212.

33. Park Y. J., K. S. Hwang, J. E. Song, J. L. Ong, H. R. Rawls, Growth of calcium phosphate on poling treated ferroelectric BaTiO$_3$ ceramics, *Biomater*, 2002, 23, 3859–3864.

34. Hwang K. S., J. E. Song, H. S. Yang, Y. J. Park, J. L. Ong, H. Rawls, Effect of poling conditions on growth of calcium phosphate crystal in ferroelectric BaTiO$_3$ ceramics, *J. Mater. Sci. - Mater. Med.*, 2002, 13, 133–138.

35. Chen X. M., B. Yang, A new approach for toughening of ceramics, *Mater. Lett.*, 1997, 33, 237–240.

36. Dubey A. K., B. Basu, K. Balani, R. Guo, A. S. Bhalla, Dielectric and pyroelectric properties of HAp-BaTiO$_3$ composites, *Ferroelectr.*, 2011, 423(1), 63–76.

37. Dubey A. K., B. Basu, K. Balani, R. Guo, A. S. Bhalla, Multifunctionality of perovskites BaTiO$_3$ and CaTiO$_3$ in a composite with hydroxyapatite as orthopedic implant materials, *Int. Ferroelectr.*, 2011, 131(1), 119–126.

38. Beloti M. M., P. T. de Oliveira, R. Gimenes, M. A. Zaghete, M. J. Bertolini, A. L. Rosa, In vitro biocompatibility of a novel membrane of the composite poly (vinylidene-trifluoroethylene)/barium titanate, *J. Biomed. Mater. Res.*, 2006, 79A (2), 282–288.

39. Dubey A. K., G. Tripathi, B. Basu, Characterization of hydroxyapatite perovskite (CaTiO$_3$) composites: Phase evaluation and cellular response, *J. Biomed. Mater. Res. B Appl. Biomater.*, 2010, 95, 320–329.

40. Webster T. J., C. Ergun, R. H. Doremus, W. A. Lanford, Increased osteoblast adhesion on titanium-coated hydroxylapatite that forms CaTiO$_3$, *J Biomed Mater Res A*, 2003, 67, 975–980.

41. Mallik P. K., B. Basu, Better early osteogenesis of electroconductive hydroxyapatite–calcium titanate composites in a rabbit animal model, *J. Biomed. Mater. Res. Part A.*, 2013, 102(3), 842–851.

42. Dubey A. K., P. K. Mallik, S. Kundu, B. Basu, Dielectric and electrical conductivity properties of multi-stage spark plasma sintered HA–CaTiO$_3$ composites and comparison with conventionally sintered materials, *J. Eur. Ceram. Soc.*, 2013, 33(15), 3445–3453.

43. Kay M. I., R. A. Young, A. S. Posner, Crystal structure of hydroxyapatite, *Nature*, 1964, 204, 1050–1052.

44. Elliott J. C., P. E. Mackie, R. A. Young, Monoclinic hydroxyapatite, *Science*, 1973, 180, 1055–1057.

45. Orlovskii V. P., V. S. Komlev, S. M. Barinov, Hydroxyapatite and hydroxyapatite-based ceramics, *Inorganic Materials*, 2002, 38, 973.

46. Ma G., X. Y. Liu, Hydroxyapatite: Hexagonal or monoclinic?, *Crystal Growth Des.*, 2009, 9(7), 2991–2994.

47. Young R. A., W. E. Brown, In biological mineralization and demineralization, ed. G. H. Nancollas, in *Dahlem Konferenzen*, Nancollas, Berlin, 1982, pp. 101–141.

48. Haverty D., S. A. M. Tofail, K. T. Stanton, J. B. McMonagle, Structure and stability of hydroxyapatite: Density functional calculation and Rietveld analysis, *Phys. Rev. B*, 2005, 71, 094103.

49. Horiuchi N., M. Nakamura, A. Nagai, K. Katayama, K. Yamashita, Proton conduction related electrical dipole and space polarization in hydroxyapatite, *J. Appl. Phys.*, 2012, 112, 0749011–0749017.

50. Yamashita K., K. Kitagaki, T. Umegaki, Thermal instability and proton conductivity of ceramic hydroxyapatite at high temperatures, *J. Am. Ceram. Soc.*, 1995, 78, 1191–1197.

51. Kobayashi K., Y. Doshida, Y. Mizuno, C. A. Randall, A route forwards to narrow the performance gap between PZT and lead-free piezoelectric ceramic with low oxygen partial pressure processed (Na$_{0.5}$K$_{0.5}$)NbO$_3$, *J. Am. Ceram. Soc.*, 2012, 95(9), 2928–2933.

52. Hatano K., K. Kobayashi, T. Hagiwara, H. Shimizu, Y. Doshida, Y. Mizuno, Polarization system and phase transition on (Li,Na,K)NbO$_3$ ceramics, *Jpn. J. Appl. Phys.*, 2010, 49, 09MD11.

53. Kobayashi K., C. A. Randall, M. Ryu, Y. Doshida, Y. Mizuno, New opportunity in alkali niobate ceramics processed in low oxygen partial pressure, *Proceedings of ISAF-ECAPD-PFM 2012*, Aveiro, 2012, pp. 1–4.

54. Orlovskii V. P., N. A. Zakharov, A. A. Ivanov, Structural transition and dielectric characteristics of high-purity hydroxyapatite, *Inorg. Mater.*, 1996, 32(6), 654–656.

55. Zakharov N. A., V. P. Orlovskii, Dielectric characteristics of biocompatible Ca$_{10}$(PO$_4$)$_6$(OH)$_2$ ceramics, *Tech. Phys. Lett.*, 2001, 27(8), 629–631.

56. Zakharov N. A., An analysis of the phase transitions in biocompatible Ca$_{10}$(PO$_4$)$_6$(OH)$_2$, *Tech. Phys. Lett.*, 2001, 27(12), 1035–1037.

57. Rani R., S. Sharma, R. Rai, A. L. Kholkin, Investigation of dielectric and electrical properties of Mn doped sodium potassium niobate ceramic system using impedance spectroscopy, *J. Appl. Phys.*, 2011, 110, 104102.

58. Suda H., M. Yashima, M. Kakihana, M. Yoshimura, Monoclinic. tautm. Hexagonal phase transition in hydroxyapatite studied by X-ray powder diffraction and differential scanning calorimeter techniques, *J. Phys. Chem.*, 1995, 99(17), 6752–6754.

59. Gittings J. P., C. R. Bowen, I. G. Turner, F. Baxter, J. Chaudhuri, Characterisation of ferroelectric-calcium phosphate composites and ceramics, *J. Eur. Ceram. Soc.*, 2007, 27, 4187.

60. Harihara B., B. Venkataraman, K. B. R. Varma, Frequency-dependent dielectric characteristics of ferroelctric SrBi2Nb2O9 ceramics, *Solid State Ionics*, 2004, 167, 197.

61. Marzec E., A comparison of dielectric relaxation of bone and keratin, *Bioelectroch. Bioener.*, 1998, 46, 29.

62. Miklavcic D., N. Pavselj, F. X. Hart, Electric properties of tissues, *Wiley Encyclopedia of Biomedical Engineering*, ed. M. Akay, 2006, 1–12.

63. Gabriel C., S. Gabriel, E. Corthout, The dielectric properties of biological tissues: I. literature survey, *Phys. Med. Biol.*, 1996, 41, 2231.

64. Fois M., A. Lamure, M. J. Fauran, C. Lacabanne. Dielectric properties of bone and its main mineral component, *10th International Symposium on Electrets, IEEE*, 1999, pp. 217–220.

65. Shames M. H., L. S. Lavine, Physical bases for bioelectric effects in mineralized tissues, *Clin. Orthop.*, 1964, 35, 177.

66. Williams P. A., S. Saha, The electrical and dielectric properties of human bone tissue and their relationship with density and bone mineral content, *Ann. Biomed. Eng.*, 1996, 24, 222.

67. Horiuchi N., J. Endo, N. Wada, K. Nozaki, M. Nakamura, A. Nagai, K. Katayama, K. Yamashita, Polarization-induced surface charges in hydroxyapatite ceramics, *J. Appl. Phys.*, 2013, 113, 134905.

68. Maiti G. C., F. Freund, Influence of fluorine substitution on the proton conductivity of hydroxyapatite, *J. Chem. Soc. Dalton Trans.*, 1981, 6, 949.

69. Reinish G. B., A. S. Nowick, Effect of moisture on electrical properties of bone, *J. Electrochem. Soc.*, 1976, 123, 1451.

70. Gittings J. P., C. R. Bowen, A. C. E. Dent, I. G. Turner, F. R. Baxter, J. B. Chaudhuri, Electrical characterization of hydroxyapatite-based bioceramics. *Acta-Biomater.*, 2009, 5(2), 743–754.

71. Royce B. S., Field-induced transport mechanisms in hydroxyapatite, *Annals New York Acad. Sci.*, 1974, 238, 131–138.

72. Yashima M., Y. Yonehara, H. Fujimori, Experimental visualization of chemical bonding and structural disorder in hydroxyapatite through charge and nuclear-density analysis, *J. Phys. Chem. C*, 2011, 115(50), 25077–25087.

73. Yamashita K., H. Owada, T. Umegaki, T. Kanazawa, T. Futagami, Ionic conduction in apatite solid solutions, *Solid State Ionics*, 1988, 28–30, 660–663.

74. Yamashita K., K. Kitagaki, T. Umegaki, Thermal-instability and proton conductivity of ceramic hydroxyapatite at high-temperatures, *J. Am. Ceram. Soc*, 1995, 78, 1191–1197.

75. Laghzizil A., N. Elherch, A. Bouhaouss, G. Lorente, T. Coradin, J. Livage, Electrical behaviour of hydroxyapatites $M_{10}(PO_4)_6(OH)_2$ (M = Ca, Pb, Ba), *Mater. Res. Bull.*, 2001, 36, 953–962.

76. Laghzizil A., N. Elherch, A. Bouhaouss, G. Lorente, J. Macquete, Comparison of electrical properties between fluoroapatite and hydroxyapatite materials, *J. Solid State Chem.*, 2001, 156, 57–60.

77. Nagai M., T. Nishino, Surface conduction of porous hydroxyapatite ceramics at elevated-temperatures, *Solid State Ionics*, 1988, 28–30, 1456–1461.

78. Jonscher A. K., The "Universal" dielectric response, *Nature*, 1977, 267, 673–679.

79. Vainas B., D. P. Almond, J. Luo, R. Stevens, An evaluation of random R-C networks for modelling the bulk ac electrical response of ionic conductors, *Solid State Ionics*, 1999, 126, 65–80.

80. Almond D. P., B. Vainas, The dielectric properties of random R-C networks as an explanation of the "universal" power law dielectric response of solids, *J. Phys. Condensed Matter*, 1999, 11, 9081–9093.

81. Bowen C. R., D. P. Almond, Modelling the "universal" dielectric response in heterogeneous materials using microstructural electrical networks, Mater. *Sci. Tech.*, 2006, 22, 719–724.

82. Sindhu M., N. Ahlawat, S. Sanghi, A. Agarwal, R. Dahiya, N. Ahlawat, Rietveld refinement and impedance spectroscopy of calcium titanate. *Curr. Appl. Phys.*, 2012, 12(6), 1429–1435.

83. Prasad K., S. Bhagat, K. Amarnath, S. N. Choudhary, K. L. Yadav, (a), Electrical conduction in $Ba(Bi_{0.5}Nb_{0.5})O_3$ ceramic: Impedance spectroscopy analysis, *Mater. Sci. - Poland.*, 2010, 28(1), 317–325.

84. Ortega N., A. Kumar, P. Bhattacharya, S. B. Majumder, R. S. Katiyar, Impedance spectroscopy of multiferroic $PbZr_xTi_{1-x}O_3/CoFe_2O_4$ layered thin films, *Phys. Rev. B.*, 2008, 77, 014111.

85. West A. R., D. C. Sinclair, N. Hirose, Characterization of electrical materials, especially ferroelectrics, by impedance spectroscopy, *J. Electroceram.*, 1997, 1(1), 65–71.

86. Jorcin J.-B., M. E. Orazem, N. Pébère, B. Tribollet, CPE analysis by local electrochemical impedance spectroscopy, *Electrochim. Acta*, 2006, 51(8), 1473–1479.

87. Guo X., W. Sigle, J. Maier, Blocking grain boundaries in Yttria-doped and undoped Ceria ceramics of high purity, *J. Am. Ceram. Soc.*, 2004, 86(1), 77–87.

88. Hodge I. M., M. D. Ingram, A. R. West, Impedance and modulus spectroscopy of polycrystalline solid electrolytes, *J. Electroanal. Chem. Interf. Electrochem.*, 1976, 74(2), 125–143.

89. Alkoy E. M., A. Berksoy-Yavuz, Electrical properties and impedance spectroscopy of pure and copper-oxide-added potassium sodium niobate ceramics, *IEEE Trans. Ultrason. Ferroelectr. Freq. Control.*, 2012, 59(10), 2121.

16

Design, Prototyping, and Performance Qualification of Thermal Protection Systems for Hypersonic Space Vehicles

Hypersonic air breathing vehicles are next-generation high-speed vehicles which enable travel at speeds greater than 5 times the speed of sound (Mach 5). Such vehicles are classified as air breathing engines as they travel within the earth's atmosphere and uses the air containing oxygen as an oxidizer for burning fuel to propel the vehicle at high speeds using a specialized class of engine, called Scramjet engine. High-speed travel results in shock waves, which rise the temperature and pressure of the air around the vehicle structure. The heating rates are typically proportional to the cube root of the vehicle velocity. Especially at the leading edges of the vehicles, stagnation (i.e., the flow velocity is brought to rest) results in a significant increase in the temperature and pressure. Typically a Mach 7 vehicle flying at an altitude of 30 km results in a freestream stagnation temperature of 2400 K and a freestream stagnation pressure of 5 MPa. Conventional materials cannot withstand such extreme loads, in particular, for long duration operation and requires a thermal protection system (TPS). In this regard, ultra-high-temperature ceramics (UHTCs) are recommended for such applications.

For sharp leading edge of hypersonic cruise vehicles, ZrB_2-based UHTCs are considered potential materials owing to their superior thermomechanical properties. While discussing many of the published results from the author's research group, an emphasis will be placed as how to couple experiments and computational analysis in interdisciplinary research. Another important aspect is the assessment of performance-limiting properties. The thermo-oxidative-structural stability of ZrB_2-SiC composites under arc jet flow, assessed using an integrated computational experimental approach, is also discussed.

16.1 Introduction

Endo-atmospheric hypersonic vehicles experience extreme pressure and heat flux arising from shock waves and viscous dissipation. The functional and operational requirements of hypersonic space vehicles demand the use of materials, which have an excellent resistance to evaporation, erosion, and oxidation as well as maintain mechanical integrity. UHTCs are suitable material choice for such applications owing to their superior melting point, hardness, chemical, and wear resistance.[1,2] ZrB_2 is a candidate material for extreme environment aerospace applications, as they possess low density for weight saving, high thermal conductivity for better heat dissipation, and high electrical conductivity for aerodynamic shaping/machining of ZrB_2s.[1,3]

Despite these advantages, ZrB_2 are difficult to process due to high melting point, low self-diffusion coefficient, and the presence of an oxide layer on the powder particles.[1] To enhance the processability and improve mechanical properties, SiC and metal additives[4–10] are added to ZrB_2. Recently, the research group of one of the authors of this book has successfully developed ZrB_2–SiC-based ceramics with Ti and $TiSi_2$[4,11–16] with an aim to achieve higher densification at lower sintering temperature along with a good combination of strength and toughness.

Translation of concept to technology in development of UHTC for hypersonic application, as shown in Figure 16.1, has been emphasized widely in published literature References [17–20,22–26]. The initial stage involves material processing using advanced sintering techniques, with optimization of sintering parameters, followed by critical investigation to understand the microstructure–property correlations. Subsequent stage encompasses numerous performance qualification tests, using plasma arc jets,[5,6] plasma wind tunnels,[7] and oxyacetylene torch flames,[8] in order to evaluate the high-temperature response of the ceramics under simulated aero-thermodynamic environment faced by the TPS of hypersonic vehicles. Also, there is a need to develop integrated computational experimental framework, which can aid in understanding vital physical phenomena, without the burden of repeated experimentations. Moreover, such computational analysis can be used to design, analyze, and optimize the flight worthy TPS. The final stages involve the demonstration of dimensional scalability of

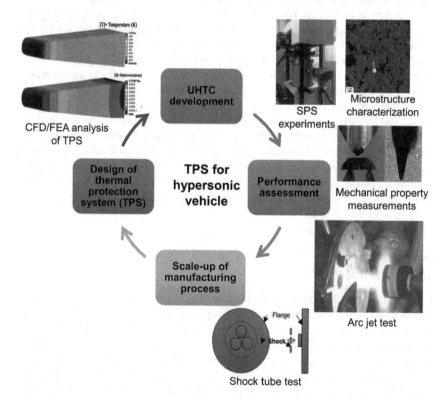

FIGURE 16.1   Schematic illustration showing the concept to technology development for UHTC-based TPS for hypersonic vehicles.

the ceramics from laboratory scale to industry scale and fabrication of component level airframe segments using advanced machining techniques.

16.2 Laboratory-Scale Development of UHTCs

Against the backdrop of the challenges and opportunities related to the development of difficult-to-sinter ZrB_2-based ceramics, this section now presents some of the author's recent published work on ZrB_2–SiC-based ceramics. As far as the materials processing is concerned, the ball-milled powder mix of high-purity ceramic powders of ZrB_2, SiC, and Ti corresponding to three different compositions, namely, ZrB_2-18 wt.%SiC-$x$wt.%Ti ($x = 0$, 10, and 20), are consolidated via spark plasma sintering using three-stage heating schedules with a final holding at 1500°C.

Microstructural features of spark plasma-sintered ZrB_2–SiC and ZrB_2–SiC–Ti is shown in Figure 16.2a,b. The backscattered SEM micrograph of ZrB_2–SiC reveals the presence of a ZrB_2 matrix (light grey contrast) with a homogeneous dispersion of SiC (dark grey contrast), in addition to few WC (bright contrast) from milling media. In case of ZrB_2–SiC–Ti, the microstructural features are significantly modified, showing the presence of additional phases, as indicated by EDS spectra in Figure 16.2. Also, the addition of Ti leads to irregular-shaped ZrB_2 grains with equiaxed and elongated SiC grains at discrete locations. The comparison between the microstructures with and without Ti addition reveals a finer microstructure in ZrB_2–SiC–Ti, as Ti aids in controlled grain growth during multistage spark plasma sintering at 1500°C. This could be phenomenologically described using coupled grain growth theory,[21] wherein a correlative topological relationship develops between the grain boundaries and interphases.

As the microstructure is quite complex with the presence of multiple phase contrasts, EDS spectra can be put for use to get some insights into the phases. A close examination of the micrograph (Figure 16.2b) with EDS suggests the presence of Zr–Si and Ti–Si phases in the interphase boundaries of ZrB_2 and SiC grains. In addition to microstructural observations, the X-ray diffractogram (not shown) indicated the presence of sintering reaction products such as ZrC, TiC, and TiB_2. However, precise identification of each of the sintering reaction products requires finer scale microstructural studies using transmission electron microscopy.

16.3 Performance-Limiting Property Assessment Using Arc Jet Testing

One of the widely used ground facilities for assessing the performance of the TPS material to simulated hypersonic reentry conditions is the arc jet test facility. Spark plasma-sintered ZrB_2–SiC and ZrB_2–SiC–Ti ceramic composites are subjected to arc jet flow to evaluate the oxidation and ablation behavior. Arc jet consists of a DC powered plasma generator with a maximum power output of 320 kW, followed by a supersonic nozzle and sample holder. During operation, the air + argon (2%) mixture is introduced into a plasma generator, operated at 101 kW, with a mass flow rate being 3.6 g/s. The plasma-heated air–argon mixture is expanded in the nozzle to reach Mach 1.4, which then interacts with the specimen, as shown in Figure 16.3.

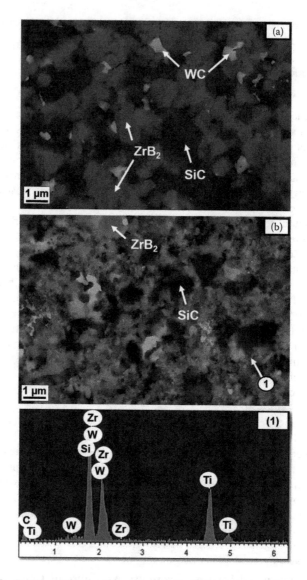

FIGURE 16.2 Scanning electron micrographs of ZrB$_2$-SiC with (a) 0 wt.%Ti and (b) 20 wt.%Ti, consolidated by spark plasma sintering at 1500°C, with EDS spectra (1). (Reproduced with permission from Ref. 27).

Ceramic composites are mounted on a phenolic cork holder and instrumented with a K-type thermocouple at the back wall to monitor the time history of the temperature. The standoff distance between the nozzle exit and sample surface is adjusted using a remote-controlled movement mechanism to achieve the desired heat flux of 2.5 MW/m². It is to be mentioned that, during the test initiation, the sample is guarded from the flame for about 20 seconds, using a water-cooled shield, in order to avoid the plasma starting transients. After stabilization, the shield is rapidly removed thereby exposing the ceramic to a desired heat flux for about 30-second test duration, followed

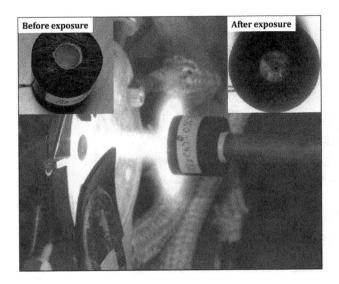

FIGURE 16.3   Typical experimental set up used in conducting arc jet testing of ultrahigh temperature ceramics showing the flame interaction with the ceramics in the phenolic cork holder assembly. The inset figures on the left and right show the sample before and after exposure to arc jet flow, respectively. (Reproduced with permission from Ref. 28).

by power shut off. Multiple tests are performed for each ceramic composition to assess the performance reliably.

16.3.1 Transient Thermal and Coupled Thermostructural Analysis

In order to assess the material response to arc jet heating, it is imperative to know the temperature at the front wall and the spatio-temporal evolution of temperature all over the ceramic composite. In this regard, a two-staged computational approach has been adopted wherein, transient thermal analysis (TTA) is initially conducted to obtain the temperature distribution, followed by coupled thermostructural analysis, wherein the temperature obtained from TTA is mapped to the structural domain to evaluate the displacement and stress.

In TTA, the ceramic composite along with the phenolic holder is modelled and discretized using high-quality second-order finite elements for better accuracy. Thermal contact is provided in the interface region between the ceramic and phenolic holder to enable heat conduction. A heat flux of 2.5 MW/m^2 is imposed on the front face of the finite-element model (FEM). Radiation to ambient atmosphere has been modeled with temperature-dependent emissivity, as ZrB_2–SiC exhibits a non-catalytic behavior at high temperature.[9] As the scanning electron micrographs of arc jet tested specimen shows significant oxide layer formation (Figure 16.4), it is vital to capture these effects, for reliable prediction of temperature. In this regard, the thickness of different oxide layers is carefully interpreted from SEM images and the corresponding material properties are assigned. The computations are carried for the experimental duration of 30 s and the back wall temperature is compared with the experimental measured temperature for validation. Post validation, the front wall temperature as well as the temperature distribution in the sample can be used for understanding the

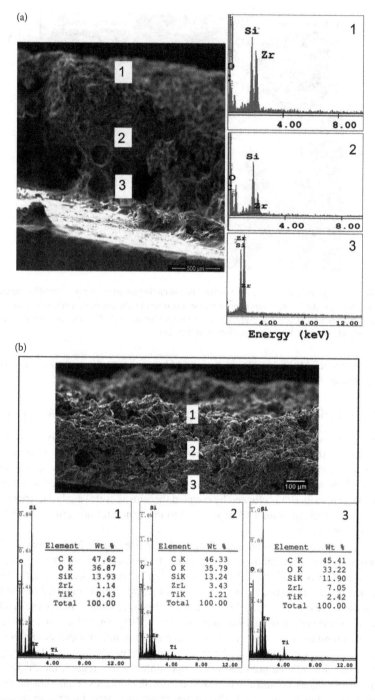

FIGURE 16.4 Cross-sectional micrograph of arc jet exposed ZrB_2–SiC showing three regions, along with elemental composition using SEM-EDS analysis: oxide layer 1 is rich in SiO_2, while oxide layer 2 is rich in ZrO_2 and oxide layer 3 shows ZrB_2 depleted with ZrO_2. (Reproduced with permission from Ref. 28).

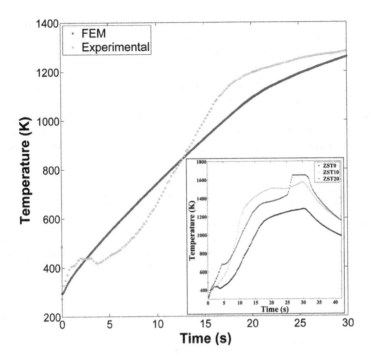

FIGURE 16.5   Time history of temperature at the back-wall of the arc jet exposed ZrB_2–SiC-based ceramics, obtained from computational and experimental methods for 30 s duration. The inset in the figure shows the experimentally obtained full time history of the back-wall temperature for three ceramic compositions. Comparison of experimentally measured and finite-element modeling estimated temperature arc jet testing of ZST0 (experimentally measured back wall temperatures of ZrB_2–SiC–Ti samples as a function of time as shown in the inset for each of the composition ZST0, ZST10, and ZST20). (Reproduced with permission from Ref. 28).

thermo-oxidative structural stability of ZrB_2-based ceramics. Further, the temperature obtained from transient thermal analysis is imported to the coupled thermo-structural analysis (CTSA) at various time steps. For CTSA, a pressure of 1.2 bar is applied to the front wall and the displacement along the rear face of the phenolic holder is constrained in all degrees of freedom to obtain the displacement and stress distribution.

Figure 16.5 shows the temporal evolution of temperature at the back wall, obtained from transient thermal analysis and experimental measurement for the case of ZST0 (ZrB_2–SiC–0Ti) composition. It can be noted that the temperature trends obtained from both techniques matches well, with minimal variations in the range of 5–10 K, which validates the computational framework and establishes its robustness in temperature prediction. From the validated computational model, the temperature at the arc jet exposed front wall is found to be 1942 K. While the front wall shows uniform distribution of temperature, the back wall temperature distribution is non-uniform, with the center region experiencing the maximum temperature which drops radially toward the circumference. The through-thickness thermal gradient is estimated to be 6.67×10^5 K/m in the oxide layer of ZST0, which experiences 1350–1950 K. In case of the virgin material, the temperature is lower than the oxide layer and lies in the range of 1156–1350 K, owing to conductive heat loss to the phenolic holder from the ceramic surface.

The inset of Figure 16.5 shows the experimentally measured back wall temperatures in arc-jet tested ZST0, ZST10, (ZrB_2–SiC–10Ti) and ZST20 (ZrB_2–SiC–20Ti) ceramics. A rapid increase in the temperature above 1000 K is noted within the first 15 seconds, following which the rate of increase is minimal. This is reasoned to the evolution of the oxide layer containing silica and zirconia, having lower thermal conductivity as compared to the virgin material. For ZST20, the temperature trend shows the attainment of steady state at ~26 seconds, which indicates thermal equilibrium between the arc jet flow (heat entry) and radiation (heat exit). After arc jet exposure, the temperature for all the ceramic surfaces drops qualitatively similar.

16.3.2 Thermodynamic Feasibility of Oxidation Reactions

Understanding the oxidation mechanisms is vital to assess the thermo-oxidative structural stability of UHTCs. Based on the phase assemblage studied carried out using X-ray diffraction, arc jet tested ZrB_2–SiC–Ti shows the formation of multiple phases, such as, SiO_2, ZrO_2, TiO_2, $ZrSi_2$, $TiSi_2$, ZrC and B_4C. Thermodynamically feasible oxidation pathways for the formation of these phases are as follows.

$$ZrB_2 + SiC + 2.5O_2(g) = ZrC + B_2O_3 + SiO_2 \tag{16.1}$$

$$2ZrB_2 + SiC + 3O_2(g) = B_4C + 2ZrO_2 + SiO_2 \tag{16.2}$$

$$SiC + 2O_2(g) = SiO_2 + CO_2(g) \tag{16.3}$$

$$ZrB_2 + 3Ti + 2SiC = TiB_2 + 2TiC + ZrSi_2 \tag{16.4}$$

$$O_2(g) + 2SiC + TiO_2 = TiSi_2 + 2CO_2(g) \tag{16.5}$$

$$ZrO_2 + SiO_2 = ZrSiO_4 \tag{16.6}$$

$$ZrO_2 + TiO_2 = ZrTiO_4 \tag{16.7}$$

$$SiC + 1.5O_2(g) = SiO_2(l) + CO(g) \tag{16.8}$$

$$SiC + O_2(g) = SiO(g) + CO(g) \tag{16.9}$$

Figure 16.6 shows the Gibbs free energy change (ΔG) with respect to temperature for the aforementioned reactions (16.1)–(16.9), calculated using commercial software, HSC Chemistry 6. It is clearly notable that all the reactions have negative ΔG values over the temperature range of 300–1942 K. This indicates the thermodynamic and kinetic feasibility during exposure to arc-jet flow thereby confirming the proposed oxidation mechanism for ZrB_2–SiC–Ti.

Thus, the integrated experimental computation approach using arc jet testing establishes the thermo-oxidative-structural stability of ZrB_2-based ceramics for hypersonic vehicle applications. It is is found that addition of Ti enhances the ablation and oxidation resistance under simulated aerodynamic conditions.

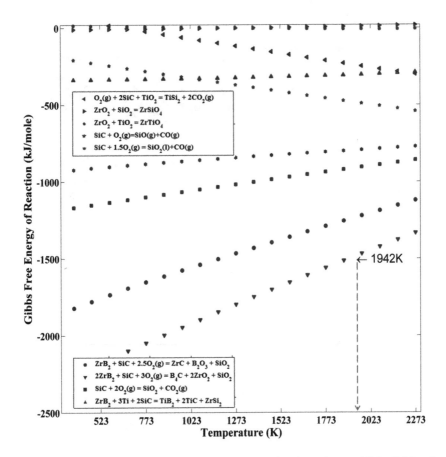

FIGURE 16.6 Thermodynamic feasibility of oxidation reactions for arc jet tested ZrB_2–SiC-based UHTCs. (Reproduced with permission from Ref. 28).

16.4 Thermo-Structural Design of Thermal Protection System

Having discussed the development of laboratory-scale UHTCs and their performance qualification, this section details on the thermostructural design of ZrB_2–SiC-based TPS for the sharp leading edges of hypersonic vehicles, using computational analysis.

The design of TPS involves making a geometric model of the three-layered TPS as shown in Figure 16.7. The top layer consists of the UHTC, followed by an insulating phenolic cork and titanium-based metallic structure. The length of the leading edge is 80 mm, and the height being 22 mm, with a nose tip radius of 4 mm. Design analysis of the TPS is carried out in two stages. In the first stage, computational fluid dynamics (CFD) simulations are performed to obtain the spatially varying thermal and pressure loads faced by the leading edge exposed to hypersonic flow. Following CFD analysis, the second stage involves the execution of coupled thermostructural analysis (CTSA), wherein the heat flux and pressure obtained from CFD is mapped

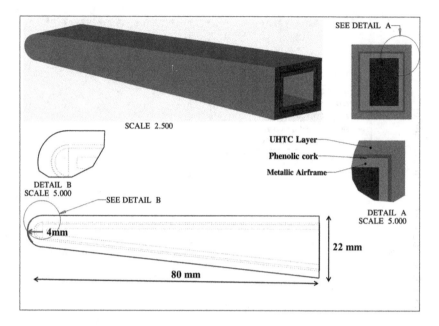

FIGURE 16.7 Geometry of the multilayered thermal protection showing the dimensional details and material layering. (Reproduced with permission from Ref. 29).

to obtain the displacement and stress fields. Such analysis is iteratively conducted to optimize the design for minimum weight while ensuring other functional and operation requirements are met with a desired factor of safety. This process is schematically depicted in Figure 16.8.

16.4.1 CFD Analysis: Hypersonic Flow around Leading Edge

CFD simulations are used for the determination of flow field around the leading edge and obtain the thermal and mechanical loads under real flight-like conditions, namely, varying Mach number, altitude, and durations. Based on the external surface of the leading edge, the fluid domain is modelled and meshed, with a dense mesh near the solid surface to accurately capture the boundary layer effects. The hypersonic flow is initialized uniformly in the fluid domain by specifying the Mach number and the free stream conditions corresponding to the altitude, which is 7 and 30 km, respectively. Based on the CFD analysis, the contours of Mach number, velocity vector, fluid density, and static pressure are obtained, as shown in Figure 16.9. Based on these contours, it is clearly seen that bow shock is formed ahead of the leading edge due to the radius of curvature provided at the nose tip. The presence of the boundary layer all along the surfaces of the leading edge can be noted from the velocity vector contour, including the flow separation at the base. Density and static pressure distribution show a sharp increase due to shock formation. Importantly, the stagnation zone forming around the nose of leading edge results in a pressure and heat flux as high as 72.8 kPa and 2.11 MW/m^2, respectively.

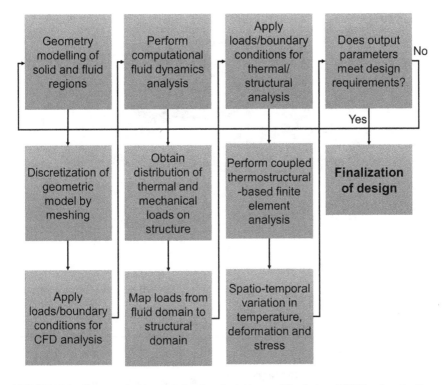

FIGURE 16.8   Framework of computational methodology used for design of TPS for sharp leading edge of hypersonic vehicles, involving CFD and finite-element-based coupled thermostructural analysis (CTSA).

16.4.2 Finite-Element-Based Coupled Thermostructural Analysis

This section summarises the results of the thermostructural analysis using ANSYS Mechanical 15.[13–15,26] A structured mesh with second-order elements is used for discretization of the solid domain, and contact elements are used at the interfaces of the different layers. Care is taken to ensure that the mesh quality is superior considering the aspect ratio, Jacobian and skewness, especially at the leading edge with a curved profile. Across the TPS thickness, multiple mesh elements have been placed to predict the temperature variations across material interfaces. After discretization, the temperature-dependent spatially varying pressure and heat flux calculated from CFD analysis is mapped from fluid to structural domain. Temperature distribution in the structure is first determined using transient thermal analysis, considering the loads from CFD. Then, the transient volumetric temperatures are mapped to structural domain from thermal domain. This is followed by coupled thermo-structural analysis, wherein the internal pressure of 1 atmosphere (equivalent to chamber internal pressure during the cruise phase) is applied on the internal surface of the leading edge, with its rear surface constrained in all degrees of freedom to simulate fixed boundary conditions. Also, the acceleration due to gravity and body loads due to the acceleration of the hypersonic cruise vehicle are considered.

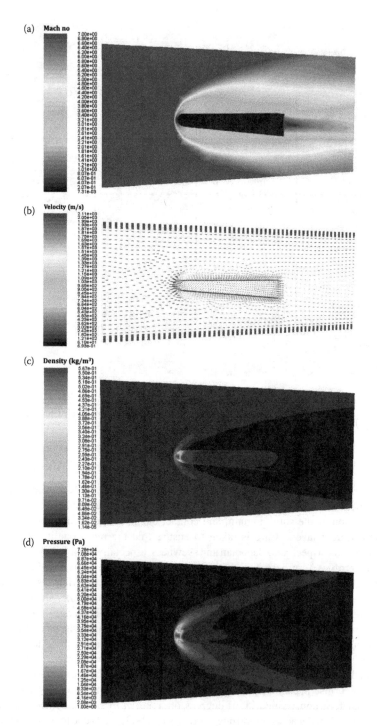

FIGURE 16.9 CFD-analysed contours of (a) Mach number; (b) velocity; (c) fluid density and (d) static pressure, around leading edge of hypersonic space vehicle to hypersonic flow. (Reproduced with permission from Ref. 29).

The governing equation in the matrix form, solved in a coupled thermostructural analysis, is shown in Figure 16.8, along with the description of the each terms. Upon solving this, the displacements are obtained, followed by the estimation of secondary derived parameters such as strain $\{\varepsilon\}$ and stress $\{\sigma\}$.

Post-processed results of von Mises stress and maximum principal stress from CTSA are shown in Figure 16.10. From the Von Mises stress contour shown in Figure 16.10a for the metallic layers, it can be seen that the variations in stress along the length is almost similar for the top and bottom surface, which is due to nearly equal lengths of these surfaces. Also, the maximum stress occurs at the trailing edge location, which is reasoned to the mechanically constrained boundary condition. It can also be noted

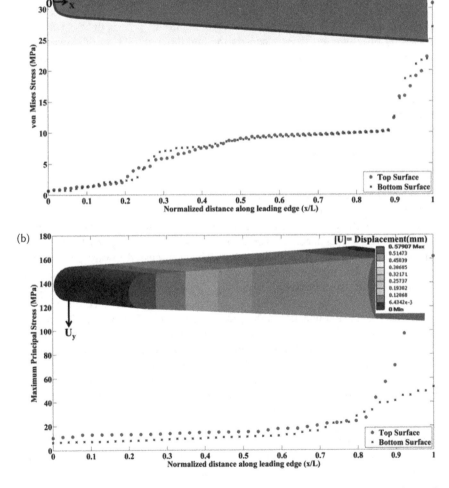

FIGURE 16.10 (a) Variation of von Mises stress along the length of the leading edge; and (b) maximum principal stress variation, with inset showing the resultant displacement. (Reproduced with permission from Ref. 29).

that the variations in stress along the length is nonlinear, which arises due to the temperature-dependent material properties coupled with spatially varying pressure and heat flux loading. In case of maximum principal stress, shown in Figure 16.10b (with inset showing the displacement contour), the top surface of the ZrB_2–SiC layer clearly is at a higher state of stress, especially at the trailing edge while the frontal edge does not show any difference.

For a displacement constrained design, it is important to maintain displacement within certain limits. In this regard, the resultant displacement of the leading edge is calculated as $[U] = \sqrt{U_x^2 + U_y^2 + U_z^2}$, where U_x, U_y, and U_z are X, Y, and Z displacement components. As shown in the inset of Figure 16.10b, it is clear that the maximum displacement is around 0.58 mm at nose tip, while it is zero at the trailing edge, which is provided with fixed boundary conditions. The overall displacement behavior is found to behave as per the applied load and boundary conditions, and the maximum values are within the aerodynamic shape requirements of the leading edge.

In addition to the temperature, von Mises stress, maximum principal stress, and displacement, interfacial shear stress are important design criteria to be checked for possibility of failure at the interfaces in the multilayered TPS. In this regard, the interfacial shear stress between the ceramic–phenolic cork layer and the phenolic cork-metal layer has been calculated. The numerical model shows that the maximum values are 43 and 2.5 MPa, respectively, which are within the adhesion strength of the material interfaces. This establishes the feasibility of the multilayered TPS concept from a thermo-structural perspective.

16.5 Closure

This chapter provided insights into various stages involved in the design, development, and testing of a TPS for sharp leading edges of hypersonic space vehicles. The development of ZrB_2–SiC-based ceramics using the multistage spark plasma sintering technique has been presented. The thermo-oxidative-structural stability of the developed ceramics under the arc jet environment has been discussed toward performance qualification of the ceramics for desired hypersonic applications. Further, the emphasis for an integrated computational experimental approach with realistic conditions is laid to minimize the number of experiments toward performance qualification experiments and better understand the underlying physics. Moreover, the importance of a computational framework toward the design of an advanced multilayered TPS is highlighted. The designed three-layered TPS is found to be satisfying the design requirements as required for a Mach 7 hypersonic cruise flight for 250 seconds.

However, these tasks do not complete the required technology and manufacturing readiness levels toward the realization of flight worthy airframe components. The complete development cycle of UHTCs for hypersonic application is given in Figure 16.11. As a logical progression, the dimensional scalability of the ceramics from laboratory-scale dimensions to flight hardware dimensions has to be proven, with isotropic properties. Toward this end, the multistage hot-pressing technique in the temperature range of 1500–1700°C has been employed to develop ZrB_2–SiC-based ceramics with a diameter up to $\phi100$ mm and thickness of ~5 mm. The achieved scalability is shown in Figure 16.12. Following demonstration of dimensional scalability, another critical step is to machine the ceramic discs to precise and complex 3D shapes

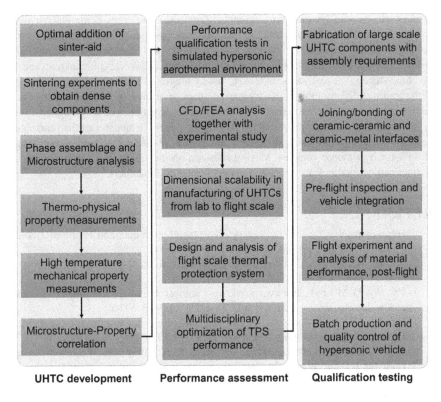

UHTC development   **Performance assessment**   **Qualification testing**

FIGURE 16.11 Various stages involved in the realization of UHTC-based integrated TPS for hypersonic application.

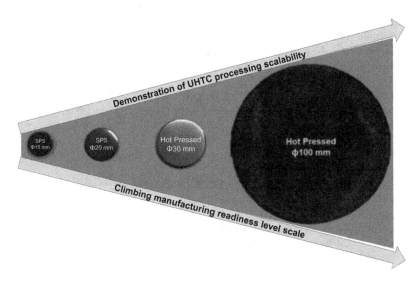

FIGURE 16.12 Demonstration of dimensional scalability of ZrB_2–SiC-based ceramics. (Adapted from 27).

as put forth by aerodynamic requirements. ZrB$_2$-based ceramics being electrically conductive can be machined using the electric discharge machining technique. From the flight vehicle perspective, joining of the ceramics with the metallic structure as well as ceramic–ceramic joints with a suitable filler material is of significant relevance. The final stage in the realization of UHTC-based integrated TPS involves flight testing of the leading edge component in a hypersonic flow environment. Post-processing and recovery of the material essentially determines the performance of the designed TPS in real time application. In summary, this chapter clearly demonstrates the marriage between the computational and experimental work toward design and prototyping of TPS for hypersonic vehicles. More discussion on the results summarized in this chapter can be found in the published papers of one of the authors.[5,14,26,27,28,29,30]

REFERENCES

1. Fahrenholtz W. G. et al., Refractory diborides of zirconium and hafnium, *J. Am. Ceram. Soc.*, 2007, 90(5), 1347–1364.
2. Zimmermann J. W. et al., Thermophysical properties of ZrB$_2$ and ZrB$_2$–SiC ceramics, *J. Am. Ceram. Soc.*, 2008, 91(5), 1405–1411.
3. Van Wie D. M. et al., The hypersonic environment: Required operating conditions and design challenges. *J. Mater. Sci.*, 2004, 39(19), 5915–5924.
4. Upadhya K., J. M. Yang, W. P. Hoffman, Materials for ultrahigh temperature structural application, *Am. Ceram. Soc. Bull*, 1997, 76, 51–56.
5. Brahma Raju G., A. Mukhopadhyay, B. Basu, K. S. Thimmappa, Review on Ultra High Temperature Boride Ceramics, *Prog. Mater. Sci.*, 2020, 111, 100651.
6. Savino R. et al., Arc-jet testing of ultra-high-temperature ceramics, *Aerospace Sci. Technol.*, 2010, 14, 178–187.
7. Monteverde F., R. Savino, M. De Stefano Fumo, Plasma wind tunnel testing of ultra-high temperature ZrB$_2$-SiC composites under hypersonic re-entry conditions, *J. Eur. Ceram. Soc.*, 2010, 30, 2313–2321.
8. Han J. et al., Oxidation resistant ZrB$_2$-SiC composites at 2200°C, *Compos. Sci. Technol.*, 2008, 68, 799–806.
9. Reddy K. M., N. Kumar, B. Basu, Innovative multi-stage spark plasma sintering to obtain strong and tough ultrafine-grained ceramics, *Scr. Mater.*, 2010, 62(7), 435–438.
10. Bellosi A., F. Monteverde, D. Sciti, Fast densification of ultra-high-temperature ceramics by spark plasma sintering, *Int. J. Appl. Ceram. Technol.*, 2006, 3(1), 32–40.
11. Shen Z. et al., Spark plasma sintering of alumina, *J. Am. Ceram. Soc.*, 2002, 85(8), 1921–1927.
12. Guo S. Q. et al., Spark plasma sintering of zirconium diborides, *J. Am. Ceram. Soc.*, 2008, 91(9), 2848–2855.
13. Karthiselva N. S., B. S. Murty, S. R. Bakshi, Low temperature synthesis of dense and ultrafine grained zirconium diboride compacts by reactive spark plasma sintering, *Scr. Mater.*, 2016, 110, 78–81.
14. Purwar A. et al., Development of ZrB$_2$-SiC-Ti by multi stage spark plasma sintering at 1600°C, *J. Ceram. Soc. Jpn.*, 2016, 124(4), 393–402.
15. Gupta N. et al., Spark plasma sintering of novel ZrB$_2$-SiC-TiSi$_2$ composites with better mechanical properties, *Mater. Sci. Eng.*, 2012, A 534, 111–118.

16. Wang H. et al., Processing and mechanical properties of zirconium diboride-based ceramics prepared by spark plasma sintering, *J. Am. Ceram. Soc.*, 2007, 90(7), 1992–1997.

17. Jain D. et al., Achieving uniform microstructure and superior mechanical properties in ultrafine grained TiB_2-$TiSi_2$ composites using innovative multi stage spark plasma sintering, *Mater. Sci. Eng.*, 2010, A 528, 200–207.

18. Akin I. et al., Microstructure and densification of ZrB_2–SiC composites prepared by spark plasma sintering, *J. Eur. Ceram. Soc.*, 2009, 29(11), 2379–2385.

19. Mukhopadhyay A., T. Venkateswaran, B. Basu, Spark plasma sintering may lead to phase instability and inferior mechanical properties: A case study with TiB_2, *Scr. Mater.*, 2013, 69, 159–164.

20. Baik S., P. F. Becher, Effect of oxygen on the densification of TiB_2, *J. Am. Ceram. Soc.*, 1987, 70(8), 527–530.

21. Liu H. L. et al., Synergetic roles of ZrC and SiC in ternary ZrB_2–SiC–ZrC ceramics, *J. Eur. Ceram. Soc.*, 2015, 35(16), 4389–4397.

22. Pierson H. O., *Handbook of Refractory Carbides and Nitrides*, William Andres Publishng/Noyes, Westwood, NJ, 1996.

23. Munro R. G., Material properties of titanium diboride, *J. Res. Natl. Inst. Stand. Technol.*, 2000, 105, 709–720.

24. Ueki M., H. Endo, Mechanical properties of particulate reinforced SiC-based ceramic composites, *ISIJ Int.*, 1992, 32(8), 943–952.

25. Rezaie A., W. G. Fahrenholtz, G. E. Hilmas, Effect of hot pressing time and temperature on the microstructure and mechanical properties of ZrB_2–SiC, *J. Mater. Sci.*, 2007, 42(8), 2735–2744.

26. Gupta N., Development of TiB_2 – Ti based homogenous/ bilayered functionally graded armor materials and ZrB_2-SiC-Ti composites using spark plasma sintering, PhD thesis, IIT Kanpur, India, 2013.

27. Mukherjee R., B. Basu, Opportunities and challenges in processing and fabrication of ultra high temperature ceramics for hypersonic space vehicles: A case study with ZrB_2-SiC, *Adv. App. Ceram.*, 2018, 117(Sup 1), S2–S8.

28. Purwar A., B. Basu, Experimental and computational analysis of thermo-oxidative-structural stability of ZrB2-SiC-Ti during Arc-jet testing, *J. Am. Ceram. Soc.*, 2017, 100, 4860–4873.

29. Purwar A., B. Basu, Thermo-structural design of ZrB_2–SiC-based thermal protection system for hypersonic space vehicles. *J. Am. Ceram. Soc.*, 2017, 100(4), 1618–1633.

30. Kirupakaran G., J. Gopalan, B. Basu, Shock wave-material interaction in ZrB2-SiC based ultra high temperature ceramics for hypersonic applications, *J. Am. Ceram. Soci.*, 2019, 102, 1–14.

17

A Way Forward

This chapter discusses the everlasting need for an integrated understanding in the field of interdisciplinary sciences and research. Clearly, the challenge is to translate scientific concepts to viable technology and eventually to engineered products of societal relevance. This is further emphasized in this concluding chapter. Accordingly, the aspects concerning education and training of young researchers to better adapt such an approach have also been emphasized.

17.1 Interdisciplinary Innovation and Translational Research

Translational research, as a concept, was introduced in medicine originally, and was defined as bench-to-bedside research. It became an acronym for cross-disciplinary and cross-sectorial research, involving exchange and co-creation of new knowledge. Translational research aims to combine disciplines, resources, expertise, and techniques to empower and enrich researchers with interdisciplinary concepts, and to build a sustainable knowledge-based society. The key requirements of translational research are reorganization of academic teams with cross-disciplinary, cross-sectorial, and cross border co-operation and use of such platforms for perpetual and circular knowledge exchange between researchers and techno-socio-economic stakeholders. Therefore, innovation communities are needed to achieve an environment that allows and nurtures the integration of the best existing knowledge required to build a successful translational research ecosystem.

Going beyond the social boundaries by which knowledge domains are structured is essential for interdisciplinary innovation. In problem solving or product development, interdisciplinary working enables one to use different skills and analytic perspectives, thereby enabling one to make use of different repositories of knowledge, to frame problems, and to develop richer solutions along the path of radical innovation. In curiosity-driven research in academia, interdisciplinary working enables one to establish conjunctions of different interests and perspectives to create new insights and to foster breakthroughs by serendipity. Transdisciplinary exchange among science, industry, government, and society is needed, since scientific inventions are not automatically relevant to society. They must be developed in response to societal needs and requirements.

Interdisciplinary working is a challenge because of silos, where different disciplines have different languages, and goals, working processes, time horizons, and at times, disregard for other disciplines. To overcome these silos, the interdisciplinary team must develop shared values and culture, such as knowledge creation, creating a common ground, and valuing different perspectives and approaches.

17.2 Integrated Understanding of Interdisciplinary Sciences

The concepts of the interdisciplinary approach can be realized in the context of scientific and technological upheaval happenings. Such an approach can potentially offer a universal battleground for the problems arising from any of the disciplines. Looking at the current demanding scenarios from each and every sector, the present philosophy of individual discipline can be treated in a classical way. In a number of chapters of this book, it has been emphasized repeatedly that one has to adopt interdisciplinary approaches as well as to use techniques of multiple disciplines to solve complex research problems. This point can be better substantiated in the contexts of biomedical engineering and "green"/renewable energy harvesting-cum-storage and the associated applications, having direct social relevance.

With respect to biomedical engineering, bones and skin can heal under several different circumstances,[1] and tissues within a few organs such as the liver have some powers of regeneration.[2] However, functional regeneration of most tissues within, for example, the musculoskeletal,[3] nervous,[4] cardiovascular,[5] and urological systems[6] is rarely achieved. Thus, some acute traumatic conditions, such as spinal cord injury, degenerative diseases, such as Parkinson's disease, Alzheimer's disease and macular degeneration, and auto-immune diseases such as type I diabetes, have few possibilities of cure through regenerative engineering. To extend such discussion further, the recent advances in stem cell science[9] and developmental biology[10,11] offer exciting possibilities of stimulating tissue generation.[12] Here, the role of biocompatible materials is of paramount importance. The biomaterials or tissue engineering scaffolds can not only support stem cell fate, but also augment stem cell differentiation through specific lineages. This process could conceivably be induced through cell therapies, wherein certain target cells loaded into a scaffold could be injected into an appropriate site and given suitable signals, that cause them to express new tissues.[13] This implies that the regenerating tissue will need a guiding template, and there will also be a need for coordination, with spatio-temporal control of the molecular and physical signalling, again implying the use of a supporting template.[14] The choice of the material for such a template remains an unresolved issue, that severely limits the progress in promoting and fully adopting these technologies.[15] The above example underlines the necessity of integrating the materials and biological sciences.

In the case of harvesting and storage of energy from "green" or renewable sources, the net contribution worldwide among just the renewable was just ~25% in 2014[7,8] and nearly all the vehicles running on the streets have been burning fossil fuels, leading to severe environmental pollution. The pollution indexes recorded in major cities around the world are testimony for the same. Accordingly, technologists and scientists have an important task on their hands, which is to, not only address the potential energy crisis of the world, but also save the environment from excessive pollution created by the "non-green" and non-renewable energy sources (like fossil fuel and coal).

In the context of "green" energy harvesting and storage, as detailed in some of the preceding chapters of this book, the basic scientific concepts including physics and physical/inorganic/organic chemistry need to evolve toward materials, mechanical, and electrical engineering. This, when coupled with efficient instrumentation can lead to the development of sustainable, efficient, and cost-effective solar panels (harvesting solar energy) and advanced batteries (storage). In this context, it is fairly incredible that the translation and integration of the aforementioned knowledge bases have resulted in the enhancement of efficiency of conversion of solar to electrical energy from $\sim$20% to $\sim$30% within a span of just 5 years,[16,17] nearly doubling the worldwide power capability and enhancing the contribution to $\sim$33% toward the renewable energies within a span of just two years (i.e., 2014–2016). Additionally, within just a decade's time, the energy storage capacity of the Li-ion battery technology has nearly doubled (viz., $\sim$200–$\sim$400 Wh/kg)[18]; and promises to get enhanced even further in a very short time by "bringing back" the Li metal anode (replacement for graphitic carbon anodes) and utilizing sulfur as the cathode.[19,20] The above-mentioned method has allowed widespread installation of solar roof tops, feasibility of "lighting-up" entire town/village with the help of solar energy and having automobiles run on the streets, powered entirely by the advanced batteries (and not using a single drop of gasoline), thus contributing big time toward not only addressing the energy concerns of the present/future human society, but also saving the environment.

17.3 Challenges in Translational Research

The interdisciplinary science-based research has been appealing to the engineering community, particularly in view of its projected societal impact. If the research is conducted within the rigid framework and boundary of a specific engineering discipline, say, mechanical engineering, then the impact also would be limited. However, the research and development would have much more impact, if the computational fluid dynamics are used to solve complex blood flow problems through arteries, as part of cardiovascular treatment. On a similar note, successful development of an improved energy storage device, such as advanced batteries, is possible only when efficient instrumentation approaches for the fabrication use novel concepts and insights drawn from electrochemistry and materials science. Hence, any research and development program to solve such and related problems demands an active contribution from engineers, materials scientists, electrochemists, clinicians, and/or biologists, as the case may be.

Such a translational approach can be further realized with more extensive discussion in the context of biomedical engineering. In the context of addressing human diseases, there is an immediate need to manufacture medical devices and implants, which should help in the repair and replacement of diseased and damaged parts of the human skeleton, heart, bone, and teeth, thereby restoring the function of the otherwise functionally compromised structures.

In particular, the implant for load-bearing applications should possess excellent biocompatibility, superior corrosion resistance in the body environment, an excellent combination of high strength and low modulus (matching with host bone), high fatigue and wear resistance, and osteointegration with tissues. For example, musculoskeletal disorders continue to be the most widespread problem, particularly with an increasing rate of trauma and diseases such as osteoporosis and osteoarthritis, which most often occur in an aging population. Other major clinical challenges include various cardiovascular and neurological diseases. Hence, in order to have a long-term clinical performance, along with the intent to create bio-interactive devices, there is an immediate need to bring together engineers, biologists, and clinicians to translate newer design strategies and manufacturing approaches to develop patient-specific biomedical implants and devices.

17.4 Examples of Multi-Institutional Translational Research in Healthcare Domain

The readers would greatly appreciate if we present brief details of some of the recently concluded multidisciplinary translational research programs, being led by one of the authors of this book. The first example is Indo-US Public-Private Networked R&D Center on Biomaterials for Healthcare, which involved 25 Indian and US young researchers. Both IIT Kanpur and Brown university led this bilateral program. The outcome of multiple bilateral projects led to the development of biomaterials for orthopedics, corneal, dermal, cartilage, and cardiovascular tissue engineering applications. Some notable achievements include (a) understanding genotoxicity and gene profiling of osteoblast cells treated with nanobioceramic composites, (b) development of HA-based electroconductive biocomposites for bone tissue engineering, and (c) PLGA-CNF based composites for synthetic heart patches, which received wide media attention.[23-37]

Another example is the Indo-UK Biomaterials center on glass ceramics for osteoporosis. This center was led by IIT Kanpur and the collaborating institutes were SCTIMST, India, and three UK Universities (Birmingham, Warwick, and Kent). Altogether six co-PIs and 15 researchers took part in this interdisciplinary research consortium. Clinically, osteoporosis is one of the most common health risks faced by the vast majority of the aging population across the globe, especially by post-menopausal women. The occurrence of this systemic skeletal disease will eventually make the bone weak, porous, and fragile. The survival of implants in such patients is very low, as osteoporosis affects the process of osseointegration and such patients otherwise have to undergo revision surgery. The biocompatability study as part of Indo-UK center confirmed the neobone formation at the interface of Sr-containing biomaterials, qualitatively using histological analysis and quantitatively using micro-computed tomography.[38] The osteoconductive properties are comparable to those of a commercially available bioactive implant (HA-based bioglass). This project led to the development of novel glass ceramic implants that were found to be biocompatible *in vivo*, with regard to the local effects after implantation, besides training a number of young researchers from India and UK.

17.5 Examples of Multi-Institutional Translational Research in Energy Sector

One of the most recent examples of a large multi-institutional interdisciplinary research program in the domain of solar energy is the SERIIUS, an acronym for the Solar Energy Research Institute for India and the United States. This program was co-led by IISc, Bangalore and NREL, Colorado, USA. In line with the discussion in two chapters of this book, the relevance for this program in the area of concentrated solar powder thrust area can be highlighted here. Using both conventional metallurgical processing approaches as well as coating deposition techniques, Cu-based intermetallic alloy-based bulk and coatings were developed for solar reflector applications. In particular, flexible glass-coated intermetallic alloy reflectors demonstrated good self-cleaning capability and excellent resistance toward environmental degradation.[39–41] Such innovation is promising in terms of both coating composition as well as scalable processing strategy for large-scale development for parabolic trough mirrors. In parallel, SERIIUS researchers developed spectrally selective novel ceramic absorber coatings for efficient renewable energy utilization in concentrated solar power systems.[42–46] In particular, multilayer absorber coatings of $W/WAlN/WAlON/Al_2O_3$ and $TiB_2/TiB_2(N)/Si_3N_4$ with high absorptance (0.96) in solar irradiation range and low emittance (0.08) in the infrared region were fabricated. Both the coatings have interesting structural configurations in terms of the gradient in the metallic properties and the refractive index from the substrate to top surface. In addition to excellent thermal stability at 500°C for 150 h in air, the recently conducted accelerated thermal shock testing (30 cycles at 450°C) on these coatings under a simulated solar field environment at Sandia National Laboratory indicated reliable service durability of 20 years.

Another ongoing interdisciplinary research program is the National Centre for Clean Coal Technology, recently established at Indian Institute of Science. The mandate of this center is to develop the most efficient ways to resolve the various demands that are made on energy technologies, and to make the power conversion systems more efficient. For thermal power conversion systems, this means operation of the generator at higher temperature. Supercritical steam power plants are now standard for the newer power plants, and the materials required for such power plants are standardized globally. However the operating temperature, in such cases, does not exceed 650°C. In order to obtain higher efficiency, several countries including India are proposing supercritical power plants that operate at 700°C or higher. These plants are expected to reach efficiency between 47% and 50%. The advanced ultra-supercritical steam power plant (AUSC) proposed by India has an operating temperature of ~710°C. There is a further push to increase the power plant operating temperature to 740°C. A large number of researchers from mechanical engineering, energy science, materials science, and electrical engineering are working together in this research program.

17.6 Impact of Translational Research

The above discussion signifies that research on biomedical implants and energy harvesting/storage devices has a significant societal impact. This impact is better utilized if the fundamental research pursued in the laboratory can be translated into products or devices for human healthcare and efficient energy harvesting/storage. Clearly, such translational research requires meaningful and close interaction of the active researchers with the scientists, engineers, clinicians, and also entrepreneurs. In this context, it must also be noted that more entrepreneurs should be encouraged to establish start-ups in the biomedical devices sector. This will enable "bench to bedside" translation into biomedically relevant products.

The impact of translational biomedical research would be felt more if there were significant research involvement of clinicians and engineers from industries. Unfortunately, the number of such programs being pursued in many of the developing nations, including in India, is not very substantial. In case of healthcare, such absence of much translational research leading to product commercialization results in the healthcare expenses remaining significantly high, because of expensive imported devices, which are often beyond the reach of financially challenged citizens. On a similar note, the development of indigenous advanced Li-ion batteries, having improved performance with translation of novel concepts to actual technology, would in turn allow the development and commercialization of electric vehicles, which the common man can better afford in a country, like India. It may be restated that the above is extremely important in the light of the rapidly increasing pollution levels in Indian cities. Overall, the above indicates, without any doubt, that translational research programs need to be given more importance and initiatives need to be taken by the researchers in academia and national laboratories, duly supported by the engineers in the concerned industries.

> The transfer of new technologies to entrepreneurs for commercialization of biomedical devices involves the assessment of Technology Readiness Level (TRL) and competitive cost-structure analysis of any newly manufactured device with respect to similar products currently sold worldwide.

With reference to medical devices, the advances in the field of materials and manufacturing enable the construction of high-quality electronic and optoelectronic devices, which can readily integrate with the soft, curvilinear surfaces of the human body. The resulting capabilities create new opportunities for studying disease states, improving surgical procedures, monitoring health/wellness, establishing human machine interfaces, and performing other functions. In particular, the usefulness of such devices in terms of their integration with the brain, the heart, and the skin needs to be established further through extensive clinical trials. With the advent of new manufacturing technologies, the fabrication of custom-made biomaterials is being attempted by various research groups.

> Typically, rapid prototyping allows implant prototypes to be produced in a wide range of materials with remarkable precision in a couple of hours.

17.7 Research Training of Next-Generation Researchers

Another important challenge is the lack of widespread academic programs in interdisciplinary subjects and approaches, as well as of effective training of young researchers in such interdisciplinary domains. As a result, many young researchers do not have much idea about the relevant concepts of interdisciplinary sciences and their significant societal impact. While the topic can be difficult to be taught in classrooms as an academic course, we believe that the approaches discussed in this book can provide some way forward to think in this direction. Eventually, this will lead to the creation of the academic research programs in this area. It has been well perceived that any research area can only grow, when courses on that subject are being taught as an undergraduate or graduate level course in academic institutions.

Another challenge is the creation and maintenance of a large research infrastructure, spanning over the spectrum of difference science disciplines. Also, it is perceived that managing the facilities in physical/chemical/engineering sciences can be less challenging, when compared to maintaining cell biology facility and battery or solar module fabrication lines. Considering the requirement of distinctly different skillsets/ expertise, the set-up and maintenance of cell lines or a culture lab by non-biologists has often been a major challenge. On a similar note, identifying the essential parameters required and the issues that might arise during battery/module fabrications are not straightforward for an engineer/technician without a sound knowledge of the science and technology involved.

An extensive network of institutes/universities/national labs needs to be developed for training and dissemination of knowledge in the field of interdisciplinary engineering sciences. The cumulative outcome will have a greater societal impact.

The safety aspects here are also very important and demand dedicated training. For example, a very high level of safety is to be ensured in any laboratory or fabrication unit, which handles viruses and other harmful chemicals, including HF. All these challenges make the field of interdisciplinary engineering sciences even more interesting, considering the opportunities to explore a new horizon at the crossroads of multiple science and engineering disciplines.

The integration of concepts of multiple disciplines has profoundly impacted the developments in various sectors of society in the recent past. In view of the fact that the interdisciplinary nature of research and development has become the root cause for the evolution of modern technology for quantum improvement of our living standard, and so on, the introduction of such courses as part of curriculum becomes necessary.

17.8 Educational Impact

In the last few decades, it has been widely observed that more and more researchers with undergraduate or core education in particular science or engineering disciplines have significantly contributed to the field of interdisciplinary engineering sciences, often at the intersection with widely disconnected disciplines. In view of the growing importance of the field of interdisciplinary science, a number of universities around the world have already started new interdisciplinary academic programs for graduate students under the broad discipline name "Biological Engineering" or "Bio-engineering" or "Biosystems Science and Engineering" or "Medical Science and Technology" or "Clinical Engineering" or "Energy Science and Engineering." Such academic programs provide a useful platform for researchers from multiple disciplines to interact and jointly supervise PhD students.

Considering the ever-increasing role of engineering in medicine and public health, it is recommended that engineering education should include the elements of physiology (human anatomy) and other branches of medical sciences. Similarly, medical education should integrate engineering principles and this requires significant revision of the academic curriculum of undergraduate and postgraduate degrees in medical universities around the world. Several approaches are suggested to accomplish this, which include the introduction of MD–PhD programs and training of a complementary team of engineers and clinicians together in academia. While the first approach has been introduced in some US universities, this certainly does not exist in other parts of the world. Concerning the second approach, the junior undergraduate students from medical institutions can be involved in summer or winter workshops, and bioengineering immersion programs in academic institutions. Similarly, "clinical immersion" internship programs can be introduced in medical institutions for undergraduate students of engineering disciplines. All these programs are expected to inculcate curiosity and interest on the part of young researchers to cross the border of their own discipline. At this point, it is worthwhile to mention that a new biology approach for the twenty-first century is perceived to be the one that depends on effective integration within biology and closer collaboration of biologists with natural scientists and engineers to develop solutions for key societal needs.

Since interdisciplinary research programs involve scientists from multiple disciplines, it is important to share contributions from different people in research publication or invention disclosure or project reports. Sharing of contributions is relatively easier when researchers from the same research groups are involved. Certainly, it is more challenging when interdisciplinary research involves researchers from different universities, medical institutions, national laboratories, industries, and companies across the country or continent.

It will be worthwhile to mention here that interdisciplinary researchers should be conscious of the ethical and intellectual property (IP) aspect. Concerning IP aspects, any study or research project involving companies needs to comply with the non-disclosure agreement (NDA), which should include the general terms and conditions of working together as well as brief mention of major IPs to be shared with Academic Institutions/National Labs. Instead of an NDA, often a memorandum of understanding (MOU) is signed in such cases. Depending on the existing laws of the land, both the documents may have legal implications and therefore, a researcher should be aware of the same. It is important that researchers from academia should respect various pieces of confidential information on the process/product of the companies and those should not be disseminated in any public forum.

Here, it is also instructive to mention that the researcher should obtain approval from the Institutional biosafety committee, before conducting any experiment using cells, bacteria, virus, or live animals. For example, research in the field of biomedical engineering often involves animal experiments (pre-clinical study) and human subjects (clinical trials). Prior to the start of any animal experiment, the researcher should obtain necessary approval from the Institutional Ethical committee, in the case of non-primate studies and from National committees in the case of primate studies. The approval becomes more difficult in case of studies involving large animals (such as sheep, goat, monkey, etc.). Once a pre-clinical study is successfully completed and a positive outcome is recorded, clinical trials are conducted on those devices or implant materials. This again requires approval from the human ethics committee in universities or medical institutions. All the approved clinical trials are to be documented in the clinical registry data. In the case of use of stem cells in any study, additional approval from the Institutional committee on stem cells is necessary.

Overall, the interdisciplinary approach facilitates the cross-talk between various disciplines, which has the potential to further enrich the educational system. As time goes on, the interdisciplinary studies will become more and more demanding at higher levels of education. However, such educational programs should be designed in a way to establish a sturdy correlation and interaction between various disciplines.

The necessity of interdisciplinary education and research has been realized by the major funding agencies such as National Institute of Health (NIH)/NationalScience Foundation (NSF), USA, Department of Biotechnology (DBT) and Department of Science and Technology (DST), India. Interdisciplinary education and research centers are one of the prime objectives of many of the premier institutes/universities around the globe. Apart from interdisciplinary research and education centers, inter-professional programs are also being offered to promote the interdisciplinary approach. However, such opportunities need to be even more widespread.[21]

Owing to the complexity and severity of plenty of upcoming challenges in every sector such as, food, health, energy, environment, and safety, it is extremely desirable to train the next-generation researchers for the application of in-depth disciplinary knowledge as an interdisciplinary team.[22] In this practice, it has to be ensured that the strong disciplinary base is not sacrificed. The fruitful results of interdisciplinary advancement can be realized with the input of experts having a sound background in a particular discipline.[22]

This book makes an attempt to explore the integration of various fundamental knowledge organizations, which can effectively facilitate interdisciplinary approach. A reasonably better practical aspect of interdisciplinary research can be realized by crossing the boundaries of the conventional engineering and science disciplines. Overall, such an interdisciplinary approach is the way forward toward more efficient translation of fundamental knowledge to actual technology development, as is mandated in rapid and ever-increasing terms by the needs and demands of human society.

REFERENCES

1. Solanas G., S. A. Benitah, Regenerating the skin; a task for the heterogeneous stem cell pool and surrounding niche, *Nat. Rev. Mol. Cell Biol.*, 2013, 737–748.
2. Si-Tayeb K., F. P. Lemaigre, S. A. Duncan, Organogenesis and development of the liver, *Dev. Cell*, 2010, 18(2), 175–189.
3. Milner D. J., J. A. Cameron, Muscle repair and regeneration: Stem cells, scaffolds, and the contributions of skeletal muscle to amphibian limb regeneration, *Curr. Top. Microbiol. Immunol.*, 2013, 367, 133–159.
4. Gu X., F. Ding, D. F. Williams, Neural tissue engineering options for peripheral nerve regeneration, *Biomaterials*, 2014, 35, 6143–6156.
5. Lin Z., W. T. Pu, Strategies for cardiac regeneration and repair, *Sci. Transl. Med.*, 2014, 6(239), 236rv1.
6. Ludlow J. W., R. W. Kelley, T. A. Bertram, The future of regenerative medicine; Urinary system, *Tiss Eng. Part B*, 2013, 18(3), 218–224.
7. *Renewables 2016 Global Status Reports*: REN21(2016); ISBN 78-3-9818107-0-7.
8. *Renewables 2017 Global Status Reports*: REN21(2017); ISBN 978-3-9818107-6-9.

9. Chen F. M., L. A. Wu, M. Zhang, R. Zhang, H. H. Sun, Homing of endogenous stem/ progenitor cells for *in situ* tissue regeneration; promises, strategies and translation perspectives, *Biomaterials*, 2011, 32(12), 3189–3209.

10. Williams D. F., *Essential Biomaterials Science*, Cambridge University Press, 2014, p. 640.

11. Levin M., The wisdom of the body: Future techniques and approaches to morphogenetic fields in regenerative medicine, developmental biology and cancer, *Regen. Med.*, 2011, 6(6), 667–673.

12. Hoover-Plow J., Y. Gong, Challenges for heart disease stem cell therapy, *Vasc. Health Risk Manage.*, 2012, 8(1), 99–113.

13. Ramsden C. M., M. B. Powner, A.-J. F. Carr, M. J. K. Smart, L. de Cruz, P. J. Coffey, Stem cells in retinal regeneration: Past, present and future, *Development* 2013, 140, 2576–2585.

14. Williams D. F., To engineer is to create, *Trends Biotech.*, 2006, 24, 4–8.

15. Williams D. F., On the nature of biomaterials, *Biomaterials*, 2009, 30(30), 5897–5909.

16. Green M. A., S. P. Bremner, *Nat. Mater.*, 2017, 16, 23.

17. Polman A., M. Knight, E. C. Garnett, B. Ehrler, W. C. Sinke, *Science*, 2016, 352(6283), 307.

18. Peters J. F., M. Baumann, B. Zimmermann, J. Braun, M. Weil, *Renew. Sust. Energ. Rev.*, 2017, 67, 491.

19. Mukhopadhyay A., M. K. Jangid, *Science*, 2018, 359(6383), 1463.

20. Fang R., S. Zhao, Z. Sun, D.-W. Wang, H.-M. Cheng, F. Li, *Adv. Mater.*, 2017, 29, 1606823.

21. Gill S. V., M. Vessali, J. A. Pratt, S. Watts, J. S. Pratt, P. Raghavan, J. M. DeSilva, The importance of interdisciplinary research training and community dissemination, *Clin. Transl. Sci.*, 2015, 8(5), 611–614.

22. Bililign S., The need for interdisciplinary research and education for sustainable human development to deal with global challenges, *Int. J. Afr. Dev.*, 2013, 1(1), 82–90.

23. Nhiem T., M. Aparna, M. Dhriti, S. Arvind, N. Suprabha, J. W. Thomas, Bactericidal effect of iron oxide nanoparticles on Staphylococcus aureus, *Int. J. Nanomedicine*, 2010, 2010(5), 277–283.

24. Nath S., B. Basu, K. Biswas, K. Wang, R. K. Bordia, Sintering, phase stability and properties ofcalcium phosphate-mullite composites, *J. Am. Ceram. Soc.*, 2010, 93(6), 1639–1649.

25. Vasita R., M. Gopinath, C. Mauli Agrawal, D. S. Katti, Surface hydrophilization of electrospun PLGA micro-/nano-fibers by blending with Pluronic® F-108, *Polymer*, 2010, 51, 3706–3714.

26. Dubey A. K., B. Basu, K. Balani, R. Guo, A. S. Bhalla, Dielectric and Pyroelectric Properties of HAp-BaTiO 3 Composites, *Ferroelectrics*, 2011, 423(1), 63–76.

27. Dubey A. K., B. Basu, K. Balani, R. Guo, A. S. Bhalla, Multifunctionality of Perovskites $BaTiO_3$ and $CaTiO_3$ in a Composite with Hydroxyapatite as Orthopedic Implant Materials, *Integr. Ferroelectr.*, 2011, 131(1), 119–126.

28. Stout D. A., B. Basu, T. J. Webster, Poly Lactic-Co-Glycolic Acid: Carbon Nanofiber Composites for Myocardial Tissue Engineering Applications, *Acta Biomater.*, 2011, 7, 3101–3112.

29. Basu B., D. Jain, N. Kumar, P. Choudhury, A. Bose, S. Bose, P. Bose, Processing, tensile and fracture properties of Injection Molded HDPE-Al 2 O 3 -HAp Hybrid Composites, *J. Appl. Polym. Sci.*, 2011, 121, 2500–2511.

30. Seil J. T., T. J. Webster, Spray deposition of live cells throughout the electrospinning process produces nanofibrous three-dimensional tissue scaffolds, *Inte. J. Nanomedicine*, 2011, 6, 1095–1099.

31. Gupta S., T. J. Webster, A. Sinha, Evolution of PVA gels prepared without crosslinking agents as a cell adhesive surface, *J. Mater. Sci: Mater. Med.*, 2011, 22(7), 1763–1772.

32. Rajyalakshmi A., B. Ercan, K. Balasubramanian, T. J. Webster, Reduced adhesion of macrophages on anodized titanium with select nanotube surface features, *Int. J. Nanomedicine*, 2011, 6, 1765–1771.

33. Reddy S., A. K. Dubey, B. Basu, R. Guo, A. S. Bhalla, Thermal Expansion Behavior of Biocompatible Hydroxyapatite-BaTiO 3 Composites for Bone Substitutes, *Integr. Ferroelectr.*, 2011, 131(1), 147–152.

34. Kumar A., T. J. Webster, K. Biswas, B. Basu, Flow cytometry analysis of human fetal osteoblast fate processes on spark plasma sintered Hydroxyapatite-Titanium biocomposites, *J. Biomed. Mater. Res. Part A*, 2013, 101(10), 2925–2938.

35. Jain S., T. J. Webster, A. Sharma, B. Basu, Intracellular reactive oxidative stress, cell proliferation and apoptosis of Schwann cells on carbon nanofibrous substrates, *Biomaterials*, 2013, 34, 4891–4901.

36. Dyondi D., T. J Webster, R. Banerjee, A nanoparticulate injectable hydrogel as a tissue engineering scaffold for multiple growth factor delivery for bone regeneration, *Int. J. Nanaomedicine*, 2013, 8, 47–59.

37. Gupta S., T. Greeshma, A. Sinha, B. Basu, Stiffness and wettability dependent myoblast cell compatibility of transparent Poly (vinyl alcohol) hydrogels, *J. Biomed. Mat. Res. Part B: Appl. Biomater.*, 2013, 101(2), 346–354.

38. Sabareeswaran A., B. Basu, S. J. Shenoy, Z. Jaffer, N. Saha, A. Stamboulis, Early osseointegration of a strontium containing glass ceramic in a rabbit model, *Biomaterials*, 2013, 34, 9278–9286.

39. Alex S., R. Kumar, K. Chattopadhyay, H. C. Barshilia, B. Basu, Thermally evaporated Cu–Al thin film coated flexible glass mirror for concentrated solar power applications, *Mater. Chem. Phys.*, 2019, 232, 221–228.

40. Alex S., K. Chattopadhyay, B. Basu, Tailored specular reflectance of Cu- based novel intermetallic alloys, *Sol. Energ. Mat. Sol. C.*, 2016, 149, 66–74.

41. Alex S., S. Sengupta, U. K. Pandey, B. Basu, K. Chattopadhyay, Electrodeposition of δ-phase based Cu–Sn mirror alloy from sulfate-aqueous electrolyte for solar reflector application, *Appl. Therm. Eng.*, 2016, 109B, 1003–1010.

42. Dan A., B. Basu, T. Echániz, I. González de Arrieta, G. A. López, H. C. Barshilia, Effects of environmental and operational variability on the spectrally selective properties of W/ WAlN/WAlON/Al2O3–based solar absorber coating, *Sol. Energ. Mat. Sol. C.*, 2018, 176, 157–166.

43. Dan A., A. Biswas, P. Sarkar, S. Kashyap, K. Chattopadhyay, H. C. Barshilia, B. Basu, Enhancing spectrally selective response of WWAlN/WAlON/Al2O3–based nanostructured multilayer absorber coating through graded optical constants, *Sol. Energ. Mat. Sol. C.*, 2018, 176, 157–166.

44. Dan A., J. Jyothi, H. C. Barshilia, K. Chattopadhyay, B. Basu, Spectrally selective absorber coating of WAlN/WAlON/Al2O3 for solar thermalapplications, *Sol. Energ. Mat. Sol. C.*, 2016, 157, 716–726.

45. Dan A., H. C. Barshilia, K. Chattopadhyay, B. Basu, Angular solar absorptance and thermal stability of WAlN/WAlON/Al2O3–based solar selective absorber coating, *Appl. Therm. Eng.*, 2016, 109 B, 997–1002.

46. Dan A., K. Chattopadhyay, H. C. Barshilia, B. Basu, Colored selective absorber coating with excellent durability, *Thin Solid Films*, 2016, 620, 17–22.

FURTHER READING

Langer R., D. Tirrell, Designing materials for biology and medicine, *Nature*, 2004, 428(6982), 487–492.

Mitragotri S., J. Lahann, Physical approaches to biomaterial design, *Nat. Mater.*, 2009, 8(1), 15–23.

Orive G., E. Anitua, J. L. Pedraz, D. F. Emerich, Biomaterials for promoting brain protection, repair and regeneration, *Nat. Rev. Neurosci.*, 2009, 10(9), 682–692.

Place E. S., N. D. Evans, M. M. Stevens, Complexity in biomaterials for tissue engineering. *Nat. Mater.*, 2009 Jun, 8(6), 457–470.

Wang D.-A. et al. Multifunctional chondroitin sulphate for cartilage tissue-biomaterial integration. *Nat. Mater.*, 2007, 6(5), 385–392.

Zhang S., Fabrication of novel biomaterials through molecular self-assembly, *Nat. Biotechnol.*, 2003, 21(10), 1171–1178.

Appendix

I. Multiple Choice Questions

Please select the appropriate answer from the choices provided below the questions:

1. The biocompatibility encompasses
   a. Cell type-dependent response *in vitro*
   b. Animal strain-dependent tissue response *in vivo*
   c. Blood compatibility
   d. All of the above

2. The static charge produces
   a. Electrostatic field
   b. Magnetic field
   c. Electromagnetic field
   d. None of the above

3. Moving charges produce
   a. Electric and magnetic field
   b. Only electric field
   c. Only magnetic field
   d. None of the above

4. The ionic pumps, present in the cell membrane,
   a. Maintain the concentration gradient of ions across the membrane
   b. Maintain the transmembrane potential across the cell membrane
   c. Utilize the energy inside the cell for transportation
   d. All of the above

5. The resting potential of the cell membrane is maintained through
   a. Synchronized functioning of ionic pumps, channels, carriers, and transporters
   b. Only ionic pumps
   c. Only transporters
   d. None of the above

6. Proteins embedded in the cell membrane are called
   a. Integral membrane proteins
   b. Peripheral membrane proteins
   c. Gated channels
   d. None of the above

7. Cell membrane contains open pathways to water molecules and these are called
   a. Ion-selective channels
   b. Aquaporins
   c. Carriers
   d. None of the above

8. Gated channels present in the cell membrane open and close according to external stimulation of type?
   a. Mechanical
   b. Electrical
   c. Synaptic
   d. All of the above

9. Which is/are the most prominent gated channels present in the cell membrane?
   a. Ca^{2+} channels
   b. K^+ channels
   c. Na^+ channels
   d. All of the above

10. The phenomenon of simultaneous force of diffusion and electric field acting on the ions present across the cell membrane is called
    a. Fick's law of diffusion
    b. Electrodiffusion
    c. Both (a) and (b)
    d. None of the above

11. Energy utilized to move the ions against their concentration gradient comes from
    a. Cytoplasm
    b. Extracellular matrix
    c. ATP hydrolysis
    d. None of the above

12. Pumps through which a net transfer of charge takes place are called
    a. Electrogenic
    b. Na-K pump
    c. Both (a) and (b)
    d. Ca^{2+} pumps

13. If the membrane potential is equivalent to the equilibrium potential of any ion X, then which statement is correct.
    a. Ion X would be impermeable across the membrane
    b. Ion X would be the only ion permeable across the membrane
    c. Ion X would be the only ion permeable across the membrane and its net flux across it would be zero
    d. None of the above

14. Endogenous electric field in living cells arises from
    a. Conductive nature of cell membrane
    b. Presence of various inorganic ions across the membrane
    c. Both (a) and (b)
    d. None of the above

15. Lipids in the cell membrane are
    a. Non-conductive
    b. Conductive
    c. Semiconductive
    d. None of the above

16. Proteins in the cell membrane
    a. Non-conductive
    b. Conductive
    c. Semiconductive
    d. None of the above

17. The equivalent circuit of an extracellular matrix is composed of
    a. Combination of resistance and capacitance
    b. Only resistance
    c. Only capacitance
    d. None of the above

18. Nuclear membrane is composed of
    a. Single bilipid layer
    b. Double bilipid layer
    c. Double bilipid layer with pores
    d. None of the above

19. The effect of E-field on the intracellular membranes is called,
    a. Extracellular electromanipulation
    b. Intracellular electromanipulation
    c. Electroporation
    d. None of the above

20. The loss of semipermeability of cell membrane through the application of E-field is called
    a. Electroporation
    b. Necrosis
    c. Electromanipulation
    d. All of the above

21. The voltage-gated channels can be represented by
    a. Combination of resistance and capacitance
    b. Combination of resistance, and battery
    c. Only resistance
    d. Only capacitance

22. The critical membrane potential is the potential of the cell membrane at which
    a. Cell membrane losses its semipermeability
    b. Cell membrane becomes conductive in nature
    c. Both (a) and (b)
    d. None of the above

23. Human bone is an example of
    a. Functionally graded material
    b. Single phase material
    c. Two phase material
    d. None of the above

24. Functional performance of any material system can be improved without any effect on its chemistry by means of development of
    a. Composite material
    b. Functionally graded material
    c. Multiphase material
    d. None of the above

25. One of the critical issues in developing functionally graded material of multicomponent system is
    a. Mismatch of thermal coefficient of expansion
    b. Surface chemistry
    c. Processing temperature
    d. None of the above

26. NKN, $BaTiO_3$ and $CaTiO_3$ have the common crystal structure, according to the notation
    a. ABO_2
    b. ABO_3
    c. Both (a) and (b)
    d. None of the above

27. Which is the most efficient piezoelectric material?
    a. NKN
    b. $BaTiO_3$
    c. $CaTiO_3$
    d. Hydroxyapatite

28. Electrets are the materials, which on application of an E-field at elevated temperatures exhibits
    a. Spontaneous polarization
    b. Permanent polarization
    c. Remnant polarization
    d. None of the above

29. The Curie temperature of the material is the temperature at which
    a. Crystal structure changes
    b. Domains are aligned
    c. Defects are created
    d. None of the above

30. For a perfect dielectric material the phase difference between the applied voltage and current is
    a. 45°
    b. 90°
    c. 180°
    d. None of the above

31. The dipolar and space charge polarization can be differentiated on the basis of
    a. Maxwell Wagner relaxation process
    b. Thermally stimulated depolarization current
    c. Both (a) and (b)
    d. None of the above

32. Dipolar polarization exhibits in molecules which possess,
    a. Permanent dipole moment
    b. Temporary dipole moment
    c. Ionic/covalent bonding
    d. None of the above

33. The space charge polarization in functionally graded materials is associated with
    a. Grain boundaries
    b. Interfacial regions
    c. Both (a) and (b)
    d. None of the above

34. During high-temperature processing, the perovskite such as calcium titanate/barium titanate develops
    a. Titanium vacancies
    b. Oxygen vacancies
    c. Calcium/barium vacancies
    d. None of the above

35. It is not advisable to use ceramics for which one of the following applications?
    a. Tiles for rocket nozzles
    b. Cutting tool inserts
    c. Cathodes for Li-ion batteries
    d. Body of an aircraft

36. Which of the following may best classify as "ultra-high temperature ceramics"?
    a. Al_2O_3
    b. ZrO_2
    c. ZrB_2
    d. SiC

37. Which of the following advanced sintering techniques may lead to near-theoretical densification with the least total duration?
    a. Hot pressing
    b. Hot isostatic pressing
    c. Microwave sintering
    d. Spark plasma sintering

38. In which of the following advanced sintering techniques there is greater chance of inhomogeneous densification from edge to the core, especially for insulating ceramics?
    a. Hot pressing
    b. Hot isostatic pressing
    c. Microwave sintering
    d. Spark plasma sintering

39. Processing of cemented carbides and cermet (such as WC-Co or TiB_2-Co) involves
    a. Solid state sintering
    b. Liquid phase sintering
    c. Melting and casting
    d. Vitrification

40. Which of the following indenters are typically used to determine fracture toughness of ceramics?
    a. Rockwell
    b. Vickers
    c. Brinell
    d. None of the above

41. For amorphous ceramics (glasses), reducing the overall dimension usually leads to an improvement in the fracture strength, because
    a. Yield strength increases
    b. Dislocation pile-up lengths decreases and hence fracture toughness increases
    c. Probability of finding critical sized flaw decreases
    d. Of quantum confinement effects

42. The typically used parameter called fracture toughness corresponds to which of the following loading types, with respect to the pre-existing flaws?
    a. Mode I.
    b. Mode II.
    c. Mode III.
    d. Independent of the loading type.

43. Yttria-stabilized tetragonal zirconia polycrystals (Y-TZP) is one of the tougher ceramics, primarily because of which of the following phenomenon?
    a. Transformation toughening.
    b. Residual stress.

    c. Ductile metal bridging.

    d. Fiber reinforcement.

44. Which of the following test configurations is usually used to get a reliable estimate of the bulk fracture strength of polycrystalline ceramics?

    a. Tensile.

    b. Compressive.

    c. Fatigue.

    d. Bend.

45. Based on the standard reduction potentials (E°), as presented below, which of the following options is correct for an electrochemical cell made from Co and Ag as electrode materials?

$$Co^{3+} + e^- \leftrightarrow Co^{2+} \quad E° = +1.81 \text{ V}$$
$$Ag^+ + e^- \leftrightarrow Ag \quad E° = +0.80 \text{ V}$$

    a. The standard EMF of the cell is 2.61 V

    b. Co is anode and Ag is cathode

    c. Co is cathode and Ag is anode

    d. One of either (a and c) or (b and c)

46. What would be the standard Gibbs free energy change for the forward (spontaneous) reaction for the cell corresponding to the reactions in question 11?

    a. −97,465 J/mol

    b. +97,465 J/mol

    c. −251,865 J/mol

    d. +251,865 J/mol

47. Which of the following energy storage applications stores energy in the form of electrical charges?

    a. Battery

    b. Capacitor

    c. Pumped hydro

    d. Flywheel

48. Which of the following electrochemical energy storage/conversion devices has the highest energy density?

    a. Batteries

    b. Fuel cells

    c. Capacitors

    d. Supercapacitors

49. Which of the following rechargeable battery systems has the highest energy density?

    a. Pb-acid

    b. Ni-Cd

    c. Ni-MH

    d. Li-ion

50. A EDLC electrode material would be called mesoporous carbon if it contains pores of sizes between
    a. 0.5–1.5 nm
    b. 15–45 nm
    c. 75–90 nm
    d. 2–50 μm

51. With respect to the basic equation for capacitance (C) (viz., $C = A\varepsilon_0\varepsilon_r/d$), which of the following statements are correct, with respect to favorably tuning the parameters in the case of activated carbon?
    a. Only A increases
    b. Only d decreases
    c. A increases and d decreases
    d. Both A and d increases

52. Bonding between Li and host in anode material, with respect to that for the corresponding cathode material, is likely to be;
    a. Stronger
    b. Weaker
    c. Of similar strength
    d. Bonding strength has no relevance

53. A graphitic carbon thin film electrode has a thickness of 100 nm and uniformly covers 90% area of one of the surfaces (i.e., faces) of a circular current collector foil of diameter 2.54 cm and height 0.25 cm. If the density of graphite is 2.23 g/cm³ and the theoretical Li-capacity is 372 mAh/g, what would be the current equivalent to C/5?
    a. 189.15 μA
    b. 33.63 μA
    c. 7.57 μA
    d. 1.68 μA

54. On which of the factors, as below, does the equilibrium electrochemical potential of Li-insertion in a host electrode lattice (based on transition metal oxide) does not depend?
    a. Energy level of the d orbital of the concerned transition metal ion.
    b. Crystal structure of the electrode material.
    c. The energy of site where the guest Li-ion gets hosted.
    d. The current density at which the electrochemical cycling is conducted.

55. The filling of molecular orbital with electrons takes place according to
    a. Hund's rule of maximum multiplicity
    b. Pauli exclusion principle
    c. The Aufbau principle
    d. All the above

56. According to MO (or molecular orbital) theory, the shape and size of a MO, which is formed due to combining of atomic orbitals (AO) depends on
    a. Orientation of the AOs
    b. Numbers of the AOs
    c. Shape and size of the AOs
    d. All the above

57. In the case of the formation of a double bond between two constituent atoms, when one of the bonds is a σ (or sigma) bond, with the other being a π (or pi) bond, overlap of what is responsible for the formation of the π bond?
    a. *s* orbitals
    b. *p* orbitals
    c. *sp* hybrid orbitals
    d. *sp²* hybrid orbitals

58. Directional preference is not shown by the s-orbital due to it
    a. Being the first orbital.
    b. Being the smallest orbital in terms of distance away from the nucleus.
    c. Being compulsorily present in every atom.
    d. Possessing spherical symmetry.

59. N_2 and O_2 have the following bond orders,
    a. +3 and +2, respectively.
    b. +2 and +3, respectively.
    c. +3 and +1, respectively.
    d. +2 and +1, respectively.

60. The energy source that currently supplies the maximum portion of the world's total energy demand is
    a. Coal
    b. Oil
    c. Natural gas
    d. Uranium

61. Which of the following is likely to increase greenhouse gas concentrations in the atmosphere?
    a. Using natural gas instead of coal to generate electrical energy
    b. Incineration of waste to generate electrical energy
    c. Increased use of wind turbines to generate electrical energy
    d. Carbon dioxide capture and storage at the power station

62. The unit of energy density of a fuel is
    a. J/m^2
    b. J/m^3
    c. J/kg
    d. kg/J

63. Which one is not correct?
    a. Generators convert mechanical energy into electrical energy
    b. Nuclear reactors convert mass into energy
    c. Chemical energy is a form of potential energy
    d. Thermal energy and solar energy are the same

64. The solar energy radiated from the sun is in the form of _____
    a. Ultraviolet radiation
    b. Infrared radiation
    c. Electromagnetic waves
    d. Transverse waves

65. Solar radiation received at any point of earth is called _____
    a. Insulation
    b. Beam radiation
    c. Diffuse radiation
    d. Infrared rays

66. The solar radiation lies from
    a. $0.2–2.5\ \mu m$
    b. $0.38–0.78\ \mu m$
    c. $0–0.38\ \mu m$
    d. $0.5–0.8\ \mu m$

67. How long does it take for sunlight to travel from the sun to earth?
    a. 8 s
    b. 8 min
    c. 0.8 min
    d. 8 h

68. More energy from the sunlight falls on the Earth in _____ than the total energy used by the entire population of the planet in one year.
    a. 1 h
    b. 5 h
    c. 10 h
    d. 24 h

69. Which of the following area is preferred for solar power plants
    a. Coastal areas
    b. Hot arid zones
    c. Mountain tops
    d. High rainfall zones

70. Global warming reduces the ice and snow cover on Earth. Which of the following correctly describes the changes in albedo and rate of energy absorption by Earth?

    Albedo: The proportion of the incident light that is reflected by the surface of the earth.

	Albedo	Rate of Energy Absorption
a.	Increase	Increase
b.	Decrease	Increase
c.	Increase	No change

Reason: Rate of energy conservation law ($A + R + T = 1$, $T = 0$ for earth, $A + R = 1$).

71. Which of the following is a practical use for solar energy?
    a. Heating swimming pool
    b. Cooking rice
    c. Powering road signs
    d. All of the above

72. Direct solar energy is used for
    a. Water heating
    b. Distillation
    c. Drying
    d. All of the above

73. Which of the following is true about solar energy?
    a. It is not difficult to produce photovoltaic cells
    b. Solar energy can currently replace all of the energy created by fossil fuels
    c. Most solar panels convert more than 25% of the light that strikes them

74. The annual variations of solar power incident per unit area at a particular point on the Earth's surface is mainly due to the change in the
    a. Distance between the Earth and the Sun.
    b. Angle at which the solar rays hit the surface of the Earth.
    c. Average albedo of the Earth.
    d. Average cloud cover of the Earth.

75. What is the most common way that solar energy is converted into electricity?
    a. Concentrating solar power plants assembled around the world
    b. Photovoltaic (PV) cells that use semiconductor technology
    c. Chemical reactions performed in laboratories that harness energy from the Sun
    d. Through the atmosphere as sunlight crosses the earth's ozone layer

76. When sunlight is incident on a solar cell an electric current is produced. This is due to
    a. A temperature gradient within the cell
    b. Very long wavelength infrared radiation
    c. Very short ultraviolet radiation
    d. The photoelectric effect

77. Which of these following is not utilized to harness solar energy?
    a. Gas
    b. Mirror
    c. PV cells
    d. All of the above

78. What is a disadvantage of solar energy?
    a. The amount of sunlight that arrives on the earth's surface is not constant
    b. A large surface area is required to collect the sun's energy at a useful rate
    c. Some toxic materials and chemicals are used in the manufacturing process of PV cells
    d. All of the above

79. For satellites the source of energy is
    a. Solar cell
    b. Fuel cells
    c. Edison cells
    d. Any of the above

80. In PV, the region where the electrons and holes diffused across the junction is called
    a. Depletion junction
    b. Depletion region
    c. Depletion space
    d. Depletion boundary

81. A module in a solar panel refers to
    a. Series arrangement of solar cells
    b. Parallel arrangement of solar cells
    c. Series and parallel arrangement of solar cells
    d. None of the above

82. The maximum theoretical efficiencies of solar cells could be around
    a. 79%
    b. 60%
    c. 48%
    d. 1%

83. The efficiency of most of the commercial solar cell is about
    a. 25%
    b. 15%
    c. 40%
    d. 60%

84. The output of the solar cell is of the order
    a. 0.5 W
    b. 1.0 W
    c. 5.0 W
    d. 10.25 W

85. What is the purpose of an inverter in an active solar system?
    a. Convert alternating current (AC) to direct current (DC)
    b. Convert direct current (DC) to alternating current (AC)
    c. Change the flow of electrons from positive to negative
    d. Change the flow of electrons from negative to positive

86. What is the major element found in solar cell?
    a. Ge
    b. Si
    c. K
    d. C

87. One disadvantage of using photovoltaic cells to power a domestic water heater is that
    a. Solar energy is a renewable source of energy
    b. The power radiated by the Sun varies significantly depending on the weather
    c. Large area of photovoltaic cells would be needed
    d. Photovoltaic cells contain CFCs, which contribute to the greenhouse effect

88. A liquid flat plate collector is usually held tilted in a fixed position, facing _____ if located in the Northern Hemisphere.
    a. North
    b. South
    c. East
    d. West

89. The collection efficiency of a flat plate collector can be improved by
    a. Putting a selective coating on the plate
    b. Evacuating the space above the absorber plate
    c. Both (a) and (b)
    d. None of the above

90. Maximum efficiency is obtained using a
    a. Flat plate collector
    b. Evacuated tube collector
    c. Line focusing collector
    d. Paraboloid dish collector

91. The following type of energy is stored as latent heat
    a. Thermal energy
    b. Chemical energy
    c. Electrical energy
    d. Mechanical energy

92. Which of the following type of collector is used for low-temperature systems?
    a. Flat plate collector
    b. Line focusing parabolic collector
    c. Paraboloid dish collector
    d. All of the above

93. Reflecting mirrors used for exploiting solar energy are called _____.
    a. Mantle
    b. Ponds
    c. Diffusers
    d. Heliostats

94. For paraboloid dish concept, the tracking of the sun can be done by rotating about
    a. One axes
    b. Two axes
    c. Three axes
    d. None of the above

95. Heat engines
    a. Produce more work output than energy input
    b. Take in thermal energy at a low temperature and exhaust it at high temperature
    c. Convert heat into mechanical energy
    d. Can be close to 100% efficient

96. Two different objects that have different temperatures are in thermal contact with one another. It is the temperatures of the two objects that determines,
    a. The amount of internal energy in each object
    b. The process by which thermal energy is transferred
    c. The specific heat capacity of each object
    d. The direction of transfer of thermal energy between the objects

97. A generator takes in an amount E_k of kinetic energy. An amount W of useful electrical energy is produced. An amount Q of thermal energy is lost due to the moving parts of the generator. The law of conservation of energy and the efficiency of the generator are given by which of the following?

    Law of conservation of energy, Efficiency
    a. $E_k = W + Q$, W/E_k
    b. $E_k = W + Q$, W/Q
    c. $E_k = W - Q$, W/Q
    d. $E_k = W - Q$, $W/(E_k - Q)$

98. The following is (are) laws of blackbody radiation.
    a. Plank's law
    b. Stefan–Boltzmann law
    c. Both (a) and (b)
    d. None of the above

99. The relative intensities of emitted wavelengths of a perfect black body depends on
    a. The surface area of the black body
    b. The temperature of the black body
    c. The radiation per square meter
    d. The radiation per second

100. The diagram shows the variation with wavelength of the power per unit wavelength (I) radiated from an area of 1 m² of two different bodies. Which of the following is a correct comparison of the temperature and of the emissivity of two bodies?

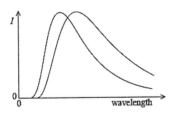

	Temperature	Emissivity
a.	Same	Same
b.	Same	Different
c.	Different	Same
d.	Different	Different

101. The average temperature of the surface of the Sun is about 20 times more than the average surface temperature of the Earth. The average power per unit area radiated by the Earth is R. The average power per unit area radiated by the Sun is
    a. 20 R
    b. 400 R
    c. 8000 R
    d. 160,000 R

102. Absorption of solar radiations at earth's surface occurs due to the presence of
    a. Ozone
    b. Water vapors
    c. Carbon dioxide
    d. All of the above

103. Global radiation can be represented as
    a. Direct radiation – diffuse radiation
    b. Direct radiation + diffuse radiation
    c. Direct radiation/diffuse radiation
    d. Diffuse radiation/direct radiation

104. The emissivity of a thin film depends on
    a. Substrate
    b. Roughness
    c. Temperature
    d. All of the above

105. What is sputtered yield?
    a. The number of sputtered atoms per unit solid angle
    b. The number of sputtered atoms per incident particle

c. The number of sputtered atoms per minute

d. None of the above

106. Which does not come under physical vapor deposition?

a. Thermal evaporation

b. Sputtering

c. Dip coating

d. None of the above

107. Metals are deposited by

a. DC sputtering

b. RF sputtering

c. Reactive sputtering

d. All of the above

108. A perfectly blackbody sphere is at a steady temperature of 473 K and is enclosed in a container at absolute zero temperature. It radiates thermal energy at a rate of 300 J/s.

I. If the temperature of the sphere is increased to 946 K it radiates heat at a rate of

a. 300 W

b. 1000 W

c. 3200 W

d. 4800 W

II. If the radius of the sphere is doubled it radiates heat at a rate of

a. 300 W

b. 1200 W

c. 3200 W

d. 4800 W

III. If the enclosure is at 473 K the net rate of heat loss would be

a. 0 W

b. 300 W

c. 1200 W

d. 100,000 W

Reason:

I. Stefan–Boltzmann law states that the total radiant heat energy emitted from a surface is proportional to the fourth power of its absolute temperature.

II. The same as the rate is only dependent on the temperature of the black body.

III. There is no thermal energy transfer and so no net rate of heat loss.

109. Each square meter of the Sun's surface emits S joules per second. The radius of the Sun is r, and the Sun is at a mean distance R from the Earth. Which of the following gives the solar power incident per unit area of the top layer of the Earth's atmosphere?

a. $(r/R)S$

b. $(r/R)^2S$

c. $(R/r)S$

d. $(R/r)^2S$

Reason: $4\pi R^2 x = 4\pi r^2 S$ Hence, $x = (r/R)^2S$

110. For sharp leading edge of Mach 7 hypersonic cruise vehicle flying at altitude of 30 km, what is the temperature of the operating environment?

a. 273 K

b. 1073 K

c. 1873 K

d. 773 K

111. Ultra-high-temperature ceramics for aerospace applications are qualified for their material performance using the following:

a. Arcjet facilities

b. Hypersonic shock tubes/tunnels

c. Scramjet facilities

d. All of the above

112. Coupled thermo-structural analysis can predict

a. Von Mises stress

b. Relative displacement

c. Temperature

d. All of the above

113. Highest heating rate can be accomplished by,

a. SPS

b. Hot pressing

c. Pressureless sintering

d. Both (b) and (c)

114. Arc jet facilities can run for a duration of

a. A few seconds

b. A few minutes

c. A few hours

d. A few milliseconds

115. Experimental test duration in a high pressure shock tube test is,

a. A few seconds

b. A few minutes

c. A few hours

d. A few milliseconds

116. Typical SPS duration is of the order of

a. A few seconds

b. A few minutes

c. A few hours

d. A few milliseconds

117. If a surface is opaque the relation between reflectance (*R*), transmittance (*T*), and absorptance (*A*) for the surface is
    a. $A + R + T = 1$
    b. $R + T = 1$
    c. $R = 1 - A$
    d. $R = 1 + A$

118. Which of the following is a practical use for solar energy?
    a. Heating swimming pool
    b. Cooking rice
    c. Powering road signs
    d. All of the above

119. What is the disadvantage of solar energy?
    a. The amount of sunlight that arrives on the earth's surface is not constant
    b. A large surface area is required to collect the sun's energy at a useful rate
    c. Some toxic materials and chemicals are used in the manufacturing process of PV cells
    d. All of the above

120. Among the following, elastic modulus is the highest for
    a. Al_2O_3
    b. Al-alloys
    c. Ti-alloys
    d. Steel

II. Fill in the Blanks

1. UHTC stands for _____.
2. SPS is the abbreviated form of _____.
3. Typical shock Mach number achievable in free piston shock tube is in the range of _____.
4. Computational fluid dynamics based conjugate heat-transfer analysis coupled _____ and _____.
5. An example of transition metal boride is _____.
6. Compatibility of a material with the blood is referred to as _____.
7. For bone mineralization the most suitable Ca and P ratio in HA is _____.
8. Two major components of natural bone are _____.
9. HA is a form of _____.
10. Nucleus can be visualized under microscope using _____ staining agent.
11. α-helix and β-sheet represent the _____ formats of Proteins.

12. Transmembrane proteins are attached to _____.
13. Triple helix structure is exhibited by _____.
14. Smallest structure-function unit of natural bone is _____.
15. Hypertrophy means _____.
16. Apoptosis is a cell fate process that represents _____.
17. Stochiometric formula of hydroxyapatite is _____.
18. Ratio of stress and strain that a material can tolerate without permanent deformation is defined by its _____.

III. Analytical Problems

1. If 500 J of work is done on a system, which, in turn, gives off 200 J of heat, what is the change in internal energy in the entire process?

2. Consider the full cell reaction $A + X^{2+}Y^{2-} \leftrightarrow X + A^{2+}Y^{2-}$. If, under standard conditions, the net free energy change for the above overall reaction is -200 kJ, estimate the EMF of the cell.

3. Consider an electrochemical cell (having all the contents at standard states) is connected to an external resistor during discharge, with the cell and the resistor being placed in separate calorimeter compartments (under 1 atm pressure and at 298 K; i.e., all under standard conditions). Assume that the useful work done by the cell during discharge is only obtained as heat energy due to the passage of finite current across the resistor of resistance R. In the situation, when $R \rightarrow \infty$, the heat evolved from the compartment having only the cell is -5000 J, while that from the compartment having only R is $-20,000$ J. What would be the standard enthalpy change for the cell reaction? What would be the net heat evolved if both the cell and the resistor are placed in the same calorimeter compartment? (Partly adapted from *Electrochemical Methods: Fundamentals and Applications*; by Allen J. Bard and Larry R. Faulkner; 2nd edition; John Wiley & Sons, Inc.)

4. If the net current drawn from an electrochemical cell is 500 mA, what would be the net rate of reduction of Fe^{2+} to metallic Fe (in mol/s/cm^2) at the cathode of 5 cm^2 in area?

5. Consider the electrode reaction $A + e^- \leftrightarrow B$. Under the conditions that $C^*_R = C^*_O = 1$ mM, $k^0 = 10^{\sim7}$ cm/s, $\alpha = 0.5$ (where the symbols have their usual meanings) and area of the electrode is 1 cm^2:

    a. Calculate the exchange current (i_0) in µA.

    b. Draw a current density (i)—overpotential (η) curve for this reaction for anodic and cathodic currents up to 600 µA/cm^2. Mass-transfer effects need to be neglected.

    c. Draw log $|i|$ vs. η curves for the current ranges in (b). Indicate the values for the slope and Y-axis intercept. (Partly adapted from *Electrochemical Methods: Fundamentals and Applications*; by Allen J. Bard and Larry R. Faulkner; 2nd edition; John Wiley & Sons, Inc.)

6. Considering a full cell reaction, $2H_2 + O_2 \rightarrow 2H_2O$, where the reactants, namely, O_2 (g) is supplied to the cathode and H_2 (g) is supplied to the anode, if the corresponding cell delivers a current of 0.5 A for 10.0 min at a potential of 0.6 V, (a) how much electrical energy (in J) would be provided by the cell? (b) If this particular cell operated at 60.0% efficiency, what amount of hydrogen gas (H_2), in mole, would be consumed by the cell?

7. Consider a typical electrical double-layer capacitor (EDLC) consisting of conductors dipped in water containing some dissolved salts, acting as dielectric with a dielectric constant of 80. If the radius of the cation of the dissolved salt is 0.2 nm, the radius of water molecule is 0.1 nm and the distance between the negative electrode surface and the center of the outer Helmholtz plane is 0.6 nm, estimate the double-layer specific capacitance (per unit area) of the electrode (permittivity of free space = 8.854×10^{-12} F/m).

8. If the maximum amount of Li that can be hosted by Si corresponds to $Li_{4.4}Si$, given the molar mass of Si being 28 g, what would be the theoretical capacity and a current equivalent to 4C for an electrode consisting of 7 mg of Si.

9. For a thin film electrode fabricated via electrodeposition of Sn onto a 230 μm thick quartz substrate of 1 inch diameter (and coated with 50 nm thick Ni), with the Sn covering the entire surface uniformly upon electrodeposition up to a thickness of 200 nm. Estimate the currents equivalent to C/5 and 5C for such an electrode (given specific capacity of Sn ~1000 mAh/g and density of Sn ~ 7 g/cm³). At which of the above currents, it is expected for the electrode to reach closer to the theoretical capacity during electrochemical lithiation?

10. For a 24 kWh, 384 V "battery pack," which is made of 48 equivalent modules in series, which has cells of ~24 Ah capacity and 4 V as the operating voltage, please indicate what the modules are made of and what is the energy density of each module.

11. The following plots, (a) and (b), are results obtained during galvanostatic cycling of thin film electrode (anode material for Li-ion full cell) in a Li "half-cell". (The scales of both the plots can be assumed to be consistent with each other.) Please identify, citing reasons, which one of them is for the 1st cycle and which is for the 10th cycle. Present rough estimations (showing steps/logics properly) for the Columbic efficiencies in both the cases. What could be the contributing factor(s) toward such observations? What are the cathode and anode materials for the cell under consideration here?

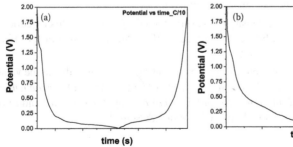
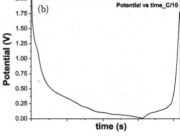

12. A thermal power station is 20% efficient and generates useful electrical power at 1000 MW. The fossil fuel used has an energy density of 50 MJ/kg. What was the mass of fuel in kg consumed every second?

13. In the model, the intensity radiated from the ground equals the intensity radiated from the atmosphere toward the ground. The temperature of ground is T_g and is assumed to radiate as a black body. The temperature of atmosphere is T_a and has an emissivity ε. What is the ratio of T_g/T_a as a function of ε?

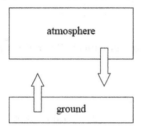

14. A home in Phoenix (Arizona) requires 62 kWh of heat on a winter day to maintain a constant indoor temperature of 20°C. (a) How much collector surface area does it need for an all-solar heating system that has a 20% efficiency? Please consider that the average solar radiation in winter in Phoenix is about 6.5 kWh/m²/day.

15. How much collector area would a 1000-MW solar farm require if the individual efficiencies of the collector system, turbine, and generator are 30, 25, and 90%, respectively? Let us assume that the average incident solar radiation at the proposed site of the plant is 200 W/m².

IV. Short Answer Type Questions

1. Explain the heating of water with the help of plasmonic materials using two examples.

2. Schematically explain the following:
   a. Different types of selective surfaces for solar thermal application
   b. Working principle of magnetron sputtering
   c. Sol–gel process
   d. Optical polarization
   e. Three absorption mechanisms in graded multilayer absorber

3. Schematically illustrate the following:
   a. Two different methods of toughening mechanisms (one each from process zone and bridging zone)
   b. Fracture toughness measurement of ceramic implants using indentation cracking and SEVNB technique
   c. Difference between hardness and strength (definition and measurement technique) of ceramics
   d. Glide of a positive edge dislocation in metal and stresses around the defect

    e. The differences between polymer, ceramic, and ductile metals with respect to tensile stress–strain response

    f. Different modes of loading and displacements at crack faces in ceramics (mode I, II, and III)

    g. Evolution of cone cracks in alumina implant under indentation loading

    h. Difference in mechanisms of bulk and surface erosion of polymers

    i. Difference between precipitation hardening and dispersion strengthening

    j. Stress–strain response and the crack propagation in ceramics under uniaxial compression

4. Distinguish with suitable graphical illustrations and examples (wherever applicable):

    a. Scaffold vs. implant

    b. Cytotoxicity vs. genotoxicity

    c. Crystallinity in metals vs. crystallinity in polymers

    d. Hardness vs. strength (in case of metals)

    e. Strengthening vs. toughening mechanism

    f. *In vitro* vs. *in vivo* tests

    g. Hot pressing vs. spark plasma sintering

    h. Cell apoptosis vs. cell necrosis

5. How does cell spreading on material surface occur? Explain the steps of cell material interaction with diagram?

6. Define tissue. With help of a schematic, explain the formation of myotubes in the context of muscle tissue.

7. Explain briefly the guidelines to be followed for the following property measurements in case of bioceramic implants: (a) hardness, (b) fracture toughness, and (c) strength.

V. Descriptive Questions

1. Present a typical plot, with proper labeling, that you would usually obtain from a fairly standard cyclic voltammetry experiment, where the associated electrochemical reactions are reversible. Why is there a gap between the oxidation and reduction peaks and what would cause the gap to get minimized?

2. What is a Ragone plot. Present a typical "Ragone plot," showing the relative positions of different electrical and electrochemical energy storage and conversion devices. The plot may be qualitative in nature (i.e., axes may have arbitrary values, if any, but with acceptable units).

3. What are the potential safety hazards and the scientific causes for them in the present generation Li-ion batteries?

4. It is usually found that Li-ion battery anodes comprised of 50 nm sized Si particles as active materials perform better than anodes comprising of 500 nm sized Si particles as active materials. Please indicate which performance(s) are being talked about here and why it/they is/are usually superior for the

former. Is any drawback expected for the 50 nm sized particles, as compared to the 500 nm sized particles?

5. Please state with proper reasoning, which among the two electrodes, viz., $LiNi^{(2+)}_{1/3}Co^{(3+)}_{1/3}Mn^{(4+)}_{1/3}O_2$ and $LiMnO_2$ (as the composition in the discharged cell), is expected to show better cyclic stability.

6. What is photovoltaics (solar electricity) or "PV"? How can we get electricity from the sun? Explain with a neat diagram the working of solar cell.

7. What's the major difference between PV and concentrated solar power systems? Discuss about different types of concentrated solar power systems. Also, explain about the concentration factors for each system.

8. Explain the principle of solar water heating systems?

9. Explain with a schematic how ellipsometry data can be analyzed? Give the examples of materials which can be analyzed using ellipsometry.

10. Define solar absorptance, thermal emittance, and spectral selectivity. Why there should be a shape edge around 1.5 μm for spectrally selective absorbers? Why thermal stability is an important aspect for the absorbers used in CSP systems?

11. The (eukaryotic) cell–material interaction is the central theme of the biomaterials science.

    a. With suitable sketches, explain the stages of cell–material interaction, *in vitro.*

    b. Comment on how the material related specific properties would influence such interaction.

    c. Describe how would you grow eukaryotic cells on a synthetic biomaterial substrate (general cell culture protocol) and mention various equipment and conditions at each stage?

    d. What are the microscopic techniques that you would use to study cell morphological changes?

    e. What are the various cell morphological parameters that one would quantify to assess the change in cell morphology at various timepoints in culture?

12. On a stress–strain plot, schematically illustrate the difference among the cytoskeletal proteins (actin, tubulin, and intermediary filaments) in terms of elastic modulus and deformation behavior.

13. What are the differences between adhesion of eukaryotic and prokaryotic cells?

14. What are the different cellular adaptation processes? Describe, with sketches, various cell fate processes, that a cell is expected to experience while growing on a biomaterial.

I. Key Answers

1. (d) All of the above
2. (a) Electrostatic field
3. (a) Electric and magnetic field
4. (d) All of the above

5. (a) Synchronized functioning of ionic pumps, channels, carriers, and transporters
6. (a) Integral membrane proteins
7. (b) Aquaporins
8. (d) All of the above
9. (c) Na^+ channels
10. (b) Electrodiffusion
11. (c) ATP hydrolysis
12. (c) Both (a) and (b)
13. (c) Ion X would be the only ion permeable across the membrane and its net flux across it would be zero
14. (b) Presence of inorganic ions across the cell membrane
15. (a) Nonconductive
16. (b) Conductive
17. (a) Combination of resistance and capacitance
18. (c) Double lipid bilayer with pores
19. (b) Intracellular electromanipulation
20. (a) Electroporation
21. (b) Combination of resistance, and battery
22. (c) Both (a) and (b)
23. (a) Functionally graded material
24. (b) Functionally graded material
25. (a) Mismatch of thermal coefficient of expansion
26. (b) ABO_3
27. (a) NKN
28. (b) Permanent polarization
29. (a) Crystal structure changes
30. (b) $90°$
31. (a) Maxwell Wagner relaxation process
32. (a) Permanent dipole moment
33. (c) Both (a) and (b)
34. (b) Oxygen vacancies
35. (d) Body of an aircraft
36. (c) ZrB_2
37. (d) Spark plasma sintering
38. (d) Spark plasma sintering
39. (b) Liquid phase sintering
40. (b) Vickers
41. (c) Probability of finding critical sized flaw decreases
42. (a) Mode I.
43. (a) Transformation toughening.
44. (d) Bend.
45. (c) Co is cathode and Ag is anode

46. (a) −97,465 J/mol
47. (b) Capacitor
48. (b) Fuel cells
49. (d) Li-ion
50. (b) 15–45 nm
51. (c) *A* increases and *d* decreases
52. (b) Weaker
53. (c) 7.57 µA
54. (d) The current density at which the electrochemical cycling is conducted.
55. (d) All the above
56. (d) All the above
57. (b) *p* orbitals
58. (d) Possessing spherical symmetry.
59. (a) +3 and +2, respectively.
60. (b) Oil
61. (b) Incineration of waste to generate electrical energy
62. (b) J/m^3
63. (d) Thermal energy and solar energy are the same
64. (c) Electromagnetic waves
65. (a) Insulation
66. (a) 0.2–2.5 µm
67. (b) 8 min
68. (a) 1 h
69. (b) Hot arid zones
70. (b) Decrease, Increase
71. (d) All of the above
72. (d) All of the above
73. (a) It is not difficult to produce photovoltaic cells.
74. (b) Angle at which the solar rays hit the surface of the Earth.
75. (b) Photovoltaic (PV) cells that use semiconductor technology
76. (d) The photoelectric effect
77. (a) Gas
78. (d) All of the above
79. (a) Solar cell
80. (b) Depletion region
81. (c) Series and parallel arrangement of solar cells
82. (c) 48%
83. (b) 15%
84. (b) 1.0 W
85. (b) Convert direct current (DC) to alternating current (AC)
86. (b) Si
87. (c) Large area of photovoltaic cells would be needed

88. (b) South
89. (c) Both (a) and (b)
90. (d) Paraboloid dish collector
91. (a) Thermal energy
92. (a) Flat plate collector
93. (d) Heliostats
94. (b) Two axes
95. (c) Convert heat into mechanical energy
96. (d) The direction of transfer of thermal energy between the objects
97. (a) $E_k = W + Q$, W/E_k
98. (c) Both (a) and (b)
99. (b) The temperature of the black body
100. (d) Different, Different
101. (d) 160,000 R
102. (d) All of the above
103. (b) Direct radiation + diffuse radiation
104. (d) All of the above
105. (b) The number of sputtered atoms per incident particle
106. (c) Dip coating
107. (a) DC sputtering
108. I. (d) 4800 W
     II. (a) 300 W
     III. (a) 0 W
109. (b) $(r/R)^2 S$
110. (c) 1873 K
111. (d) All of the above
112. (d) All of the above
113. (a) SPS
114. (c) A few hours
115. (d) A few milliseconds
116. (b) A few minutes
117. (c) $R = 1 - A$
118. (d) All of the above
119. (d) All of the above
120. (a) Al_2O_3

II. Answers for Fill in the Blanks

1. Ultra-high temperature ceramics
2. Spark plasma sintering

3. 5–10

4. Conduction in solid and convection/radiation in fluid

5. ZrB_2 (or) TiB_2 (or) NbB_2

6. Hemocompatibility

7. 1.67

8. Collagen and HA

9. Calcium phosphate

10. DAPI

11. Secondary

12. Plasma membrane

13. Collagen

14. Osteon

15. Increase in cell size

16. Programmed cell death

17. $Ca_{10}(PO_4)_6 (OH)_2$

18. Elastic modulus

III. Answers for Analytical Questions

1. As per the first law of thermodynamics, the work done by the system tends to lower the energy (Δw), whereas the absorption of heat (Δq) raises the energy. Accordingly, the change in internal energy ΔU, of a system is given by the heat supplied minus the work done by the system; as per $\Delta U = \Delta q - \Delta w$.

   Therefore, ΔU of change in internal energy of the system $= -200$ J $+ 500$ J $= 300$ J.

2. Under standard conditions (as per Nernst equation), the ΔG^0 for a cell is $= -nF(\text{EMF})$, where n is the number of electrons transferred per mole of the product formed (which is 2 here) and F is Faraday's constant ($\sim$96,500 C/mol).

   Rearranging the above equation, the EMF $= -\Delta G^0/nF = 200,000/(2 \times 96,500) \sim 1$ V.

3. It may be noted that discharging a cell through an extremely high resistance, especially when R $\to \infty$, can be assumed to be a thermodynamic reversible process. Accordingly, the heat that is evolved when the cell is undergoing such a thermodynamically reversible process can be as termed as $\Delta q_{\text{reversible}}$.

   Hence, under standard conditions, according to the 2nd law of thermodynamics, $\Delta q_{\text{reversible}} = T\Delta S^0 = -5000$ J.

   The heat dissipated just at the resistance is the electrical energy (here, dissipated as heat) obtained from the reversible work done by the cell. Hence, $\Delta G^0 = -20,000$ J.

   Since, $\Delta G^0 = \Delta H^0 - T\Delta S^0$, the standard enthalpy change (or ΔH^0) of the cell $= -20,000 + (-5000) = -25,000$ J.

   If both the cell and the resistor were placed in the same compartment, the net heat evolved would have been same as the ΔH^0, which is $= -25,000$ J.

4. Since the rate of electrochemical reaction (R; in mol/s/cm²) is dependent on the passage of current (i), which is the rate of passage of the reactant electrons, $R = i/nFA$; where i is the current (here 500 mA), n is 2 (for two electron transfer reaction; that is, $Fe^{2+} + 2e^- \rightarrow Fe$) and $F = 96{,}500$ C/mol. Hence, R would be 5.2×10^{-7} mol/s/cm².

5. (a) i_0, per unit area $= Fk^\circ C^* = 96{,}500 \times 10^{-7} \times 1 \times 10^{-3} \times 10^{-3} = 96{,}500 \times 10^{-13}$ A/cm² $= 96{,}500 \times 10^{-7}$ μA/cm² $= 9.65 \times 10^{-3}$ μA/cm². For an electrode of surface area 1 cm², the exchange current $i_0 = 9.65 \times 10^{-3}$ μA.

(b) The relation to be used for estimating the overpotential as a function of current density is: $i_{net} = i_0[\exp\{-\alpha f\eta\} - \exp\{(1-\alpha)f\eta\}]$, which is valid when i_0 is very small and mass transfer effects are neglected.

The curve cannot bend toward horizontal direction because mass transfer effects have been neglected.

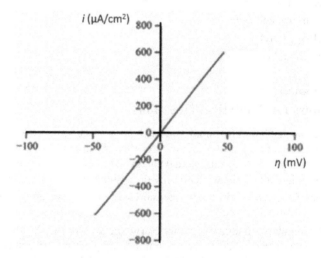

(c) This is Tafel plot, as per $\eta = (RT/\alpha F)\ln i_0 - (RT/\alpha F)\ln i$; based on the concepts of Tafel behavior and Equations 7.14 and 7.15.

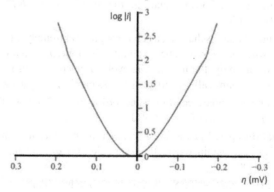

In terms of log–log plot, the slope is either $-\alpha F/2.3\,RT$ or $(1-\alpha)F/2.3\,RT = -0.3 \times 96{,}500/2.3 \times 8.314 \times 298 = -5$ or 11.9.

Since the electrode area is 1 cm^2, Y-axis intercept $= \log(i_0) = \log(9.65 \times 10^{-3}) = -2$ (in μA).

6. (a) Energy delivered $=$ current $\times$ time $\times$ voltage $= 0.5 \times 10 \times 60 \times 0.6 = 180$ J.

   (b) Anode reaction: $H_2 \rightarrow 2H^+ + 2e^-$; 2 moles of e$^-$ from 1 mole of H$_2$; no. of moles of e$^-$ transferred $= 0.5 \times 10 \times 60/96{,}500 = 0.003$.

   With 100% efficiency, 0.0015 moles of H$_2$ would have been needed. However, with 60% efficiency, no. of moles of H$_2$ needed is $0.0015/0.6 = 0.0025$ moles of H$_2$.

7. $1/C_{net} = 1/C_{IHP} + 1/C_{OHP}$

   C_{IHP}/negative surface (per unit area) $= (8.854 \times 10^{-12} \times 80)/(0.1 \times 10^{-9}) \sim 7.08$ F/m^2

   C_{OHP}/IHP (per unit area) $= (8.854 \times 10^{-12} \times 80)/[0.5 \times 10^{-9}] \sim 1.77$ F/m^2

   Hence; $1/C_{net} = 1/C_{IHP} + 1/C_{OHP}$
   $$= 0.71 \text{ m}^2/\text{F}$$
   $$C_{net} \sim 1.41 \text{ F/m}^2$$

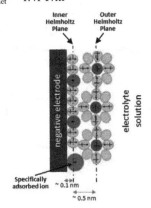

Helmholtz model of electrochemical double layer at electrode/electrolyte interface; crossed arrows represent dipoles

8. Theoretical specific capacity of insertion electrode (in mAh/g) $= nF/3.6$ M.

   In the case of Si, $n = 4.4$ (i.e., the number of Li hosted peer Si atom).

   Hence, the theoretical specific capacity for Si electrode is $(4.4 \times 96{,}500)/(3.6 \times 28) = 4200$ mAh/g (approximately).

9. The volume of the active Sn electrode film is $= \pi \times d^2 \times h/4 = 3.142 \times (2.54)^2 \times 200 \ (\times 10^{-7})/4 = 10.13 \times 10^{-5}$ cm^3.

   Hence, the mass of the active Sn film electrode is $7 \times 0.13 \times 10^{-5} = 0.71$ mg.

   Hence, the net capacity of the active Sn film electrode is $1000 \times (0.71 \times 10^{-3}) = 0.71$ mAh.

   Hence, the C-rate for the electrode is 0.71 mA, with C/5 equivalent to $0.71/5 = 0.142$ mA and 5C equivalent to $0.71 \times 5 = 3.55$ mA.

   The electrode has a better chance of delivering a capacity closer to its theoretical capacity at the lower current, that is, C/5 here. This is because, the Li-alloying/de-alloying in the bulk Sn is expected to get rate limited or kinetically hindered at the higher current density, but with the potential quickly moving toward the cut-off potential due to IR component of the cell.

10. Parallel arrangement of cells lead to an increase in the overall capacity, whereas series arrangement leads to an increase in overall voltage.

    If 48 modules (in series) can make the entire battery pack having a net voltage of 384 V, then each module will be having a voltage of 384/48 = 8 V. Since each cell has a voltage of 4, this necessitates placing 2 cells in series in each module. However, since the capacity of each module in series (leading to no enhancement in the capacity) has to be 24,000 (Wh)/384(V) ∼ 60 Ah, 2 cells (of 30 Ah) also need to be placed in parallel in each module. Accordingly, the energy density of each module is 0.48 kWh.

    Now, this can allow the formation of a battery pack of ∼24 kWh (i.e., 384 V × 60 Ah). This implies that 4 (in each module) × 48 (modules) = 192 cells are needed to make the concerned battery pack.

11. The plot (b) is for 1st cycle and plot (a) is for 10th cycle.

    This is apparent due to the greater irreversibility between discharging (lithiation) and charging (delithiation) in plot (b), which is more pronounced in the 1st cycle.

    Coulombic efficiency (CE) = time taken during delithiation or charge/ time taken during lithiation or discharge. Since the time axis in both the plots are in the same scale, approximately, for plot (a) CE ∼ 5/6 ∼ 83%; for plot (b) CE ∼ 2/6 ∼ 3%.

    The lower CE, that is, greater irreversibility in plot (b) is due to irreversible surface reactions, including SEI layer formation (which leads to irreversible consumption of Li-ions and charge).

    The cathode and anode materials for the cell under consideration here are graphitic carbon and Li metal, respectively.

12. Ans: 100 kg

    Total input energy: 5000 W (100%)

    That is supplied by fuel and Watt = J/s

13. Ans. $\varepsilon^{1/4}$

    Reason: $1. \sigma T_g^4 = \varepsilon \sigma T_a^4$

14. Ans: The daily quantity of thermal energy obtained using collectors will be

    $$\text{Thermal energy} = \frac{6.5 [\text{kWh(solar)}]}{\text{m}^2 \, \text{day}} \frac{20}{100} = 1.3 \, \text{kWh/m}^2 \, \text{day}$$

    This means that for every square meter of collector surface area, 1.3 kWh of heat are produced every day. Therefore, the required collector surface area is obtained as follows:

    Collector surface area: 62 kWh day^{-1}/1.3 kWh/m^2 day = 48 m^2

15. Ans: Solar energy converted to thermal energy/m^2 is

    $200 \times 30\% \times 25\% \times 90\% = 13.5 \, \text{W/m}^2$

    Collector area = $1000 \times 10^6 \, \text{W}/13.5 \, \text{W/m}^2 = 7.4 \times 10^7 \, \text{m}^2$

    J Kg^{-1} * x kg = 5000 W hence, x = 100 kg

IV. Answers for Some of the Descriptive Questions

1.

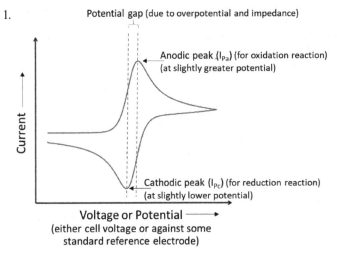

The voltage gap between the oxidation and reduction peaks (i.e., oxidation peak being at a higher voltage than the corresponding reduction peak) is primarily associated with the overpotential(s) needed beyond the equilibrium potential for the corresponding redox reaction(s) to take place. The overall impedance of the cell also contributes toward the voltage difference between the oxidation and reduction peaks. The greater the overpotential needed and greater the impedance, the greater is the gap between the oxidation and reduction peaks.

2. Ragone plot is a map, which ranks the different energy storage and conversion technologies, as per the ranges of power density and energy density achieved by them.

The following is a Ragone plot, which includes only the major electrochemical energy storage and conversion techniques.

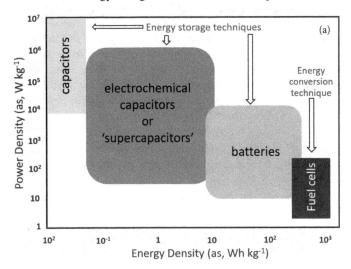

3. The potential safety hazards are:
   - Li-plating on the graphite electrode, followed by formation of Li dendrites, which may lead to short circuiting. This is due to the potential corresponding to Li-intercalation in graphite being only slightly greater than that of Li-plating, with overpotential and polarization (especially at higher current densities) often leading to the intercalation potential being almost same as the Li-plating potential.
   - Overcharging of the Li-T_M-oxide based cathode may lead to electrochemical oxidation of the electrolyte and also gaseous O_2 evolution from the surface of the cathode.
   - The flammable organic electrolyte can lead to fire hazards in case of short circuiting or even excess heating of the cell, especially at higher current densities.

4. The performances impacted are: rate capability of electrode and cyclic stability of electrode.

   The nanoparticles show improved rate capability due to lesser transport distance for Li through the bulk and also more contact area between electrode and electrolyte.

   The nanoparticles are better in terms of cyclic stability due to lesser stress induced degradation resulting from more uniform Li-distribution from the surface to the core arising from more contact area between electrolyte and particle surface for a given volume and also reduced Li transport distance through the particle bulk (from surface to core). Improved mechanical integrity in the case of nanoparticles also helps. This is particularly important in the case of Si because of the huge volume expansion upon Li-insertion, as compared to many other anode materials.

5. The electrode (cathode for Li-ion cell) based on $LiNi^{(+2)}_{1/3}Co^{(3+)}_{1/3}Mn^{(4+)}_{1/3}O_2$ is expected to show greater cyclic stability, as compared to the electrode based on $LiMnO_2$.

   This is because, Mn is in (+3) stage in the later ($LiMnO_2$), as opposed to the (+4) state in the $LiNi^{(+2)}_{1/3}Co^{(3+)}_{1/3}Mn^{(4+)}_{1/3}O_2$. Mn in (+3) state is the primary issue here because in the (+3) state, high spin Mn-ion undergoes Jahn–Teller distortion, which causes structural instability of the electrode material and hence cyclic instability during electrochemical cycling. Furthermore, Mn^{3+} has a tendency for disproportionation into Mn^{2+} and Mn^{4+}, with the former being susceptible to dissolving in the organic electrolyte used. Dissolution of Mn^{2+} leads to Mn-loss from the cathode and also accrued formation of SEI layer on the surface of the graphitic carbon anode upon getting deposited on the surface of the same, causing irreversible Li-loss and increase in impedance.

   In fact, $LiNi^{(+2)}_{1/3}Co^{(3+)}_{1/3}Mn^{(4+)}_{1/3}O_2$ is one of the most promising cathode materials for the present generation Li-ion cells, unlike $LiMnO_2$.

6. The word itself helps to explain how photovoltaic (PV) or solar electric technologies work. The word has two parts: photo, a stem derived from the Greek phos, which means light, and volt, a measurement unit named for

Alessandro Volta a pioneer in the study of electricity. Therefore, photovoltaics could literally be translated as light-electricity.

When certain semiconducting materials, such as certain kinds of silicon, are exposed to sunlight, they release small amounts of electricity. This process is known as the photoelectric effect. The photoelectric effect refers to the emission, or ejection, of electrons from the surface of a metal in response to light. It is the basic physical process in which a solar electric or photovoltaic (PV) cell converts sunlight to electricity.

Sunlight is made up of photons, or particles of solar energy. Photons contain various amounts of energy, corresponding to the different wavelengths of the solar spectrum. When photons strike a PV cell, they may be reflected or absorbed, or they may pass right through. Only the absorbed photons generate electricity. When this happens, the energy of the photon is transferred to an electron in an atom of the PV cell (which is actually a semiconductor).

With its newfound energy, the electron escapes from its normal position in an atom of the semiconductor material and becomes part of the current in an electrical circuit. By leaving its position, the electron causes a hole to form. Special electrical properties of the PV cell—a built-in electric field—provide the voltage needed to drive the current through an external load (such as a light bulb).

7. Photovoltaic (PV) systems convert sunlight directly to electricity by means of PV cells made of semiconductor materials.

    Concentrating solar power (CSP) systems concentrate the sun's energy using reflective devices such as troughs or mirror panels to produce heat that is then used to generate electricity.

8. Every solar water-heating system features a solar collector that faces the sun to absorb the sun's heat energy. This collector can either heat water directly or heat a "working fluid" that is then used to heat the water. In active solar water-heating systems, a pumping mechanism moves the heated water through the building. In passive solar water-heating systems, the water moves by natural convection. In almost all cases, solar water-heating systems work in tandem with conventional gas or electric water-heating systems; the conventional systems operate as needed to ensure a reliable supply of heated water.

Index

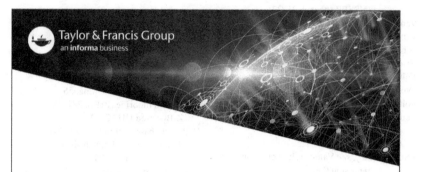